Nutrition During Pregnancy and Lactation

Nutrition During Pregnancy and Lactation

Implications for Maternal and Infant Health

Special Issue Editor

Leanne M. Redman

MDPI • Basel • Beijing • Wuhan • Barcelona • Belgrade

MDPI

Special Issue Editor
Leanne M. Redman
Pennington Biomedical Research Center
USA

Editorial Office
MDPI
St. Alban-Anlage 66
4052 Basel, Switzerland

This is a reprint of articles from the Special Issue published online in the open access journal *Nutrients* (ISSN 2072-6643) in 2019 (available at: https://www.mdpi.com/journal/nutrients/special_issues/ Pregnancy_Lactation_Infant_Health).

For citation purposes, cite each article independently as indicated on the article page online and as indicated below:

LastName, A.A.; LastName, B.B.; LastName, C.C. Article Title. *Journal Name* **Year**, *Article Number, Page Range.*

ISBN 978-3-03928-054-4 (Pbk)
ISBN 978-3-03928-055-1 (PDF)

Contents

About the Special Issue Editor

Leanne M. Redman is the Founder and Director of the Reproductive Endocrinology and Women's Health Laboratory. She is the Principal Investigator of multiple research studies currently enrolling subjects at Pennington Biomedical. Her research studies are comprised both of clinical and translational science research and involve investigations into women's health, weight management, lifestyle intervention, endocrinology, and energy metabolism. She is an internationally renowned expert in human metabolic phenotyping and brings more than 15 years of clinical research experience in phenotyping human subjects, including pregnant women, in an effort to understand the mechanisms of obesity (weight gain and weight loss) as well as to develop and test inventions for effective treatment and prevention. Her research involves the controlled manipulation of diet and/or physical activity or the administration of pharmaceutical agents to alter body energy stores and therefore body weight. In these studies, she uses an array of sophisticated methodologies (doubly labeled water, whole-room indirect calorimetry, DXA, and whole body MRI) to derive estimates of energy intake and energy expenditure to understand the role of these factors on body-weight regulation.

Preface to "Nutrition During Pregnancy and Lactation"

Pregnancy is a viewed as a window to future health. With the birth of the developmental origins of human adult disease hypothesis, research and clinical practice has turned its attention to the influence of maternal factors such as health and lifestyle surrounding pregnancy as a means to understand and prevent the inter-generational inheritance of chronic disease susceptibility. Outcomes during pregnancy have long-lasting impacts on both women on children. Moreover, nutrition early in life can influence growth and the establishment of lifelong eating habits and behaviors.

Maternal nutrition is probably one of the most well described factors known to directly impact fetal development and infant health. For example, inadequate folate intake in mothers who gave birth to children with neural tube defects led to studies on folate supplementation, widespread food fortification programs and clinically to the routine prescription of vitamin and mineral supplements to pregnant women. In the modern world, pregnancy and lactation is now plagued by new challenges brought about by poor quality diets irrespective of their energy content. Dubbed the double burden of malnutrition, the maternal diet can influence the healthy progression of pregnancy. For women, the maternal diet in pregnancy can influence the likelihood of gestational diabetes and gestational hypertension disorders. For children, a mother's diet can influence size at birth and lifelong progression for obesity, type 2 diabetes and cardiovascular disease. New research is emerging on the unique role the maternal diet can have on breastmilk, influencing the nutritive and non-nutritive components which not only impacts normal growth but susceptibility to allergies and asthma.

Efforts have been made to improve the quality and quantity of the maternal diet in pregnancy and during lactation to alter the downstream implications on maternal and child health. Approaches while varied most often times result in an improvement in diet quality yet studies vary in their impacts on adverse pregnancy outcomes and child health. New research that investigates the influence of specific dietary components, maternal eating attitudes and behaviors and the interactions with the gut microbiome is needed to advance our understanding of maternal nutrition during pregnancy and lactation to optimize health outcomes of women and children.

This Special Issue on "Nutrition during Pregnancy and Lactation: Implications for Maternal and Infant Health" is intended to highlight new epidemiological, mechanistic and interventional studies that investigate maternal nutrition around the pregnancy period on maternal and infant outcomes. Submissions may include original research, narrative reviews, and systematic reviews and meta-analyses.

Leanne M. Redman
Special Issue Editor

nutrients

MDPI

Article

Dysregulation of Neuronal Genes by Fetal-Neonatal Iron Deficiency Anemia Is Associated with Altered DNA Methylation in the Rat Hippocampus

Yu-Chin Lien [1], David E Condon [1], Michael K Georgieff [2], Rebecca A Simmons [1,3,*] and Phu V Tran [2,*]

[1] Center for Research on Reproduction and Women's Health, Perelman School of Medicine, The University of Pennsylvania, Philadelphia, PA 19104, USA; ylien@pennmedicine.upenn.edu (Y.-C.L.); dec986@gmail.com (D.E.C.)
[2] Department of Pediatrics, University of Minnesota School of Medicine, Minneapolis, MN 55455, USA; georg001@umn.edu
[3] Children's Hospital of Philadelphia, Philadelphia, PA 19104, USA
* Correspondence: rsimmons@pennmedicine.upenn.edu (R.A.S.); tranx271@umn.edu (P.V.T.); Tel.: +1-215-662-3269 (R.A.S.); Tel.: +1-612-626-7964 (P.V.T.)

Received: 17 April 2019; Accepted: 22 May 2019; Published: 27 May 2019

Abstract: Early-life iron deficiency results in long-term abnormalities in cognitive function and affective behavior in adulthood. In preclinical models, these effects have been associated with long-term dysregulation of key neuronal genes. While limited evidence suggests histone methylation as an epigenetic mechanism underlying gene dysregulation, the role of DNA methylation remains unknown. To determine whether DNA methylation is a potential mechanism by which early-life iron deficiency induces gene dysregulation, we performed whole genome bisulfite sequencing to identify loci with altered DNA methylation in the postnatal day (P) 15 iron-deficient (ID) rat hippocampus, a time point at which the highest level of hippocampal iron deficiency is concurrent with peak iron demand for axonal and dendritic growth. We identified 229 differentially methylated loci and they were mapped within 108 genes. Among them, 63 and 45 genes showed significantly increased and decreased DNA methylation in the P15 ID hippocampus, respectively. To establish a correlation between differentially methylated loci and gene dysregulation, the methylome data were compared to our published P15 hippocampal transcriptome. Both datasets showed alteration of similar functional networks regulating nervous system development and cell-to-cell signaling that are critical for learning and behavior. Collectively, the present findings support a role for DNA methylation in neural gene dysregulation following early-life iron deficiency.

Keywords: hippocampus; DNA methylation; DNA sequencing; iron; neurobiology; transcriptome; micronutrient deficiency; neuroplasticity

1. Introduction

Fetal and neonatal (early-life) iron deficiency with or without anemia affects more than 30% of pregnant women and preschool age children worldwide, and results in long-term cognitive and behavioral abnormalities [1–8]. We have previously investigated the effects of early-life iron deficiency using a rat model, whereby pups were made iron-deficient (ID) from gestational day 2 through postnatal day (P) 7 by providing pregnant and nursing dams with an ID diet, after which they were rescued with an iron-sufficient (IS) diet. This model of maternal-fetal iron deficiency results in a 50% reduction in brain iron concentration by P7 [9], the age at which rat brain development approximates that of a full-term human newborn [10,11]. The deficit in brain iron content is similar to the degree of brain iron deficiency observed in full-term newborn humans [12,13]. Similar cognitive and behavioral abnormalities are observed in our rat model [14–16] and are accompanied by abnormal neuronal morphology [17,18] and glutamatergic neurotransmission [19] in the hippocampus. Iron treatment starting at P7 resolves brain iron deficiency by P56 [20]. Despite this resolution, the formerly iron-deficient (FID) rats show persistent cognitive impairment accompanied by abnormal neuronal morphology [17,18], glutamatergic neurotransmission [19], and lower expression of genes critical for neural plasticity in the hippocampus [21–23]. The persistent dysregulation of hippocampal gene expression in the adult FID rat hippocampus [22] suggests a possible role for epigenetic regulation. Indeed, in a previous study we showed that early-life iron deficiency induced epigenetic modifications at the *Bdnf* locus, a critically important gene coding for a growth factor that regulates brain development and adult synaptic plasticity [24]. As such, comprehensive genome-wide analyses of DNA and histone methylation remain uninvestigated as iron is a critical cofactor for DNA and histone modifying proteins, such as the ten-eleven translocation (TET) enzymes and the Jumonji C-terminal domain (JmjC) family of histone demethylases [25,26].

DNA methylation is essential for neuronal differentiation and maturation in the developing central nervous system and plays a critical role in learning and memory in the adult brain [27]. Altered DNA methylation patterns are associated with many neurological and psychiatric disorders [27]. While DNA methylation at promoter regions is relatively well studied and strongly associated with transcriptional silencing [28], methylation in intergenic regions and gene bodies has been less characterized and may have different functions [27]. Whole genome bisulfite sequencing (WGBS) is the most comprehensive method to analyze 5-methyl cytosine (5mC) at a single-nucleotide resolution [29]. In our previously published methodological paper, a novel method to identify differentially methylated regions (DMRs), namely the Defiant program, was developed. Here, using the same WGBS dataset [30], we present the first genome-wide assessment of DNA methylation in the developing postnatal day 15 (P15) rat hippocampus, during a period of peak iron deficiency and robust axonal growth and dendritic branching [18,31]. In addition to confirming previously reported individual genes and loci that were altered epigenetically due to iron deficiency, we identify novel loci critical to neural function that are epigenetically modified by early-life iron deficiency.

2. Materials and Methods

2.1. Animals

The University of Minnesota Institutional Animal Care and Use Committee approved all experiments in this study. Gestational day 2 (G2) pregnant Sprague-Dawley rats were purchased from Charles Rivers (Wilmington, MA). Rats were kept in a 12 h:12 h light:dark cycle with ad lib food and water. Fetal-neonatal iron deficiency was induced by dietary manipulation as described previously [32]. In brief, pregnant dams were given a purified ID diet (4 mg Fe/kg, TD 80396, Harlan Teklad, Madison, WI) from G2 to P7, at which time nursing dams were given a purified iron sufficient (IS) diet (200 mg Fe/kg, TD 01583, Harlan Teklad) to generate ID pups. Both ID and IS diets were similar in all contents with the exception of iron (ferric citrate) content. Control IS pups were generated from pregnant dams maintained on a purified IS diet. All litters were culled to eight pups with six males and two females at birth. Only male offspring were used in experiments.

2.2. Hippocampal Dissection

P15 male rats were sacrificed by an intraperitoneal injection of Pentobarbital (100 mg/kg). The brains were removed and bisected along the midline on an ice-cold metal block. Each hippocampus was dissected and immediately flash-frozen in liquid N_2 and stored at $-80\,°C$.

2.3. Whole Genome Bisulfite Sequencing and Library Preparation

Genomic DNA from IS and ID hippocampi was isolated using an AllPrep DNA/RNA Mini Kit (Qiagen). WGBS was performed using a previously published protocol [33]. Briefly, 1 µg of genomic DNA was fragmented into ~300 bp fragments using a M220 Covaris Ultrasonicator (Covaris, Woburn, MA, USA). Sequencing libraries were generated using a NEBNext genomic sequencing kit (New England Biolabs, Ipswich, MA, USA) and ligated with Illumina methylated paired end adaptors. Libraries were bisulfite-converted using an Imprint DNA modification kit (MilliporeSigma, St. Louis, MO, USA), and the size of 300–600 bp was selected using the Pippin Prep DNA size selection system (Sage Science, Beverly, MA, USA). Libraries were then amplified using Pfu-Turbo Cx Hotstart DNA polymerase (Agilent Technologies, Santa Clara, CA, USA). Paired-end libraries were sequenced to 100 bp on an Illumina hiSeq2000. Three biological replicates for each group were performed in WGBS. WGBS data are available on the Gene Expression Omnibus under GSE98064.

2.4. Identification of DMRs Using the Defiant Program

DMRs were identified by our in-house developed Defiant (DMRs: Easy, Fast, Identification and ANnoTation) program based on five criteria, as described previously [30]. Briefly, adapters were trimmed from the reads using a custom C language program. Trimmed reads were aligned against the rat genome (rn4). When reads overlapped at a base, the methylation status from read 1 was used. Methylation data at the C and G in a CpG pair were merged to produce the estimate for that locus. DMRs were defined with a minimum coverage of 10 in all six samples, p-value < 0.05, and a minimum methylation percentage change of 10%. Since the Defiant program did not use a pre-defined border to identify DMRs, the p-value < 0.05 cutoff only influenced the widths and quantity of DMRs. The Benjamini–Hochberg approach was applied for multiple testing to obtain false discovery rate (FDR, q-values). Genes were assigned to the DMRs based on a promoter cutoff of 15 kb to the transcription start site, with the direction of transcription taken into account.

2.5. Bioinformatics

The knowledge-based Ingenuity Pathway Analysis® (IPA, Qiagen, Germantown, MD, USA) was employed to identify networks, canonical pathways, molecular and cellular functions, and behavioral and neurological dysfunctions using a P15 DNA methylation dataset from WGBS. The microarray dataset from a prior study [34] was also analyzed by IPA. IPA maps gene networks using an algorithm based on molecular function, cellular function, and functional group. Fisher's exact test was used to calculate the significance of the association between genes in the datasets and the analyzed pathways or functions.

3. Results

3.1. Early-Life Iron Deficiency Induced Differential DNA Methylation in the Rat Hippocampus

We performed whole genome cytosine methylation bisulfite sequencing on P15 ID (n = 3) and IS (n = 3) rat hippocampi. To determine whether iron deficiency alters the genome-wide pattern of DNA methylation in the developing hippocampus, DNA methylation at 1000 randomly selected loci were compared between ID and IS samples to generate a representative heat map. This unsupervised clustering approach showed consistent patterns of methylation across all samples, without an overall shift toward hypo- or hypermethylation in the ID group (Figure 1a). To determine whether iron deficiency induces changes in DNA methylation at a locus-specific level, a $\geq 10\%$ methylation change with p-value < 0.05 was used as an inclusion criterion [30]. We identified 229 DMRs (Figure 1b and Table S1), including 58% intergenic, 26% intronic, and 11% exonic regions (Figure 1c). Approximately 4% of DMRs were located in promoter regions. These DMRs mapped to within 15 kb of the transcription start site of 108 genes with 63 hypermethylated and 45 hypomethylated loci in ID compared to IS hippocampi (Table 1).

(a)

(b)

(c)

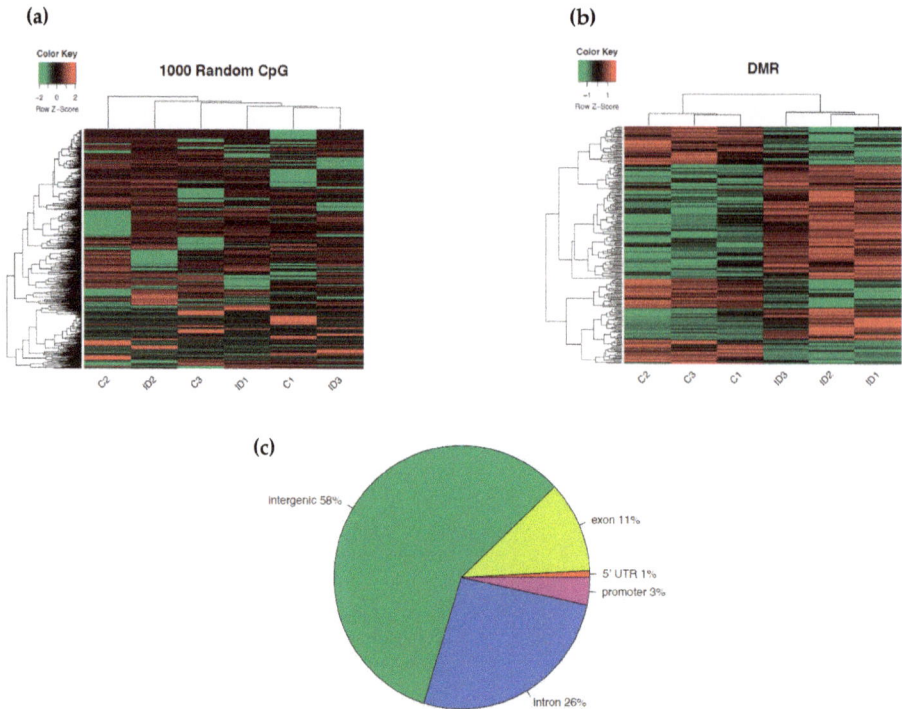

Figure 1. DNA methylome of the postnatal day (P) 15 rat hippocampus. (**a**) An unsupervised clustering heat map of 1000 randomly selected loci showing an absence of bias in global methylation between iron-sufficient (IS) and iron-deficient (ID) hippocampi. Each row in the heat map corresponds to data from a single locus. The branching dendrogram at the top corresponds to the relationships among samples. Hyper- and hypomethylation are shown on a continuum from red to green, respectively. (**b**) Heat map of differentially methylated regions (DMRs) showing significant differences in cytosine methylation between IS (labeled C1-3) and ID (labeled ID1-3) hippocampi. Each row in the heat map corresponds to data point from a single locus, whereas columns correspond to individual samples. The branching dendrogram corresponds to the relationships among samples, as determined by clustering using the 229 identified DMRs. Hyper- and hypomethylation are shown on a continuum from red to green, respectively. (**c**) Pie chart representing the location and proportion of DMRs. The gene body included exons and introns. The promoter was limited to 15 kb upstream from the transcriptional start site. The 5′-untranslated region began at the transcription start site and ended before the initiation sequence. The intergenic region is comprised of the regions not included in the above defined regions.

Table 1. CpG methylation within the 15 kb promoter region of genes in the P15 iron-deficient rat hippocampus.

Hypermethylation				Hypomethylation			
Gene Name	#CpG	DMethylation(%)	*q*-value	Gene Name	#CpG	DMethylation(%)	*q*-value
Adamts19	5	58.5	0.016	*Abhd11*	5	−36.6	0.032
Aebp1	5	10.5	0.118	*Adarb2*	6	−19.7	0.031
Ak4	6	56.0	0.016	*Arhgap28*	5	−21.2	0.026
Ankrd13a	5	26.8	0.024	*Arhgef15*	5	−41.6	0.025
Arf1	6	17.8	0.026	*Arhgef3*	6	−19.8	0.048
Arhgap31	5	36.8	0.035	*Armc8*	9	−10.7	0.041
Armc3	8	43.1	0.031	*Bag2*	6	−21.8	0.059
B4galnt3	5	31.8	0.040	*Cds1*	6	−27.8	0.047

Table 1. *Cont.*

Hypermethylation				Hypomethylation			
Gene Name	#CpG	DMethylation(%)	*q*-value	Gene Name	#CpG	DMethylation(%)	*q*-value
Bcl11b	6	44.7	0.035	*Commd1*	6	−40.4	0.016
Btbd9	5	23.6	0.039	*Dip2a*	10	−57.3	0.039
Cacna1c	5	27.9	0.032	*Dnaja2*	6	−39.0	0.037
Camk2b	9	19.3	0.032	*Dnpep*	5	−14.3	0.051
Capn12	7	21.5	0.024	*Dpf1*	10	−20.6	0.039
Chd2	7	59.7	0.031	*Dpf3*	6	−13.3	0.018
Clvs1	11	26.5	0.031	*Fkrp*	10	−77.7	0.018
Cog3	5	27.4	0.032	*Guca1a*	6	−29.0	0.032
Dgki	5	51.3	0.026	*Hint1*	8	−25.1	0.025
Ephb1	5	33.2	0.040	*Jak3*	8	−20.4	0.023
Ezr	5	51.9	0.018	*Kif26b*	6	−44.2	0.031
Fat3	9	50.1	0.018	*Klhl40*	5	−25.0	0.028
Fig4	5	47.9	0.016	*Lims2*	10	−11.6	0.016
Foxb2	5	32.6	0.039	*LOC691083*	5	−45.6	0.045
Gucy2c	9	86.6	0.016	*Mknk1*	5	−56.8	0.025
Hip1r	5	56.4	0.034	*Mobp*	5	−48.9	0.048
Iqcg	9	52.6	0.020	*Ncf1*	9	−38.4	0.032
Itsn1	5	29.8	0.032	*Pck1*	5	−48.2	0.023
Jph3	8	25.1	0.024	*Pgm3*	5	−69.0	0.024
Kank3	5	28.7	0.031	*Pon2*	6	−10.0	0.029
Kctd15	6	55.1	0.032	*Ppp1r3b*	7	−19.4	0.031
Kctd6	7	27.3	0.016	*Rasd2*	5	−22.4	0.016
Macrod1	5	35.4	0.033	*RGD735029*	5	−17.2	0.032
Map3k11	10	12.9	0.016	*Sardh*	5	−42.9	0.022
Marveld2	6	17.9	0.018	*Sh3pxd2a*	5	−39.4	0.016
Mc3r	5	42.2	0.016	*Slit3*	5	−41.5	0.029
Mib1	6	48.9	0.032	*Smyd3*	6	−28.2	0.112
Mogat1	5	27.4	0.035	*Ss18l1*	6	−18.6	0.042
Mrpl19	6	23.9	0.059	*St8sia1*	6	−20.1	0.016
Myo3b	5	17.1	0.024	*Tal1*	7	−35.1	0.024
Neto2	6	24.3	0.031	*Tmem120b*	5	−37.7	0.024
Olr987	5	21.1	0.016	*Tmem181*	5	−47.1	0.018
Pabpn1l	5	33.3	0.016	*Trrap*	5	−28.2	0.023
Pde2a	5	28.6	0.025	*Usf2*	10	−18.3	0.016
Pde6c	6	46.1	0.024	*Ush1g*	9	−17.2	0.037
Ppp1r21	15	24.9	0.030	*Ust*	5	−54.1	0.037
Prkar1b	5	40.2	0.024	*Wiz*	5	−45.3	0.034
Ptpn14	5	30.1	0.016				
Rev3l	5	16.7	0.023				
Ric8b	6	42.0	0.025				
Riok2	5	48.0	0.031				
Sbk1	6	39.5	0.026				
Scrt2	6	36.8	0.038				
Slc38a1	5	38.2	0.029				
Slc5a1	7	37.4	0.026				
Snurf	5	34.7	0.001				
Spon1	8	73.6	0.031				
Srgap2	5	28.6	0.026				
Tbc1d20	6	13.5	0.042				
Tenm2	5	42.5	0.016				
Tfap2b	5	22.6	0.035				
Tgif2	5	32.2	0.017				
Tnni1	31	29.8	0.016				
Unc93b1	5	54.8	0.020				
Usp36	21	28.8	0.043				

3.2. Early-Life Iron Deficiency Altered the Methylation Status of Genes Regulating Neuronal Development and Function

To identify potential molecular pathways disrupted in the ID hippocampus, IPA was used to map DMRs onto functional networks. The top 10 canonical pathways are shown for DMRs from ID hippocampi (Table 2). Notable pathways critical for neuronal differentiation and function include β-adrenergic signaling, axonal guidance signaling, reelin signaling, Rho family GTPase signaling, cAMP-mediated signaling, and synaptic long-term potentiation.

Table 2. Top 10 canonical pathways implicated by DMRs in the P15 iron-deficient rat hippocampus.

Ingenuity Canonical Pathways	*p*-Value	Differentially Methylated Genes in the Pathway
Nitric Oxide Signaling in the Cardiovascular System	0.002	*CACNA1C, PRKAR1B, PDE2A, GUCY2C*
Cardiac β-Adrenergic Signaling	0.005	*CACNA1C, PRKAR1B, PDE2A, PDE6C*
cAMP-Mediated Signaling	0.005	*CAMK2B, MC3R, PRKAR1B, PDE2A, PDE6C*
Axonal Guidance Signaling	0.006	*ARHGEF15, ITSN1, SLIT3, MKNK1, PRKAR1B, EPHB1, SRGAP2*
Relaxin Signaling	0.007	*PRKAR1B, PDE2A, GUCY2C, PDE6C*
Reelin Signaling in Neurons	0.010	*ARHGEF15, ARHGEF3, MAP3K11*
G-Protein Coupled Receptor Signaling	0.011	*CAMK2B, MC3R, PRKAR1B, PDE2A, PDE6C*
Protein Kinase A Signaling	0.013	*CAMK2B, PTPN14, TNNI1, PRKAR1B, PDE2A, PDE6C*
Synaptic Long-Term Potentiation	0.021	*CAMK2B, CACNA1C, PRKAR1B*
Signaling by Rho Family GTPases	0.034	*ARHGEF15, ARHGEF3, MAP3K11, EZR*

3.3. The Methylation Status of Genes Regulating Axonal Guidance Was Altered in the P15 ID Hippocampus

Neuronal connections are formed by the extension of axons to reach their synaptic targets. This process is controlled by ligands and their receptors at the axonal growth cone, which can sense attractive and repulsive guidance cues to help navigate an axon to its destination [35–39]. These guidance molecules include netrins, slits, semaphorins, and ephrins. Iron deficiency altered methylation at the genes regulating ephrin B signaling/ephrin receptor signaling (data not shown), including increased methylation at *Ephb1*, *Itsn1*, *Prkar1b*, and *Srgap2*, and decreased methylation at *Arhgef15*, *Mknk1*, and *Slit3* loci (Table 1). Decreased methylation at *Arhgef15* and *Arhgef3* and increased methylation at *Map3k11* and *Ezr* loci suggest altered Rho GTPase signaling (Table 2), which transduces guidance signals in the growth cone and regulates cytoskeletal dynamics, an important cellular process for the formation of long-term potentiation (LTP) [40], a cellular basis of learning and memory [41,42].

3.4. Differential DNA Methylation is a Potential Epigenetic Mechanism Contributing to Neural Gene Dysregulation in the P15 ID Hippocampus

To determine whether differential DNA methylation in the P15 ID hippocampus potentially contributes to neural gene dysregulation, we compared our WGBS methylomic dataset and the P15 ID hippocampal transcriptomic dataset [34]. IPA revealed that cAMP-mediated signaling, axonal guidance signaling, reelin signaling, synaptic long-term potentiation, Rho family GTPase signaling, and ephrin B signaling were among the 18 pathways that were altered in both datasets (Table 3). The top functional networks altered in the P15 ID hippocampal methylome (Table 4) were also observed in the P15 ID hippocampal transcriptome. These include cell-to-cell signaling, nervous system development and function, behavior, neurological disease, molecular transport, and lipid metabolism. The transcriptomic dataset corroborates the methylome data and further highlights the disruption of synaptic transmission (Figure 2a), neuritogenesis (Figure 2b), and movement disorders (Figure 2c,d).

Table 3. Overlapping canonical pathways of the P15 DNA methylome and P15 microarray datasets.

	Methylome Analysis		Microarray Analysis	
Ingenuity Canonical Pathways	*p*-value	Differentially Methylated Genes	*p*-value	Differentially Expressed Genes
Nitric Oxide Signaling in the Cardiovascular System	0.002	*CACNA1C, PRKAR1B, PDE2A, GUCY2C*	0.000	*ITPR2, PIK3R3, KDR, PTPN11, PRKAA1, GUCY2D, ITPR1, CAMK4, PRKAG1, PDE2A, PDGFC*
Cellular Effects of Sildenafil (Viagra)	0.004	*CACNA1C, PRKAR1B, PDE2A, GUCY2C*	0.000	*MYH3, CACNG8, ITPR2, ADCY3, GPR37, GUCY2D, ITPR1, ADCY2, PLCE1, CAMK4, PRKAG1, PDE2A*
Cardiac β-Adrenergic Signaling	0.005	*CACNA1C, PRKAR1B, PDE2A, PDE6C*	0.036	*ADCY3, PKIG, ADCY2, PRKAG1, PDE2A, PPP2R2A, PPP1R11*
cAMP-Mediated Signaling	0.005	*CAMK2B, MC3R, PRKAR1B, PDE2A, PDE6C*	0.000	*GABBR1, CHRM3, CAMK4, VIPR1, PDE2A, Htr5b, CHRM2, CNGA2, CAMK2A, GNAI3, ADCY3, HRH3, PKIG, ADCY, LHCGR, OPRM1, GRM6*
Axonal Guidance Signaling	0.006	*ARHGEF15, ITSN1, SLIT3, MKNK1, PRKAR1B, EPHB1, SRGAP2*	0.003	*CXCL12, PIK3R3, TUBB, EPHA3, ROBO1, PLCE1, DPYSL5, RTN4R, RTN4, GNAI3, FZD4, PDGFC, BAIAP2, SEMA4F, CXCR4, NRAS, CFL1, PTPN11, NTRK2, PRKAG1*
Relaxin Signaling	0.007	*PRKAR1B, PDE2A, GUCY2C, PDE6C*	0.008	*PIK3R3, ADCY3, PTPN11, GUCY2D, ADCY2, PRKAG1, PDE2A, NFKBIA, GNAI3*
Reelin Signaling in Neurons	0.010	*ARHGEF15, ARHGEF3, MAP3K11*	0.004	*PAFAH1B1, PIK3R3, PTPN11, APP, MAPT, ARHGEF9, APBB1*
G-Protein Coupled Receptor Signaling	0.011	*CAMK2B, MC3R, PRKAR1B, PDE2A, PDE6C*	0.000	*PIK3R3, GABBR1, CHRM3, CAMK4, VIPR1, PDE2A, Htr5b, NFKBIA, CHRM2, CAMK2A, GNAI3, NRAS, PDPK1, ADCY3, HRH3, PTPN11, ADCY2, PRKAG1, GRM5, LHCGR, OPRM1, GRM6*
Protein Kinase A Signaling	0.013	*CAMK2B, Ptpn14, TNNI1, PRKAR1B, PDE2A, PDE6C*	0.000	*ITPR2, PLCE1, NFKBIA, CNGA2, GNAI3, PYGB, ADCY3, PTPN11, ITPR1, ADCY2, PTPRF, TGFBR1, PPP1R1B, YWHAB, PPP1R11, DUSP12, PTPRN, CDC25A, PTPN2, PTPRO, H3F3A/H3F3B, CAMK4, PDE2A, PTPN23, CAMK2A, BAD, DUSP5, PTPN12, PRKAG1*
Breast Cancer Regulation by Stathmin1	0.017	*CAMK2B, ARHGEF15, ARHGEF3, PRKAR1B*	0.000	*ITPR2, PIK3R3, TUBB, CAMK4, PPP2R2A, CAMK2A, GNAI3, STMN1, NRAS, ADCY3, PTPN11, ITPR1, ADCY2, ARHGEF9, PRKAG1, PPP1R11*
Synaptic Long-Term Potentiation	0.021	*CAMK2B, CACNA1C, PRKAR1B*	0.000	*NRAS, ITPR2, GRINA, ITPR1, PLCE1, CAMK4, PRKAG1, GRM5, GRIN1, CAMK2A, GRM6, PPP1R11*
Gustation Pathway	0.023	*PRKAR1B, PDE2A, PDE6C*	0.000	*CACNG8, ITPR2, ADCY3, CACNB4, CACNA2D1, P2RX5, ITPR1, ADCY2, PRKAG1, PDE2A, P2RY1, CACNA1H*
Sperm Motility	0.023	*MAP3K11, PRKAR1B, PDE2A*	0.002	*ITPR2, PAFAH1B1, ITPR1, PLCE1, CAMK4, PRKAG1, PDE2A, CNGA2, CACNA1H*
GNRH Signaling	0.032	*CAMK2B, MAP3K11, PRKAR1B*	0.000	*CACNG8, ITPR2, CACNB4, CAMK4, CAMK2A, GNAI3, CACNA1H, NRAS, ADCY3, CACNA2D1, GNRHR, ITPR1, ADCY2, PRKAG1*
Signaling by Rho Family GTPases	0.034	*ARHGEF15, ARHGEF3, MAP3K11, EZR*	0.010	*BAIAP2, CFL1, RHOT2, PIK3R3, PTPN11, RHOB, CDH1, ARHGEF9, PLD1, RHOV, GNAI3, STMN1*
Molecular Mechanisms of Cancer	0.042	*CAMK2B, ARHGEF15, ARHGEF3, JAK3, PRKAR1B*	0.000	*RASGRF1, RHOT2, PIK3R3, CDC25A, CASP9, NFKBIA, CAMK2A, BAD, GNAI3, FZD4, NCSTN, NRAS, RALBP1, ADCY3, PTPN11, RHOB, HIF1A, ADCY2, CASP3, TGFBR1, CDH1, ARHGEF9, PRKAG1, RHOV*
Melatonin Signaling	0.048	*CAMK2B, PRKAR1B*	0.021	*PLCE1, CAMK4, PRKAG1, CAMK2A, GNAI3*
Ephrin B Signaling	0.049	*ITSN1, EPHB1*	0.022	*CXCL12, CXCR4, CFL1, CAP1, GNAI3*

Table 4. IPA annotated functional similarity between the DNA methylome and transcriptome of the P15 ID rat hippocampus.

Category	Diseases or Functions Annotation	*p*-value	Differentially Methylated Genes	*p*-value	Number of Genes
Cell-To-Cell Signaling	Synaptic Depression/Neurotransmission	1.65E-04	*CAMK2B, ARF1, ITSN1, DGKI, PRKAR1B, EPHB1*	1.94E-10	21
Nervous System Development and Function	Neuritogenesis/Extension of Neurites	8.40E-03	*CAMK2B, ST8SIA1, ITSN1, SS18L1, BCL11B, EZR, SLIT3, UST, EPHB1, SRGAP2*	4.92E-16	62
Behavior	Locomotion	3.09E-04	*RASD2, HINT1, MC3R, BTBD9, NCF1, CACNA1C, JPH3, FIG4, TAL1*	1.08E-13	40
	Learning	2.22E-02	*CAMK2B, NCF1, BTBD9, DGKI, CACNA1C, PRKAR1B, JPH3*	3.51E-21	57
Neurological Disease	Cell Death of Cerebral Cortex Cells	1.33E-02	*ST8SIA1, ITSN1, MAP3K11, NCF1, SH3PXD2A*	8.55E-14	32
	Movement Disorder	4.68E-02	*CAMK2B, AEBP1, CDS1, ST8SIA1, HINT1, BCL11B, TFAP2B, PDE6C, USP36, RASD2, MC3R, BTBD9, CACNA1C*	5.58E-32	117
Lipid Metabolism	Quantity of Sphingolipid/Steroid	2.73E-03	*ST8SIA1, HINT1, BCL11B, PON2*	4.24E-09	40
Molecular Transport	Quantity of Heavy Metal	1.13E-02	*ARF1, USF2, COMMD1*	4.32E-19	58
	Transport of Molecule	1.92E-02	*SLC5A1, SLC38A1*	5.41E-31	144

Integrating the P15 ID WGBS methylome and microarray datasets led to the identification of three genes, including *Pde2a*, *Mobp*, and *Cds1* (Table 5). All three genes showed differential methylation in their intronic regions. *Pde2a* (+28.6%) was hypermethylated while *Mobp* (−48.9%) and *Cds1* (−27.8%) were hypomethylated in the P15 ID hippocampus. All three genes were upregulated in the P15 ID hippocampus. While DNA methylation at gene promoters is strongly associated with gene silencing [28], DNA methylation in intronic regions may mark enhancers or repressors and can be associated with changes in gene expression [43,44]. Phosphodiesterase 2A (Pde2a) is highly expressed in the brain and metabolizes cGMP and cAMP to regulate short-term synaptic plasticity, axonal excitability, and transmitter release in the hippocampal, cortical, and striatal networks [45,46]. Myelin-associated oligodendrocyte basic protein (Mobp) is the third most abundant protein in the central nervous system (CNS), and is exclusively expressed in oligodendrocytes, the myelinating glial cells of the CNS [47]. Mobp plays a role in compacting or stabilizing the myelin sheath and regulates the morphological differentiation of oligodendrocytes [48]. CDP-diacylglycerol synthase 1 (Cds1) is a key enzyme in regulating second messenger phosphatidylinositol 4,5-bisphosphate (PIP$_2$) levels. It is localized in the endoplasmic reticulum and mitochondria [49] and is involved in the synthesis of phosphatidylglycerol and cardiolipin, an important component of the inner mitochondrial membrane [50]. Cds1 is a novel regulator of lipid droplet formation, lipid storage, and adipocyte development [51], and plays a critical role in mammalian energy storage, which is compromised in developing iron-deficient neurons [52].

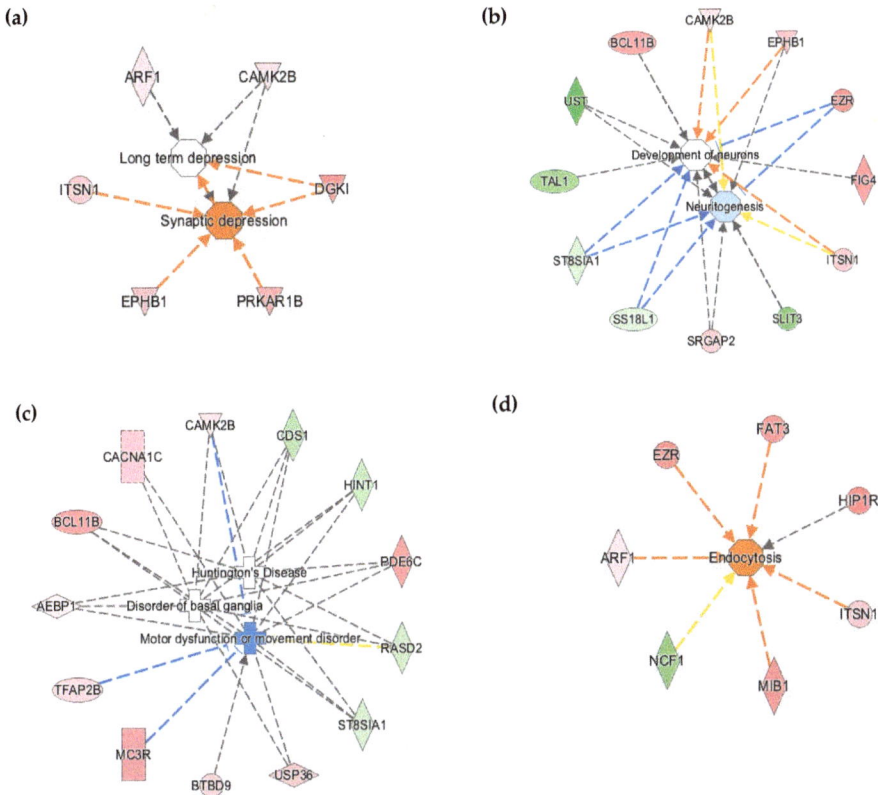

Figure 2. Ingenuity Pathway Analysis® (IPA) functional annotation of altered DNA methylation at loci that are involved in (**a**) synaptic depression, (**b**) neuritogenesis and neuronal development, (**c**) pathogenesis of neurological diseases, and (**d**) lipid metabolism and molecular transport, including endocytosis. Red-filled and green-filled shapes indicate increased and decreased methylation, respectively. Orange-red lines indicate activation; blue lines indicate inhibition; yellow lines indicate findings inconsistent with the state of downstream activity; grey lines indicate that the effect was not predicted.

Table 5. Overlapping genes of the P15 DNA methylome and microarray datasets.

Gene Name	Symbol	Δ Methylation (%)	CpGs Location	FC (ID/IS)	Location	Type(s)
Phosphodiesterase 2A	*Pde2a*	28.6	Intron 2	1.16	Plasma Membrane	enzyme
Myelin-associated oligodendrocyte basic protein	*Mobp*	−48.9	Intron 2	1.37	Cytoplasm	other
CDP-diacylglycerol synthase 1	*Cds1*	−27.8	Intron 11	1.23	Endoplasmic reticulum & mitochondria	enzyme

Δ Methylation values are means from DNA methylome, and FC (fold change) values are means from microarray.

4. Discussion

Fetal and early postnatal life iron deficiency causes long-lasting impairments in learning, memory, and socio-emotional behaviors [1,14–16,53], including an increased risk for autism, depression, and schizophrenia in humans [2,54,55]. These long-term neurobehavioral deficits occur despite early diagnosis and treatment, indicating the need for adequate iron during critical periods of brain development. In preclinical models, these effects have been ascribed in part to the persistence of abnormalities in monoamine signaling, myelination, neural metabolism, and the expression of neuroplasticity-associated proteins into adulthood [20,56–58]. The molecular mechanisms underlying these persistent changes have not been fully elucidated. The present study goes well beyond previous studies by systematically analyzing the alteration of DNA methylation induced by early-life iron deficiency using a whole genome bisulfite sequencing approach. Consistent with previous transcriptomic analysis, the changes in DNA methylation in the ID hippocampus mapped to functional networks that are important for neuronal plasticity.

DNA methylation is an important epigenetic mechanism regulating gene expression, often across the lifespan. Methylation at genomic regions has different influences on gene transcriptional activity depending on the location of DNA methylation. In the present study, differential methylation was highly enriched at intergenic regions (58%) in the ID hippocampus. This outcome is similar to our previous findings in pancreatic islets of an intrauterine growth restriction rat model [59], where approximately 65% of DMRs were located in intergenic regions, as well as to other models of early-life adverse environments [33,60–62]. These conserved intergenic regions may represent important enhancers or cis-regulatory sites in regulating gene expression [43,63,64]. Thus, these DMRs might account for a substantially fewer number of loci with DMRs compared to a number of differentially expressed genes in the microarray dataset and a small overlap between these two datasets. Our data also showed that approximately 37% of DMRs in the ID hippocampus were located in gene bodies (26% and 11% in introns and exons, respectively). DNA methylation in gene bodies is generally associated with higher gene expression in dividing cells [65], in contrast to the regulatory effect of DNA methylation in promoter regions. However, this association is not seen in non-dividing cells [27]. Although not many cells in the hippocampus are actively dividing at P15, these DNA modifications might have occurred during the period of active proliferation in the prenatal period when the pregnant dam and fetus were iron-deficient. Additional DMR analysis at a timepoint when the developing hippocampus undergoes active proliferation will provide further insight into this notion. DNA methylation in gene bodies may define the exon boundaries, regulate alternative promoters in gene bodies, and regulate mRNA splicing and alternative splicing [65–68]. Wan et al. (2013) showed that tissue-specific DMRs are preferentially located in exons and introns of protein-coding genes [69]. These biologically relevant DMRs are enriched in alternatively spliced genes and a subset of developmental genes. It is possible that the real effect of DNA methylation in the P15 ID hippocampus is within these domains. Our microarray analysis [34] was insufficient to probe such effects. Finally, iron deficiency-induced intragenic DMRs could modify potential gene enhancers [43,44,70,71]. The intragenic DMRs in the ID hippocampus may directly contribute to neural gene dysregulation by modifying the accessibility of alternative splice sites or promoters. These analyses constitute a potential direction for future study.

Our WGBS analysis of the ID hippocampus identified pertinent signaling pathways that could underlie the neurobehavioral abnormalities associated with early-life iron deficiency. The DNA methylome showed that DNA methylation at genes regulating cAMP-mediated signaling and protein kinase A signaling was significantly altered in the P15 ID rat hippocampus (Table 2). Both pathways play critical roles in regulating LTP, as well as the plasticity of axonal guidance responses [72,73]. In addition, the predicted changes to the β-adrenergic signaling and nitric oxide signaling pathways in the ID hippocampus would likely result in lower activities of cAMP, cGMP, protein kinase C (PKC), mitogen-activated protein kinase (MAPK), and N-methyl-D-aspartate (NMDA) receptors [74,75]. Likewise, altered Rho GTPase signaling could change the axonal responses to guidance cues and affect neuronal connections and LTP formation. The Rho family of GTPases plays a key role in the formation of LTP by regulating cellular processes, including axon outgrowth and growth cone dynamics [76,77]. This effect is consistent with and provides a molecular basis for our previous finding of abnormal dendritogenesis and synaptogenesis in these ID rats [17]. Our study also revealed the alteration of the reelin signaling pathway in the ID hippocampus. Reelin regulates neuronal migration and cell positioning in the developing neocortex and cerebellar cortex [78], and modulates synaptogenesis, synaptic plasticity, and LTP, which are necessary for learning and memory in the adult brain [79,80]. Reelin can bind to two lipoprotein receptors, apolipoprotein E receptor 2 (ApoER2) and very-low-density-lipoprotein receptor (VLDLR), and initiates signaling cascades, including NMDA receptor activity, that are critical for hippocampal-dependent learning and memory [79,81]. Altered reelin signaling has been implicated in the pathogenesis of schizophrenia, bipolar disorder, and autism [82,83], all which have been associated with early-life iron deficiency [2,54,55].

DNA methylation patterns are dynamically regulated by DNA methyltransferase (DNMT) and TET enzymes [84]. TET proteins are responsible for catalyzing the conversion of 5-methylcytosine (5mC) to 5-hydroxymethylcytosine (5hmC) and other oxidized methylcytosines, thereby initiating the active DNA demethylation process. TET enzymatic activity is iron-dependent [25,85]. Consequently, iron deficiency could decrease TET activity, leading to global DNA hypermethylation. However, our WGBS results did not show an overall shift toward hypermethylation, suggesting a minimal effect of potentially compromised TET activity in the P15 ID hippocampus. As such, 5hmC has been shown to be an important epigenetic modification for chromatin structure and transcriptional regulation [86–88]. TET proteins and 5hmC are highly enriched in the brain and play an important role in neuronal development and differentiation [87,89,90]. Changes to 5hmC levels are associated with neurodegenerative diseases, such as Alzheimer's disease, Huntington's disease, and Parkinson's disease [91]. Early-life iron deficiency may decrease TET activity and dynamically alter the DNA methylation pattern between 5mC and 5hmC, leading to the dysregulation of neuronal gene expression in the ID hippocampus. Due to methodological limitations, our current WGBS study could not distinguish 5mC and 5hmC levels. Investigations of loci-specific and genome-wide 5hmC alterations in the ID hippocampus are ongoing and will be presented in a future publication.

DNA methylation has been demonstrated to be an important factor in central nervous system development, the modulation of normal brain function, and the pathogenesis of neurological and psychiatric disorders [27]. Despite the limitations of our study, such as minimal biological replicates due to the high cost of WGBS and possible sex-dependent differences in ID rat brains, our current study is the first hippocampal methylome study on the developing rat brain, and provides evidence for DNA methylation as a potential epigenetic mechanism contributing to hippocampal gene dysregulation in early-life ID animals. Given that 95% of DMRs were found in the intergenic and intragenic regions, future studies will need to uncover the mechanisms by which DMRs reprogram gene regulation and mRNA splicing alteration in the ID hippocampus.

Supplementary Materials: The following is available online at http://www.mdpi.com/2072-6643/11/5/1191/s1, Supplemental Table S1: Differentially methylated regions in P15 ID rat hippocampus.

Author Contributions: Y.-C.L., P.V.T., R.A.S., and M.K.G. designed the study, which was conducted and analyzed by Y.-C.L. and P.V.T. D.E.C. developed the Defiant method for DMR analysis. This manuscript was written by Y.-C.L. together with P.V.T. and R.A.S. with input from all authors. All authors have read and approved the manuscript.

Funding: This work was supported by grant NICHD R01 HD29421-20 and NINDS R01 NS099178.

Acknowledgments: We thank Kyoung-Jae Won and Jonathan Schug for assisting in whole genome bisulfite-sequencing data analysis.

Conflicts of Interest: Y.-C. Lien, D.E. Condon, M.K. Georgieff, R.A. Simmons, and P.V. Tran have no conflict of interest.

References

1. Lozoff, B.; Beard, J.; Connor, J.; Barbara, F.; Georgieff, M.; Schallert, T. Long-lasting neural and behavioral effects of iron deficiency in infancy. *Nutr. Rev.* **2006**, *64*, S34–S91. [CrossRef] [PubMed]
2. Insel, B.J.; Schaefer, C.A.; McKeague, I.W.; Susser, E.S.; Brown, A.S. Maternal iron deficiency and the risk of schizophrenia in offspring. *Arch. Gen. Psychiatry* **2008**, *65*, 1136–1144. [CrossRef]

3. Christian, P.; Murray-Kolb, L.E.; Khatry, S.K.; Katz, J.; Schaefer, B.A.; Cole, P.M.; Leclerq, S.C.; Tielsch, J.M. Prenatal micronutrient supplementation and intellectual and motor function in early school-aged children in Nepal. *JAMA* **2010**, *304*, 2716–2723. [CrossRef]

4. Li, Q.; Yan, H.; Zeng, L.; Cheng, Y.; Liang, W.; Dang, S.; Wang, Q.; Tsuji, I. Effects of maternal multimicronutrient supplementation on the mental development of infants in rural western China: Follow-up evaluation of a double-blind, randomized, controlled trial. *Pediatrics* **2009**, *123*, e685–e692. [CrossRef] [PubMed]

5. Schmidt, M.K.; Muslimatun, S.; West, C.E.; Schultink, W.; Hautvast, J.G. Mental and psychomotor development in Indonesian infants of mothers supplemented with vitamin A in addition to iron during pregnancy. *Br. J. Nutr.* **2004**, *91*, 279–286. [CrossRef] [PubMed]

6. Chang, S.; Zeng, L.; Brouwer, I.D.; Kok, F.J.; Yan, H. Effect of Iron Deficiency Anemia in Pregnancy on Child Mental Development in Rural China. *Pediatrics* **2013**. [CrossRef]

7. Siddappa, A.M.; Georgieff, M.K.; Wewerka, S.; Worwa, C.; Nelson, C.A.; Deregnier, R.A. Iron deficiency alters auditory recognition memory in newborn infants of diabetic mothers. *Pediatr. Res.* **2004**, *55*, 1034–1041. [CrossRef]

8. Rao, R.; Georgieff, M.K. Iron in fetal and neonatal nutrition. *Semin. Fetal Neonatal Med.* **2007**, *12*, 54–63. [CrossRef]

9. Rao, R.; Tkac, I.; Townsend, E.L.; Gruetter, R.; Georgieff, M.K. Perinatal iron deficiency alters the neurochemical profile of the developing rat hippocampus. *J. Nutr.* **2003**, *133*, 3215–3221. [PubMed]

10. Avishai-Eliner, S.; Brunson, K.L.; Sandman, C.A.; Baram, T.Z. Stressed-out, or in (utero)? *Trends Neurosci.* **2002**, *25*, 518–524. [PubMed]

11. Semple, B.D.; Blomgren, K.; Gimlin, K.; Ferriero, D.M.; Noble-Haeusslein, L.J. Brain development in rodents and humans: Identifying benchmarks of maturation and vulnerability to injury across species. *Prog. Neurobiol.* **2013**, *106–107*, 1–16. [CrossRef] [PubMed]

12. Petry, C.D.; Eaton, M.A.; Wobken, J.D.; Mills, M.M.; Johnson, D.E.; Georgieff, M.K. Iron deficiency of liver, heart, and brain in newborn infants of diabetic mothers. *J. Pediatr.* **1992**, *121*, 109–114. [CrossRef]

13. Georgieff, M.K.; Petry, C.D.; Wobken, J.D.; Oyer, C.E. Liver and brain iron deficiency in newborn infants with bilateral renal agenesis (Potter's syndrome). *Pediatr. Pathol. Lab. Med.* **1996**, *16*, 509–519. [CrossRef] [PubMed]

14. Unger, E.L.; Hurst, A.R.; Georgieff, M.K.; Schallert, T.; Rao, R.; Connor, J.R.; Kaciroti, N.; Lozoff, B.; Felt, B. Behavior and monoamine deficits in prenatal and perinatal iron deficiency are not corrected by early postnatal moderate-iron or high-iron diets in rats. *J. Nutr.* **2012**, *142*, 2040–2049. [CrossRef]

15. Schmidt, A.T.; Waldow, K.J.; Grove, W.M.; Salinas, J.A.; Georgieff, M.K. Dissociating the long-term effects of fetal/neonatal iron deficiency on three types of learning in the rat. *Behav. Neurosci.* **2007**, *121*, 475–482. [CrossRef] [PubMed]

16. Felt, B.T.; Beard, J.L.; Schallert, T.; Shao, J.; Aldridge, J.W.; Connor, J.R.; Georgieff, M.K.; Lozoff, B. Persistent neurochemical and behavioral abnormalities in adulthood despite early iron supplementation for perinatal iron deficiency anemia in rats. *Behav. Brain Res.* **2006**, *171*, 261–270. [CrossRef]

17. Brunette, K.E.; Tran, P.V.; Wobken, J.D.; Carlson, E.S.; Georgieff, M.K. Gestational and neonatal iron deficiency alters apical dendrite structure of CA1 pyramidal neurons in adult rat hippocampus. *Dev. Neurosci.* **2010**, *32*, 238–248. [CrossRef]

18. Jorgenson, L.A.; Wobken, J.D.; Georgieff, M.K. Perinatal iron deficiency alters apical dendritic growth in hippocampal CA1 pyramidal neurons. *Dev. Neurosci.* **2003**, *25*, 412–420. [CrossRef] [PubMed]

19. Jorgenson, L.A.; Sun, M.; O'Connor, M.; Georgieff, M.K. Fetal iron deficiency disrupts the maturation of synaptic function and efficacy in area CA1 of the developing rat hippocampus. *Hippocampus* **2005**, *15*, 1094–1102. [CrossRef]

20. Rao, R.; Tkac, I.; Schmidt, A.T.; Georgieff, M.K. Fetal and neonatal iron deficiency causes volume loss and alters the neurochemical profile of the adult rat hippocampus. *Nutr. Neurosci.* **2011**, *14*, 59–65. [CrossRef]

21. Clardy, S.L.; Wang, X.; Zhao, W.; Liu, W.; Chase, G.A.; Beard, J.L.; True Felt, B.; Connor, J.R. Acute and chronic effects of developmental iron deficiency on mRNA expression patterns in the brain. *J. Neural Transm. Suppl.* **2006**, 173–196.

22. Tran, P.V.; Kennedy, B.C.; Pisansky, M.T.; Won, K.; Gewirtz, J.C.; Simmons, R.A.; Georgieff, M.K. Prenatal choline supplementation diminishes early-life iron deficiency induced reprogramming of molecular networks associated with behavioral abnormalities in the adult rat hippocampus. *J. Nutr.* **2016**, *146*, 484–493. [CrossRef]

23. Tran, P.V.; Carlson, E.S.; Fretham, S.J.; Georgieff, M.K. Early-life iron deficiency anemia alters neurotrophic factor expression and hippocampal neuron differentiation in male rats. *J. Nutr.* **2008**, *138*, 2495–2501. [CrossRef]

24. Tran, P.V.; Kennedy, B.C.; Lien, Y.C.; Simmons, R.A.; Georgieff, M.K. Fetal iron deficiency induces chromatin remodeling at the Bdnf locus in adult rat hippocampus. *Am. J. Physiol. Regul. Integr. Comp. Physiol.* **2015**, *308*, R276–R282. [CrossRef]

25. Tahiliani, M.; Koh, K.P.; Shen, Y.; Pastor, W.A.; Bandukwala, H.; Brudno, Y.; Agarwal, S.; Iyer, L.M.; Liu, D.R.; Aravind, L.; et al. Conversion of 5-methylcytosine to 5-hydroxymethylcytosine in mammalian DNA by MLL partner TET1. *Science* **2009**, *324*, 930–935. [CrossRef]

26. Hou, H.; Yu, H. Structural insights into histone lysine demethylation. *Curr. Opin. Struct. Biol.* **2010**, *20*, 739–748. [CrossRef]

27. Moore, L.D.; Le, T.; Fan, G. DNA methylation and its basic function. *Neuropsychopharmacology* **2013**, *38*, 23–38. [CrossRef]

28. Weber, M.; Hellmann, I.; Stadler, M.B.; Ramos, L.; Pääbo, S.; Rebhan, M.; Schübeler, D. Distribution, silencing potential and evolutionary impact of promoter DNA methylation in the human genome. *Nat. Genet.* **2007**, *39*, 457–466. [CrossRef]

29. Stevens, M.; Cheng, J.B.; Li, D.; Xie, M.; Hong, C.; Maire, C.L.; Ligon, K.L.; Hirst, M.; Marra, M.A.; Costello, J.F.; et al. Estimating absolute methylation levels at single-CpG resolution from methylation enrichment and restriction enzyme sequencing methods. *Genome Res.* **2013**, *23*, 1541–1553. [CrossRef]

30. Condon, D.E.; Tran, P.V.; Lien, Y.C.; Schug, J.; Georgieff, M.K.; Simmons, R.A.; Won, K.J. Defiant: (DMRs: Easy, fast, identification and ANnoTation) identifies differentially Methylated regions from iron-deficient rat hippocampus. *BMC Bioinf.* **2018**, *19*, 31. [CrossRef]

31. Fretham, S.J.; Carlson, E.S.; Georgieff, M.K. The role of iron in learning and memory. *Adv. Nutr.* **2011**, *2*, 112–121. [CrossRef]

32. Tran, P.V.; Fretham, S.J.; Wobken, J.; Miller, B.S.; Georgieff, M.K. Gestational-neonatal iron deficiency suppresses and iron treatment reactivates IGF signaling in developing rat hippocampus. *Am. J. Physiol. Endocrinol. Metab.* **2012**, *302*, E316–E324. [CrossRef]

33. Sheaffer, K.L.; Kim, R.; Aoki, R.; Elliott, E.N.; Schug, J.; Burger, L.; Schübeler, D.; Kaestner, K.H. DNA methylation is required for the control of stem cell differentiation in the small intestine. *Genes Dev.* **2014**, *28*, 652–664. [CrossRef]

34. Carlson, E.S.; Stead, J.D.; Neal, C.R.; Petryk, A.; Georgieff, M.K. Perinatal iron deficiency results in altered developmental expression of genes mediating energy metabolism and neuronal morphogenesis in hippocampus. *Hippocampus* **2007**, *17*, 679–691. [CrossRef]

35. Chilton, J.K. Molecular mechanisms of axon guidance. *Dev. Biol.* **2006**, *292*, 13–24. [CrossRef]

36. Dickson, B.J. Molecular mechanisms of axon guidance. *Science* **2002**, *298*, 1959–1964. [CrossRef]

37. Egea, J.; Klein, R. Bidirectional Eph-ephrin signaling during axon guidance. *Trends Cell. Biol.* **2007**, *17*, 230–238. [CrossRef]

38. Rohani, N.; Canty, L.; Luu, O.; Fagotto, F.; Winklbauer, R. EphrinB/EphB signaling controls embryonic germ layer separation by contact-induced cell detachment. *PLoS Biol.* **2011**, *9*, e1000597. [CrossRef]

39. Shamah, S.M.; Lin, M.Z.; Goldberg, J.L.; Estrach, S.; Sahin, M.; Hu, L.; Bazalakova, M.; Neve, R.L.; Corfas, G.; Debant, A.; et al. EphA receptors regulate growth cone dynamics through the novel guanine nucleotide exchange factor ephexin. *Cell* **2001**, *105*, 233–244. [CrossRef]

40. Rex, C.S.; Chen, L.Y.; Sharma, A.; Liu, J.; Babayan, A.H.; Gall, C.M.; Lynch, G. Different Rho GTPase-dependent signaling pathways initiate sequential steps in the consolidation of long-term potentiation. *J. Cell. Biol.* **2009**, *186*, 85–97. [CrossRef]

41. Malenka, R.C.; Nicoll, R.A. Long-term potentiation—A decade of progress? *Science* **1999**, *285*, 1870–1874. [CrossRef]

42. Nicoll, R.A. A Brief History of Long-Term Potentiation. *Neuron* **2017**, *93*, 281–290. [CrossRef]

43. Blattler, A.; Yao, L.; Witt, H.; Guo, Y.; Nicolet, C.M.; Berman, B.P.; Farnham, P.J. Global loss of DNA methylation uncovers intronic enhancers in genes showing expression changes. *Genome Biol.* **2014**, *15*, 469. [CrossRef]

44. Singer, M.; Kosti, I.; Pachter, L.; Mandel-Gutfreund, Y. A diverse epigenetic landscape at human exons with implication for expression. *Nucleic Acids Res.* **2015**, *43*, 3498–3508. [CrossRef]

45. Stephenson, D.T.; Coskran, T.M.; Kelly, M.P.; Kleiman, R.J.; Morton, D.; O'Neill, S.M.; Schmidt, C.J.; Weinberg, R.J.; Menniti, F.S. The distribution of phosphodiesterase 2A in the rat brain. *Neuroscience* **2012**, *226*, 145–155. [CrossRef]

46. Fernández-Fernández, D.; Rosenbrock, H.; Kroker, K.S. Inhibition of PDE2A, but not PDE9A, modulates presynaptic short-term plasticity measured by paired-pulse facilitation in the CA1 region of the hippocampus. *Synapse* **2015**, *69*, 484–496. [CrossRef]

47. Montague, P.; McCallion, A.S.; Davies, R.W.; Griffiths, I.R. Myelin-associated oligodendrocytic basic protein: A family of abundant CNS myelin proteins in search of a function. *Dev. Neurosci.* **2006**, *28*, 479–487. [CrossRef]

48. Schäfer, I.; Müller, C.; Luhmann, H.J.; White, R. MOBP levels are regulated by Fyn kinase and affect the morphological differentiation of oligodendrocytes. *J. Cell Sci.* **2016**, *129*, 930–942. [CrossRef]

49. Kuchler, K.; Daum, G.; Paltauf, F. Subcellular and submitochondrial localization of phospholipid-synthesizing enzymes in Saccharomyces cerevisiae. *J. Bacteriol.* **1986**, *165*, 901–910. [CrossRef]

50. Nowicki, M.; Müller, F.; Frentzen, M. Cardiolipin synthase of Arabidopsis thaliana. *FEBS Lett.* **2005**, *579*, 2161–2165. [CrossRef]

51. Qi, Y.; Kapterian, T.S.; Du, X.; Ma, Q.; Fei, W.; Zhang, Y.; Huang, X.; Dawes, I.W.; Yang, H. CDP-diacylglycerol synthases regulate the growth of lipid droplets and adipocyte development. *J. Lipid Res.* **2016**, *57*, 767–780. [CrossRef]

52. Bastian, T.W.; von Hohenberg, W.C.; Georgieff, M.K.; Lanier, L.M. Chronic Energy Depletion due to Iron Deficiency Impairs Dendritic Mitochondrial Motility during Hippocampal Neuron Development. *J. Neurosci.* **2019**, *39*, 802–813. [CrossRef]

53. Corapci, F.; Calatroni, A.; Kaciroti, N.; Jimenez, E.; Lozoff, B. Longitudinal evaluation of externalizing and internalizing behavior problems following iron deficiency in infancy. *J. Pediatr. Psychol.* **2010**, *35*, 296–305. [CrossRef]

54. Lukowski, A.F.; Koss, M.; Burden, M.J.; Jonides, J.; Nelson, C.A.; Kaciroti, N.; Jimenez, E.; Lozoff, B. Iron deficiency in infancy and neurocognitive functioning at 19 years: Evidence of long-term deficits in executive function and recognition memory. *Nutr. Neurosci.* **2010**, *13*, 54–70. [CrossRef]

55. Schmidt, R.J.; Tancredi, D.J.; Krakowiak, P.; Hansen, R.L.; Ozonoff, S. Maternal intake of supplemental iron and risk of autism spectrum disorder. *Am. J. Epidemiol.* **2014**, *180*, 890–900. [CrossRef]

56. Connor, J.R.; Menzies, S.L. Relationship of iron to oligodendrocytes and myelination. *Glia* **1996**, *17*, 83–93. [CrossRef]

57. Beard, J.L.; Wiesinger, J.A.; Connor, J.R. Pre- and postweaning iron deficiency alters myelination in Sprague-Dawley rats. *Dev. Neurosci.* **2003**, *25*, 308–315. [CrossRef]

58. Tran, P.V.; Fretham, S.J.; Carlson, E.S.; Georgieff, M.K. Long-term reduction of hippocampal brain-derived neurotrophic factor activity after fetal-neonatal iron deficiency in adult rats. *Pediatr. Res.* **2009**, *65*, 493–498. [CrossRef]

59. Thompson, R.F.; Fazzari, M.J.; Niu, H.; Barzilai, N.; Simmons, R.A.; Greally, J.M. Experimental intrauterine growth restriction induces alterations in DNA methylation and gene expression in pancreatic islets of rats. *J. Biol. Chem.* **2010**, *285*, 15111–15118. [CrossRef]

60. Xiao, Y.; Camarillo, C.; Ping, Y.; Arana, T.B.; Zhao, H.; Thompson, P.M.; Xu, C.; Su, B.B.; Fan, H.; Ordonez, J.; et al. The DNA methylome and transcriptome of different brain regions in schizophrenia and bipolar disorder. *PLoS ONE* **2014**, *9*, e95875. [CrossRef]

61. Li, M.; Gao, F.; Xia, Y.; Tang, Y.; Zhao, W.; Jin, C.; Luo, H.; Wang, J.; Li, Q.; Wang, Y. Filtrating colorectal cancer associated genes by integrated analyses of global DNA methylation and hydroxymethylation in cancer and normal tissue. *Sci. Rep.* **2016**, *6*, 31826. [CrossRef]

62. Radford, E.J.; Ito, M.; Shi, H.; Corish, J.A.; Yamazawa, K.; Isganaitis, E.; Seisenberger, S.; Hore, T.A.; Reik, W.; Erkek, S.; et al. In utero effects. In utero undernourishment perturbs the adult sperm methylome and intergenerational metabolism. *Science* **2014**, *345*, 1255903. [CrossRef]

63. Consortium, E.P. An integrated encyclopedia of DNA elements in the human genome. *Nature* **2012**, *489*, 57–74. [CrossRef]

64. Pennacchio, L.A.; Ahituv, N.; Moses, A.M.; Prabhakar, S.; Nobrega, M.A.; Shoukry, M.; Minovitsky, S.; Dubchak, I.; Holt, A.; Lewis, K.D.; et al. In vivo enhancer analysis of human conserved non-coding sequences. *Nature* **2006**, *444*, 499–502. [CrossRef]

65. Lev Maor, G.; Yearim, A.; Ast, G. The alternative role of DNA methylation in splicing regulation. *Trends Genet.* **2015**, *31*, 274–280. [CrossRef]

66. Maunakea, A.K.; Nagarajan, R.P.; Bilenky, M.; Ballinger, T.J.; D'Souza, C.; Fouse, S.D.; Johnson, B.E.; Hong, C.; Nielsen, C.; Zhao, Y.; et al. Conserved role of intragenic DNA methylation in regulating alternative promoters. *Nature* **2010**, *466*, 253–257. [CrossRef]

67. Oberdoerffer, S. A conserved role for intragenic DNA methylation in alternative pre-mRNA splicing. *Transcription* **2012**, *3*, 106–109. [CrossRef]

68. Gelfman, S.; Cohen, N.; Yearim, A.; Ast, G. DNA-methylation effect on cotranscriptional splicing is dependent on GC architecture of the exon-intron structure. *Genome Res.* **2013**, *23*, 789–799. [CrossRef]

69. Wan, J.; Oliver, V.F.; Zhu, H.; Zack, D.J.; Qian, J.; Merbs, S.L. Integrative analysis of tissue-specific methylation and alternative splicing identifies conserved transcription factor binding motifs. *Nucleic Acids Res.* **2013**, *41*, 8503–8514. [CrossRef]

70. Ishihara, K.; Oshimura, M.; Nakao, M. CTCF-dependent chromatin insulator is linked to epigenetic remodeling. *Mol. Cell.* **2006**, *23*, 733–742. [CrossRef]

71. Pikaart, M.J.; Recillas-Targa, F.; Felsenfeld, G. Loss of transcriptional activity of a transgene is accompanied by DNA methylation and histone deacetylation and is prevented by insulators. *Genes Dev.* **1998**, *12*, 2852–2862. [CrossRef]

72. Ming, G.L.; Song, H.J.; Berninger, B.; Holt, C.E.; Tessier-Lavigne, M.; Poo, M.M. cAMP-dependent growth cone guidance by netrin-1. *Neuron* **1997**, *19*, 1225–1235. [CrossRef]

73. Moore, S.W.; Kennedy, T.E. Protein kinase A regulates the sensitivity of spinal commissural axon turning to netrin-1 but does not switch between chemoattraction and chemorepulsion. *J. Neurosci.* **2006**, *26*, 2419–2423. [CrossRef]

74. Sweatt, J.D. Toward a molecular explanation for long-term potentiation. *Learn. Mem.* **1999**, *6*, 399–416. [CrossRef]

75. Thomas, M.J.; Moody, T.D.; Makhinson, M.; O'Dell, T.J. Activity-dependent beta-adrenergic modulation of low frequency stimulation induced LTP in the hippocampal CA1 region. *Neuron* **1996**, *17*, 475–482. [CrossRef]

76. Bito, H.; Furuyashiki, T.; Ishihara, H.; Shibasaki, Y.; Ohashi, K.; Mizuno, K.; Maekawa, M.; Ishizaki, T.; Narumiya, S. A critical role for a Rho-associated kinase, p160ROCK, in determining axon outgrowth in mammalian CNS neurons. *Neuron* **2000**, *26*, 431–441. [CrossRef]

77. Heasman, S.J.; Ridley, A.J. Mammalian Rho GTPases: New insights into their functions from in vivo studies. *Nat. Rev. Mol. Cell. Biol.* **2008**, *9*, 690–701. [CrossRef]

78. Lakatosova, S.; Ostatnikova, D. Reelin and its complex involvement in brain development and function. *Int. J. Biochem. Cell. Biol.* **2012**, *44*, 1501–1504. [CrossRef]

79. Weeber, E.J.; Beffert, U.; Jones, C.; Christian, J.M.; Forster, E.; Sweatt, J.D.; Herz, J. Reelin and ApoE receptors cooperate to enhance hippocampal synaptic plasticity and learning. *J. Biol. Chem.* **2002**, *277*, 39944–39952. [CrossRef]

80. D'Arcangelo, G. Apoer2: A reelin receptor to remember. *Neuron* **2005**, *47*, 471–473. [CrossRef]

81. Chen, Y.; Beffert, U.; Ertunc, M.; Tang, T.S.; Kavalali, E.T.; Bezprozvanny, I.; Herz, J. Reelin modulates NMDA receptor activity in cortical neurons. *J. Neurosci.* **2005**, *25*, 8209–8216. [CrossRef]

82. Fatemi, S.H.; Kroll, J.L.; Stary, J.M. Altered levels of Reelin and its isoforms in schizophrenia and mood disorders. *Neuroreport* **2001**, *12*, 3209–3215. [CrossRef]

83. Fatemi, S.H.; Stary, J.M.; Halt, A.R.; Realmuto, G.R. Dysregulation of Reelin and Bcl-2 proteins in autistic cerebellum. *J. Autism Dev. Disord.* **2001**, *31*, 529–535. [CrossRef]

84. Li, E.; Zhang, Y. DNA methylation in mammals. *Cold Spring Harb. Perspect. Biol.* **2014**, *6*, a019133. [CrossRef]

85. Zhao, B.; Yang, Y.; Wang, X.; Chong, Z.; Yin, R.; Song, S.H.; Zhao, C.; Li, C.; Huang, H.; Sun, B.F.; et al. Redox-active quinones induces genome-wide DNA methylation changes by an iron-mediated and Tet-dependent mechanism. *Nucleic Acids Res.* **2014**, *42*, 1593–1605. [CrossRef]

86. Freudenberg, J.M.; Ghosh, S.; Lackford, B.L.; Yellaboina, S.; Zheng, X.; Li, R.; Cuddapah, S.; Wade, P.A.; Hu, G.; Jothi, R. Acute depletion of Tet1-dependent 5-hydroxymethylcytosine levels impairs LIF/Stat3 signaling and results in loss of embryonic stem cell identity. *Nucleic Acids Res.* **2012**, *40*, 3364–3377. [CrossRef]

87. Zhu, X.; Girardo, D.; Govek, E.E.; John, K.; Mellén, M.; Tamayo, P.; Mesirov, J.P.; Hatten, M.E. Role of Tet1/3 Genes and Chromatin Remodeling Genes in Cerebellar Circuit Formation. *Neuron* **2016**, *89*, 100–112. [CrossRef]

88. Branco, M.R.; Ficz, G.; Reik, W. Uncovering the role of 5-hydroxymethylcytosine in the epigenome. *Nat. Rev. Genet.* **2011**, *13*, 7–13. [CrossRef]

89. Szulwach, K.E.; Li, X.; Li, Y.; Song, C.X.; Wu, H.; Dai, Q.; Irier, H.; Upadhyay, A.K.; Gearing, M.; Levey, A.I.; et al. 5-hmC-mediated epigenetic dynamics during postnatal neurodevelopment and aging. *Nat. Neurosci.* **2011**, *14*, 1607–1616. [CrossRef]

90. Hahn, M.A.; Qiu, R.; Wu, X.; Li, A.X.; Zhang, H.; Wang, J.; Jui, J.; Jin, S.G.; Jiang, Y.; Pfeifer, G.P.; et al. Dynamics of 5-hydroxymethylcytosine and chromatin marks in Mammalian neurogenesis. *Cell. Rep.* **2013**, *3*, 291–300. [CrossRef]

91. Al-Mahdawi, S.; Virmouni, S.A.; Pook, M.A. The emerging role of 5-hydroxymethylcytosine in neurodegenerative diseases. *Front. Neurosci.* **2014**, *8*, 397. [CrossRef]

nutrients

MDPI

Article

A Slow-Digesting Carbohydrate Diet during Rat Pregnancy Protects Offspring from Non-Alcoholic Fatty Liver Disease Risk through the Modulation of the Carbohydrate-Response Element and Sterol Regulatory Element Binding Proteins

Rafael Salto [1], Manuel Manzano [2], María Dolores Girón [1,*], Ainara Cano [3], Azucena Castro [3], José Dámaso Vílchez [1], Elena Cabrera [1] and José María López-Pedrosa [2]

[1] Department of Biochemistry and Molecular Biology II, School of Pharmacy, University of Granada, Campus de Cartuja, 18071 Granada, Spain; rsalto@ugr.es (R.S.); e.damaso@go.ugr.es (J.D.V.); elenacc_20@hotmail.com (E.C.)
[2] Abbott Nutrition R&D, Abbott Laboratories, 18004 Granada, Spain; manuel.manzano@abbott.com (M.M.); jose.m.lopez@abbott.com (J.M.L.-P.)
[3] OWL Metabolomics, Parque Tecnológico de Bizkaia, 48160 Deiro, Spain; acano@owlmetabolomics.com (A.C.); acastro@owlmetabolomics.com (A.C.)
* Correspondence: mgiron@ugr.es; Tel.: +34-958-246363

Received: 12 February 2019; Accepted: 11 April 2019; Published: 14 April 2019

Abstract: High-fat (HF) and rapid digestive (RD) carbohydrate diets during pregnancy promote excessive adipogenesis in offspring. This effect can be corrected by diets with similar glycemic loads, but low rates of carbohydrate digestion. However, the effects of these diets on metabolic programming in the livers of offspring, and the liver metabolism contributions to adipogenesis, remain to be addressed. In this study, pregnant insulin-resistant rats were fed high-fat diets with similar glycemic loads but different rates of carbohydrate digestion, High Fat-Rapid Digestive (HF–RD) diet or High Fat-Slow Digestive (HF–SD) diet. Offspring were fed a standard diet for 10 weeks, and the impact of these diets on the metabolic and signaling pathways involved in liver fat synthesis and storage of offspring were analyzed, including liver lipidomics, glycogen and carbohydrate and lipid metabolism key enzymes and signaling pathways. Livers from animals whose mothers were fed an HF–RD diet showed higher saturated triacylglycerol deposits with lower carbon numbers and double bond contents compared with the HF–SD group. Moreover, the HF–RD group exhibited enhanced glucose transporter 2, pyruvate kinase (PK), acetyl coenzyme A carboxylase (ACC) and fatty acid (FA) synthase expression, and a decrease in pyruvate carboxylase (PyC) expression leading to an altered liver lipid profile. These parameters were normalized in the HF–SD group. The changes in lipogenic enzyme expression were parallel to changes in AktPKB phosphorylation status and nuclear expression in carbohydrate-response element and sterol regulatory element binding proteins. In conclusion, an HF–RD diet during pregnancy translates to changes in liver signaling and metabolic pathways in offspring, enhancing liver lipid storage and synthesis, and therefore non-alcoholic fatty liver disease (NAFLD) risk. These changes can be corrected by feeding an HF–SD diet during pregnancy.

Keywords: early programming; hepatic lipogenesis; insulin-resistant pregnancy; metabolic flexibility; non-alcoholic fatty liver disease; slow digesting carbohydrates

1. Introduction

Humans have developed metabolic adaptations to promote energy storage in periods of fasting. They are especially suited to stimulating fat storage from other nutrients and carbohydrates as an evolutionary advantage for famine periods. However, alterations to this tightly regulated process have a relevant role in well-known pathological situations such as diabetes or obesity. It is also important during development, as catch-up growth in children after nourishment periods, or in the perinatal period, since it has been clearly established that a mother's nutrition effects the metabolic performance of offspring in adulthood.

Metabolic flexibility is one of the main parameters that is modified by the glycemic index (GI) of dietary carbohydrates. Metabolic flexibility is the capability of an organism to select fuel oxidation in function of the nutrient availability (or the prediction that the organism makes about the nutrient availability). A relevant situation that involves fuel selection, fat storage and metabolic flexibility takes place during pregnancy [1]. Undernourishment or high-fat (HF) diets during this period lead to severe alterations in the offspring, both in human and animal models. In rats, feeding high-fat diets during pre-mating and/or gestation causes weight gain and glucose intolerance in offspring, regardless the post-weaning diet [2,3]. Therefore, dietary alterations during pregnancy can have a strong impact on the metabolic flexibility of the offspring in adulthood.

Dietary carbohydrates are able to promote metabolic adaptations to facilitate fat storage in a coordinated way, involving short term regulatory processes mainly involving changes in hormone (insulin and glucagon) secretion and response, as well as changes in signaling pathways (for example phosphorylation and regulation of key enzymes in response to the hormone secretion). Furthermore, carbohydrates are able to promote long-term adaptations in metabolism, involving modifications of gene expression at the cell nucleus that over time sustain the channeling of glucose to fat [4,5].

In a previous work [6], we have shown that the offspring of pregnant rats fed a high-fat diet containing slow-digesting (SD) carbohydrates during gestation (HF–SD group) seemed to protect against an increase in adipose tissue mass during adolescence. The HF–SD animals had reduced body fat mass and lower levels of cholesterol and triacylglycerols (TAGs) in plasma when compared with their counterparts, who were exposed to high-fat and rapidly digestible carbohydrate diets (HF–RD group) during gestation.

In this work, we have analyzed the effects of HF–SD and HF–RD diets during pregnancy in the liver metabolism of offspring in adolescence. Our results indicate that in the HF–RD group, liver metabolism is modified to promote glucose uptake that is converted into TAGs, producing changes in the lipid profile that indicate an enhanced risk of non-alcoholic fatty liver disease (NAFLD) in the offspring. On the contrary, the offspring from the HF–SD group showed a decreased tendency of lipogenesis and lipid storage. The results highlight the relevance of liver metabolic programming during pregnancy to control body homeostasis of the offspring. The supplementation of diet during pregnancy with slow-digestive carbohydrates normalizes metabolic and signaling pathways promoting metabolic flexibility, and appears to exert a protective effect on the offspring.

2. Materials and Methods

2.1. Animal and Housing

Female and male Sprague Dawley rats (aged 10 weeks) were purchased from Charles River Laboratories (Wilmington, MA, USA). Animals were maintained on a 12-h light/12-h dark cycle at 23–24 °C with food and water available ad libitum. All experimental procedures were approved by the Animal Welfare Committee at Estación Experimental del Zaidín-CSIC (Granada, Spain) in accordance with guidelines and recommendations of Spanish legislations on animal welfare and the European Convention for the Protection of Vertebrate Animals used for Experimental and other Scientific Purposes.

2.2. Diets and Experimental Design

A detailed description of the experimental diets and design has been previously described [6]. In brief, a model of insulin resistance was used by feeding female rats with a highly palatable obesogenic lard diet (HF; 20.5% fat, 24.2% protein, 41.5% carbohydrates and 7.9% fiber per weight) prior to mating for six weeks. Control animals were fed a standard reference diet AIN93M. Rats were mated for three days and then randomly assigned to one of the experimental diets during gestation: a high-fat diet containing slow-digesting carbohydrates (HF–SD group), a high-fat diet containing rapid-digesting carbohydrates (HF–RD group) or an AIN93G diet for the reference group. Composition of the three diets used during gestation is described in Martin et al. [6] (Supplementary Table S1). All diets were prepared at Abbott Nutrition R&D facilities. At the delivery, all animals were fed AIN93G diets regardless of the diet consumed during pregnancy. On day 21 after delivery, all pups were weaned onto the AIN93M diet and housed for seven additional weeks (Figure 1). At the outcome, insulin and glucagon serum concentrations were assayed (Supplementary Table S2).

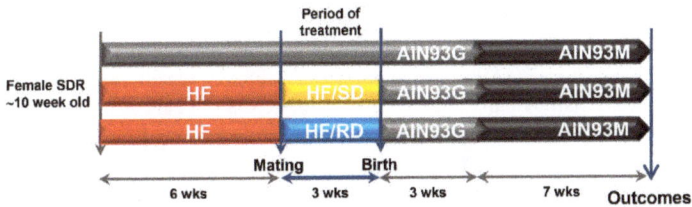

Figure 1. Experimental model. Virgin rats were assigned to one of three experimental groups: reference dams fed a standard rodent diet before mating and throughout pregnancy; dams fed a high-fat diet (HF) six weeks before mating and then an HF diet containing either carbohydrates with a high (HF–RD) or low (HF–SD) digestion rate throughout pregnancy. At delivery, all the animals were fed the standard rodent diet for the remainder of the study (10 weeks). SDR, Sprague-Dawley rats.

2.3. Liver Lipidomics Analysis

Absolute concentration of TAGs in the liver was measured using a triglycerides-LQ kit (Spinreact, Barcelona, Spain). Metabolite profiles were analyzed as previously described [7,8]. Briefly, two separate UHPLC-time-of-flight (TOF)–MS-based platforms (Agilent Technologies, Santa Clara, CA, USA) analyzing methanol and chloroform–methanol liver extracts were combined to semi-quantify lipid species. Non-esterified fatty acyls, bile acids and lysoglycerophospholipids were analyzed in the methanol extract platform. For this, methanol was added to the frozen liver tissue (30:1, *v/w*), and this mixture was homogenized with a Precellys 24 grinder (Precellys, Montigny-le-Bretonneux, France), followed by protein precipitation. The methanol used for extraction was spiked with internal standards not found in liver tissue using the same method. After brief vortex mixing, samples were incubated overnight at −20 °C. Supernatants were collected and dried after centrifugation at 16,000× *g* for 15 min. The dried extracts were resuspended in methanol, centrifuged at 16,000× *g* for 5 min and supernatants were collected and transferred to vials for UHPLC–MS (Agilent Technologies, Santa Clara, CA, USA) analysis.

The chloroform–methanol extract platform provided coverage over glycerolipids, glycerophospholipids, sterol lipids and sphingolipids. Liver tissues were homogenized in the Precellys 24 grinder by mixing with chloroform–methanol (2:1, *v/v*) and sodium chloride (50 mmol/L) (overall ratio 1:30:3, *w/v/v*), followed by protein precipitation. The extraction solvent was spiked with internal standards not detected using the same method. After brief vortex mixing, samples were incubated at −20 °C for 1 h. After centrifugation at 16,000× *g* for 15 min, the lower organic phase was collected and the solvent removed. The dried extracts were then resuspended in acetronitrile–isopropanol (1:1), centrifuged at 16,000× *g* for 5 min and, finally, supernatants were transferred to vials for UHPLC–MS analysis. Lipid nomenclature follows the LIPID MAPS convention, www.lipidmaps.org.

This combined analysis was established for rodent liver tissue by OWL Metabolomics (Derio, Spain). Data obtained with the UHPLC–MS were processed with the TargetLynx application manager for MassLynx 4.1 (Waters Corp., Milford, MA, USA). Intra- and inter-batch normalization followed the procedure published by Martinez-Herranz et al. [9]. All the calculations were performed with R v3.2.0 (R Development Core Team, Vienna, Austria, 2010).

2.4. Western Blot Analysis

Liver lysates were obtained in lysis buffer (RIPA buffer containing protease and phosphatase inhibitors). The lysate was centrifuged at 16,000× g at 4 °C for 15 min. The supernatant was transferred to a new microcentrifuge tube (1.5 mL) and sonicated for 15 s (cycle 0.5, amplitude 60%). Nuclear extracts from liver samples were prepared as described [10].

Protein concentration was determined using the bicinchoninic acid method, and 20–60 μg were loaded for western blot. The following antibodies were used: carbohydrate-responsive element-binding protein (ChREBP), glucose transporter 2 (GLUT2), sterol regulatory element-binding protein (SREBP) SREBP1 and SREBP2 from Santa Cruz Biotechnology (Santa Cruz, CA, USA). Acetyl Coenzyme A carboxylase (ACC), phosphoenolpyruvate carboxykinase (PEPCK), total and phospho-(Ser473)-PKBAkt, AMP-activated protein kinase α2 (AMPKα2) and phospho-AMPKα2 (Thr172), ERK1/2 and phospho-p44/42 Erk1/2 (Thr202/Tyr204), mammalian target of rapamycin (mTOR) and phospho-mTOR (Ser2448) were provided by Cell Signaling (Beverly, MA, USA). Pyruvate carboxylase (PyC) antibody was raised at Salto's laboratory. GAPDH (Sigma-Aldrich, Saint Louis, MO, USA) was used as a load control. Data were normalized using the values of the reference animals as 100%.

2.5. Glycogen Content

Hepatic glycogen was isolated as described by Chang and Exon [11]. Liver homogenates (10%) were made in 30 mmol/L HCl and spread evenly on pieces of filter paper (Whatman 3M chromatography paper, 2.0 × 2.0 cm) in duplicate. The papers were dropped immediately into a beaker containing 66% ethanol and stirred gently by a rotating magnet screened from the papers by a wire mesh. The papers were subsequently washed three times for 40 min in 66% ethanol. Next, they were briefly rinsed with acetone and dried under a stream of warm air. The dried filter papers were cut into four pieces and placed in a tube containing 0.4 mL of 0.2 mol/L acetate buffer, pH 4.8; 0.2 mg of amylo-α-1,4-a-1,6-glucosidase; and H_2O, to a final volume of 2 mL. The vials were incubated for 90 min at 37 °C with gentle shaking. Appropriate controls were prepared by incubating aliquots of homogenate in acetate buffer minus amyloglucosidase. Glucose concentration in the incubated samples was determined by the glucose oxidase method.

2.6. Statistical Analyses

One-way ANOVA test was applied when comparing three groups, and a Student's t test was used when comparing two groups. Homoscedasticity was checked by Barlett's test, and non-parametric tests were applied when appropriate. Differences were considered significant at $p < 0.05$.

3. Results

In this article, we have addressed the question of whether the changes owing to a high-fat/rapid-digestive carbohydrates diet during pregnancy produces metabolic adaptations in the liver of the offspring. This question is relevant, since the liver plays a central role in the metabolic flexibility of the organism and, therefore, is the key organ with the capacity to distribute nutrients to the rest of the organs and tissues. A second relevant question to be addressed was to determine if a slow-digesting carbohydrate diet during pregnancy could prevent and normalize the adverse effects of a high-fat diet on this organ in offspring.

For this purpose, liver lipidome was first analyzed to depict long-term metabolic changes induced by the different diets during pregnancy on adolescent offspring (Figure 2), and 309 lipid metabolites were individually semi-quantified. Data were calibrated with quality controls, and changes in relative abundance of the liver lipidomes were searched, comparing the HF–RD and HF–SD experimental groups (Supplementary Table S3). Results are plotted as a heat map in Figure 2a. The liver lipid profile of the HF–RD animals was significantly different to those in the HF–SD, with significant differences ($p < 0.05$) in 99 lipid metabolites. The largest differences found were detected among TAGs; adolescent HF–RD rats had higher amounts of TAGs and diglycerides (DAGs) in their livers compared with the HF–SD animals. Besides, in the HF–RD group, TAGs had fatty acids (FAs) with lower carbon numbers and double-bond content than in the HF–SD animals, showing a characteristic pattern (Figure 3b).

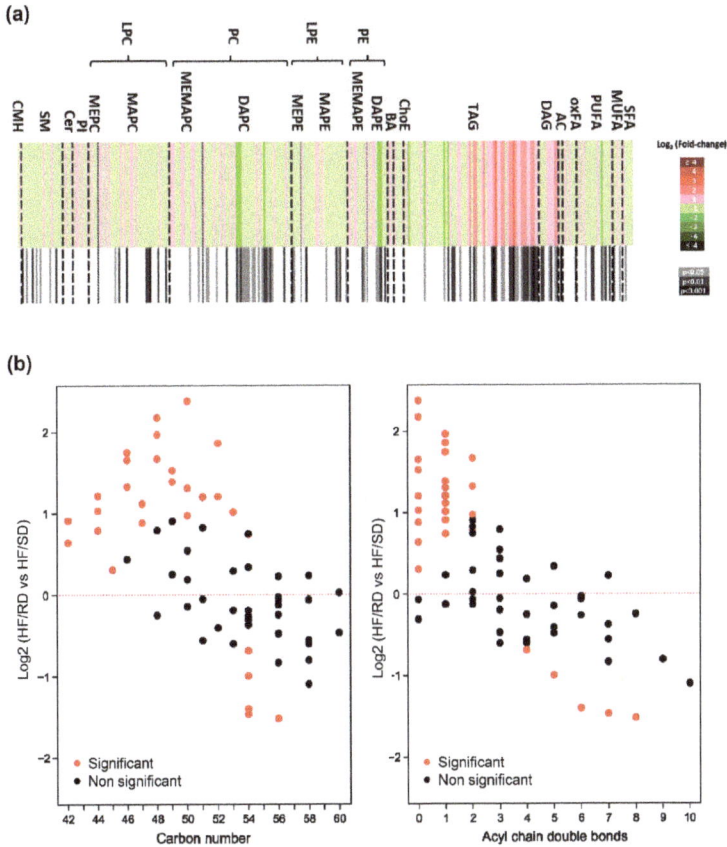

Figure 2. Rat liver lipidome. (**a**) Heat map showing relevant changes in liver lipids. Each metabolite is shown as a line whose color is defined by the sign and magnitude of the change. Adjacent column to each comparison shows the results of the *t* test (*p* value). Color scales are shown in the right: upper scale, Log2 fold-change; lower scale, statistical significance (*n* HF–SD = 7, *n* HF–RD = 8). BA, bile acids; Cer, ceramides; CMH, monohexosyl ceramide; ChoE, cholesterol esters; DAGs, diglycerides; DAPC, diacylglycerophosphocholine; DAPE, diacylglycerophosphoethanolamine; oxFA, oxidized free fatty acids; LPCs, lysophosphatidylcholines; LPEs, lysophosphatidyletanolamines; MEPC, 1-Monoetherglycerophosphocholine; MAPC, MonoacylglyceroPhosphocholine; MEMAPC, 1-ether, 2-acylglycerophosphocholine; PCs, phosphatidylcholines; PEs, phosphatidyletanolamines; PIs, phosphatidylinositols; MUFAs, monounsaturated fatty acids; PUFAs, polyunsaturated fatty acids; SFAs, saturated fatty acids; SMs, sphingomyelins; TAGs, triacylglycerols. (**b**) Correlation of HF–RD vs. HF–SD fold-change with TAG carbon number (left), and with acyl chain double bonds (right).

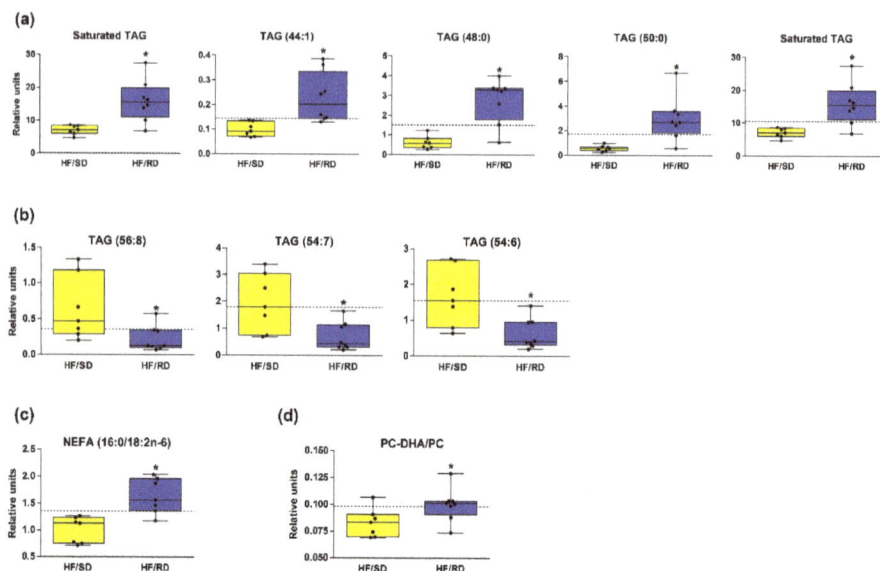

Figure 3. Relevant lipid species and ratios. (**a**) TAGs enriched in saturated fatty acids. (**b**) TAGs enriched in polyunsaturated fatty acids. (**c**) Lipogenic index. (**d**) Ratio Docosahexaenoyl Phosphatidylcholine/Phosphatidylcholine (PC–DHA/PC). The bar plots show the normalized values. The boxes range from 25%–75% percentiles; the 5% and 95% percentiles are indicated as error bars; single-data points are indicated by dots. Medians are indicated by horizontal lines within each box. Lines in all graphs show the reference group mean value. * $p < 0.05$ compared to the HF–SD group. NEFA, Non-esterified Fatyy Acids.

As observed in Figure 2b, TAGs showed a remarkable profile in the livers of the HF–RD vs. HF–SD rats. TAG concentrations in the livers of the HF–RD presented a marked, gradual increase in the concentration of TAGs with shorter acyl chains and less unsaturation when compared with the HF–SD. The concentration of saturated or monounsaturated TAGs increased in the HF–RD when compared with the HF–SD, since the length of their carbon acyl chains was generally medium (16 carbons, palmitic acid). Some individual TAG species in the HF–RD were much higher than in the HF–SD, with TAGs (48:0) (TAGs (16:0/16:0/16:0)) and TAGs (50:0) (TAGs (16:0/16:0/18:0)) being the ones that changed the most (fold change 4.53 and 5.22, respectively) (Figure 3a). On the contrary, the concentration of long carbon acyl chain polyunsaturated TAGs decreased in the HF–RD when compared with the HF–SD (Figure 3b). The length of the carbon acyl chains in the HF–SD was 18 carbons or longer (Figure 3b). The HF–SD rats showed lower levels of TAG species enriched in saturated fatty acids (SFAs) than the reference group, and equal or higher levels of unsaturated TAG species than the same group.

A hallmark of NAFLD is the accumulation of liver TAGs, driven by increased palmitate and decreased polyunsaturated fatty acids (PUFAs) [7]. The ratio of FA (16:0/18:2n-6), called the lipogenic index [12], increased 63% in the HF–RD vs. HF–SD rats (Figure 3c).

The heat map also revealed a significant decrease in some phosphatidylcholine (PC), especially diacyl-PC, and phosphatidylethanolamine (PE) species in the HF–RD compared to the HF–SD group. Phosphatidylcholines (PCs) formed via phosphatidylethanolamine (PE) N-Methyltransferase (PEMT) pathway are primarily enriched in long-chain polyunsaturated fatty acids (PUFAs), such as docosahexaenoic acid (22:6n-3). Thus, the increased PC (22:6n-3) to total PC ratio (20%, $p \leq 0.05$) found in HF–RD rats when compared with HF–SD rats (Figure 3d) suggests decreased PEMT flux in their livers, since previous studies indicated a correlation between the PC–DHA/PC ratio and the estimation of PEMT activity [8,13].

Next, we tried to elucidate if the changes observed in the lipid profile of the offspring could be related to modifications in glucose transporter expression and carbohydrate (Figure 4) and lipid metabolism (Figure 5) key enzymes. For that purpose, glucose transporter (GLUT2) was measured in the liver tissue of adolescent rats. Western blots of glyceraldehyde 3-phosphate dehydrogenase (GAPDH) were used as a load control. Our results showed that the HF–RD group had a significantly higher expression of GLUT2 compared to HF–SD rats, whose GLUT2 expression was similar to the reference group.

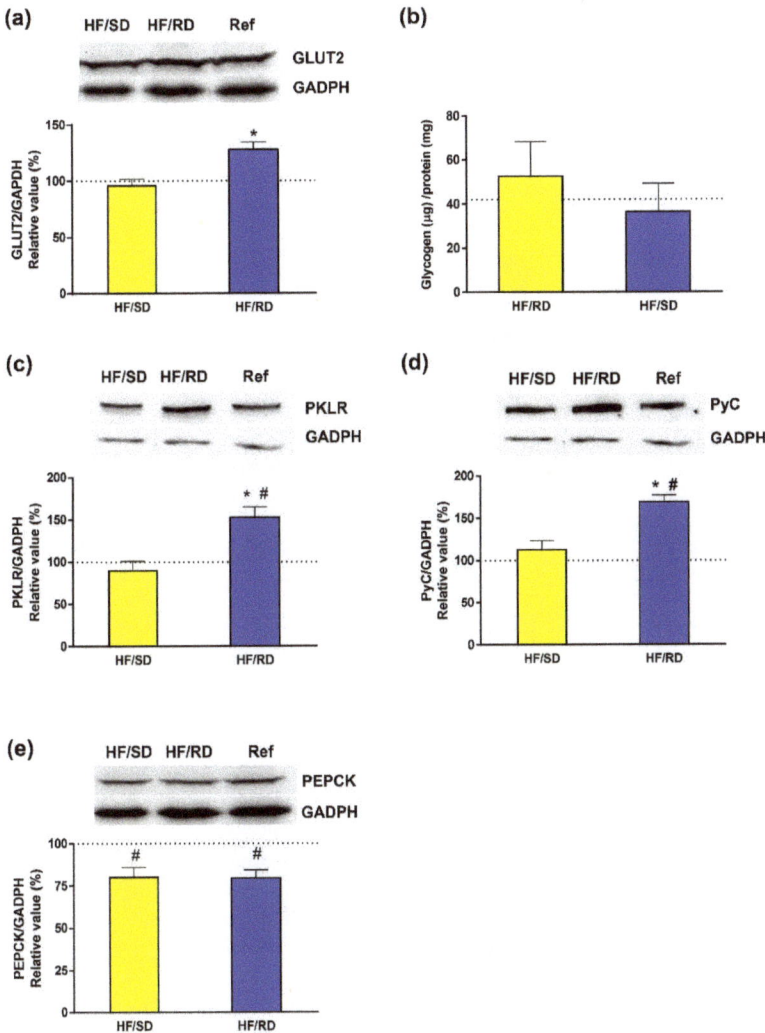

Figure 4. Liver carbohydrate metabolism. The expression of relevant transporters and enzymes has been analyzed by western blot; liver glycogen content has been assayed enzymatically. (**a**) GLUT2 transporter. (**b**) Glycogen content. (**c**) Liver pyruvate kinase isoenzyme (PKLR). (**d**) Pyruvate carboxylase (PyC). (**e**) Phosphoenol pyruvate carboxykinase (PEPCK). Results have been referred to reference rats (100% value). Results are mean ± SEM ($n = 6$ animals per group). * $p < 0.05$ compared to the HF–SD group; # $p < 0.05$ compared to the reference group.

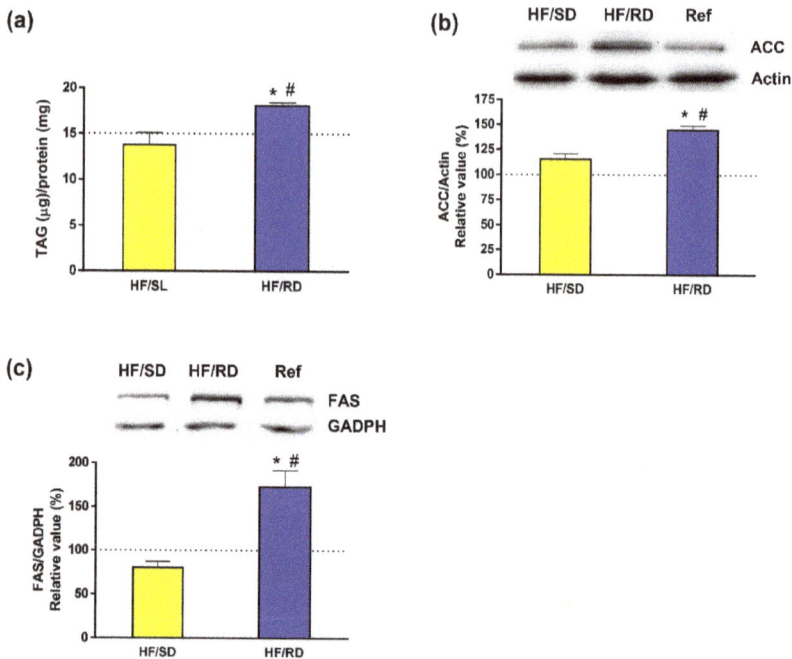

Figure 5. Liver lipogenesis. The expression of key enzymes has been analyzed by western blot, and liver triglyceride content has been assayed enzymatically. (**a**) Triacylglicerols (TAG) content. (**b**) Acetyl Coenzyme A carboxylase (AAC). (**c**) Fatty acid synthase (FAS). Results have been referred to control rats (100% value). Results are mean ± SEM (n = 6 animals per group). * $p < 0.05$ compared to the HF–SD group; # $p < 0.05$ compared to the reference group.

To address the fate of glucose in the liver of the HF–RD group, glycogen content and liver pyruvate kinase (PK) isoenzyme were determined as indicators of glycolytic flux. Glycogen content was similar in all experimental groups. In addition, glucokinase and glycogen phosphorylase activities were enzymatically assayed (Supplementary Figure S1) and no significant differences were found among experimental groups. Therefore, in the HF–RD group, the enhanced glucose uptake was not directed to glycogen, and we targeted a key enzyme of the glycolytic pathway, PK. Our results show that the expression of the PK main liver isoform (PKLR) was significantly higher in the HF–RD group with respect to the HF–SD and reference groups, having similar levels of both.

In addition, liver gluconeogenesis was analyzed in offspring by measuring the expression of pyruvate carboxylase (PyC) and phosphoenolpyruvate carboxykinase (PEPCK) (Figure 4d,e). Our results showed that the levels of PEPCK expression were similar between the two HF groups, and significantly lower compared with the reference group. More interestingly, expression of PyC in liver lysates was significantly higher in the HF–RD compared with the HF–SD and reference groups.

Next, the involvement of dietary intervention during pregnancy was analyzed in liver lipogenesis. The enzymatic measurement of total TAG content in the liver from the different experimental groups (Figure 5a) indicated that TAGs were being accumulated in the liver of the HF–RD when compared to the other two offspring groups. Furthermore, the expression of acetyl-CoA carboxylase (ACC) and fatty acid synthase (FAS), two key proteins involved in the metabolism of lipids in the liver, were measured. Both the levels of ACC (Figure 5b) and FAS (Figure 5c) were significantly higher in the HF–RD than in the HF–SD and reference groups.

To study the effects of dietary carbohydrates as regulators of signaling pathways involved in glucidic and lipid homeostasis was our next task. Therefore, we studied the phosphorylation status of kinases that are essential for the regulation of these pathways (Figure 6). We analyzed the activation of PKBAkt (Figure 6a) as a marker of insulin responsiveness in the liver [4]. Our results indicate that in the HF–SD group there was a significant increase in the phosphorylation of the kinase compared with the HF–RD and reference groups.

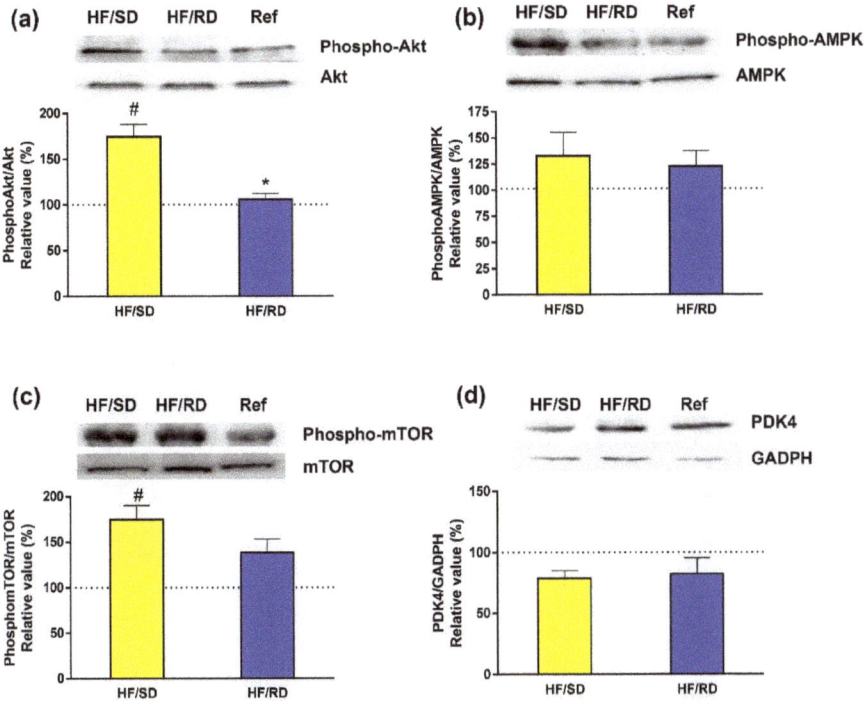

Figure 6. Signaling pathways involved in the regulation of liver metabolism. The phosphorylation of key kinases has been analyzed by western blot. (**a**) PKB–Akt, (**b**) AMPK, (**c**) mTOR and (**d**) PDK4. Results have been referred to control rats (100% value). Results are mean ± SEM (n = 6 animals per group). * $p < 0.05$ compared to the HF–SD group; # $p < 0.05$ compared to the reference group.

Next, we assayed the activation of AMPK and mTOR, since they are involved in the control of cell growth and development [14,15]. While AMPK showed similar phosphorylation levels in all groups (Figure 6b), the phosphorylation of mTOR was higher in the HF–SD group compared with the other two groups, reaching significance with respect to the reference group.

Pyruvate dehydrogenase complex (PDC) constitutes one of the main decision points in the competition between fatty acids and glucose for oxidation [1]. The expression of PDK4 as the main isoform widely distributed of PDK was also measured, revealing that there were similar levels among groups.

Finally, the expression in cell lysates and nuclear extracts of carbohydrate-response element-binding protein (ChREBP), which is regulated by dietary carbohydrates [16], was measured. Our results (Figure 7) indicate that, while the total amount of ChREBP protein was found to be similar in all experimental groups (Figure 7a), there was a significant increase in the nuclear concentration of the transcription factor in the HF–RD group (Figure 7b). Since the sterol regulatory binding protein 1 and 2 (SREBP1 and SREBP2) transcription factors act in a coordinated way with ChREBP to regulate transcription of lipogenic genes, the expression of these transcription factors was assayed in nuclear extracts from the different experimental groups (Figure 7c,d). The results indicate that SREBP1 and SREBP2 nuclear expression was significantly enhanced in the HF–RD group compared with the HF–SD group, while there was a decrease in the HF–SD group compared with the reference group.

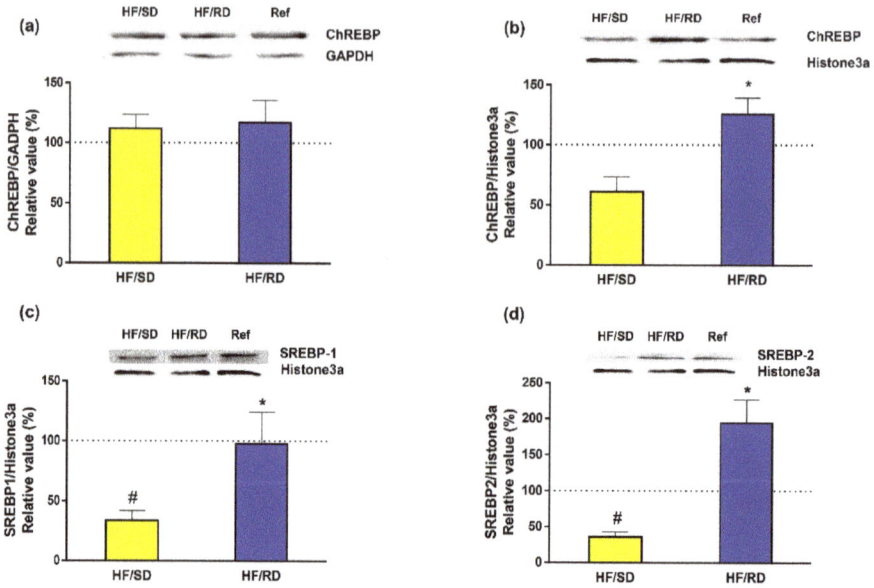

Figure 7. Carbohydrate-response element-binding protein (ChREBP) and sterol regulatory binding protein 1 and 2 (SREBP1 and SREBP2) expression in the liver. Total cellular (**a**) and nuclear (**b**) ChREBP content was analyzed by western blot. Nuclear expression of SREBP1 (**c**) and SREBP2 (**d**) was measured by western blot. Results were referred to rats (100% value). Results are mean ± SEM ($n = 6$ animals per group). * $p < 0.05$ compared to the HF–SD group; # $p < 0.05$ compared to reference group. Glyceraldehyde-3-phosphate Dehydrogenase, GADPH.

4. Discussion

Avoiding pathologies such as obesity, diabetes, metabolic syndrome derived from the use of highly processed meals and sedentary life are objectives of the WHO. These diseases are associated with NAFLD, which is the commonest cause of chronic liver disease not only in adults but also in children, particularly in Western countries. NAFLD is a clinical term that refers to excess hepatic fat (5%–10% by weight) accumulation in the absence of excessive alcohol consumption [17]. Therefore, the promotion of healthy diets rich in fiber and slow-digesting carbohydrates could be of importance in the prevention of pathologies later in life [18].

Studies in animal models have been used to mimic human physiology, and the cafeteria diet (high-fat content and rapid-digesting carbohydrates) constitutes an appropriate model to test the importance of dietary intervention in the development of obesity, adiposity, metabolic syndrome and the associated NAFLD [19].

Dietary intervention has a direct influence on normal physiological function, but also in different pathological situations as metabolic syndrome, diabetes or cardiovascular disease [20]. Additionally, there are evidences that adverse situations during early development could have a direct impact in disease risk in later periods of life [21]. In fact, this is consistent with the theory of the developmental origins of health and disease (DOHaD) that environmental factors, including nutritional status, during the stages of development through the fetal and neonatal period are related to the risk of non-communicable diseases, such as metabolic syndrome, in later life [22].

One of the key points for dietary intervention is the control of glycaemia during pregnancy, especially among obese women since there is a well-established relationship between insulin resistance and adverse consequences for both mother and offspring. Studies are beginning to use different mixtures of carbohydrates in which rapid-digesting carbohydrates (sacharose) are replaced with low glycemic index (GI) carbohydrates as a rational approach [23].

It has been well established that a high-fat diet rich in rapid-digesting carbohydrates has deep effects on liver carbohydrate and lipid metabolism. However, the knowledge of molecular bases of how these changes can be translated to offspring is scarce. Some studies have reported that high-fat diets during the period of pre-mating or gestation induce glucose intolerance in pups, and this change is independent of the post-weaning diet [2,3]. Other studies have described that when high-fiber and high-protein diets are consumed during pregnancy and lactation, there is a regulation of satiety hormones and genes involved in glucose metabolism in offspring [5]. Furthermore, our previous results indicate that a high-fat, rapid-digestive carbohydrate diet during pregnancy promotes an increase in adipose tissue mass and a concomitant alteration of carbohydrate and lipid metabolism in this tissue in the offspring [6]. More interesting, these adipose tissue alterations can be corrected when slow digestive ones substitute the rapid-digesting carbohydrates.

In an individual, a high-fat diet containing rapid-digestive carbohydrates is able to increase liver glucose uptake, glycolysis and alter glycogen metabolism. At the same time, it inhibits gluconeogenesis. Therefore, under these circumstances the liver is a net glucose consumer, rather than a glucose exporter. If we pay attention to lipid metabolism, a diet enriched in rapid-digesting carbohydrates blocks the use of fatty acids to obtain energy and enhances the conversion of glucose to TAGs. Consequently, these diets promote a hepatic lipogenic program that has local and systemic consequences, such as an increase in blood TAGs and cholesterol, enhanced TAG transport to adipose tissue and, even worse, hepatic steatosis [17].

The question that this article addresses is to confirm if these alterations leading to NAFLD are also mimicked in the offspring of animals fed these diets during pregnancy [24], and whether a slow-digestive carbohydrate supplementation to pregnant mothers has any protective effect later on the offspring to prevent this risk.

In the HF–SD diet, the sugars (sucrose and maltodextrin) as rapidly metabolized carbohydrates were substituted for a mixture of slow-digestive carbohydrates composed of isomaltulose, resistant maltodextrins and fructooligosaccharides [6]. This dietary intervention during pregnancy has been proven to prevent excess adipogenesis in offspring by modulating adipose tissue metabolism. More important is to determine if these positive effects on adipose tissue are also present in the livers of the HF–SD offspring.

The liver has a key role on metabolism homeostasis, and it has been demonstrated that maternal obesity imposes a developmental programming on offspring, affecting mainly the liver, and pre-disposing to obesity and NAFLD [17,25].

Our results confirm that a high-fat diet containing rapid-digestive carbohydrates during pregnancy promotes a predisposition to develop NAFLD in the livers of offspring. These results are in agreement with previous studies [24,25].

The liver lipid profile of the HF–RD animals was significantly different from the HF–SD rats, mainly among TAGs. Remarkably, TAG and DAG concentrations in the livers of the HF–RD adolescent rats increased, with the acyl chains in TAGs being shorter and less unsaturated than in the HF–SD (Figure 3b). It should be highlighted that some individual TAG species in the HF–RD were greatly higher than those of the other groups (Figure 3a). Interestingly, Petry et al. [26] suggest that fetal imprinted genes may influence maternal circulating clinically relevant TAG concentrations early in pregnancy. Concretely, TAGs (44:1), which increased 2.32-fold in the livers of the HF–RD when compared with the HF–SD rats in our study (Figure 3a), were found to be very strongly associated to maternal and paternal transmissions and linked to insulin resistance.

Additionally, the concentration of polyunsaturated TAGs with long carbon acyl chains decreased in the HF–RD when compared with the HF–SD (Figure 3b), with the length of their carbon acyl chains being 18 carbons or longer (Figure 3b). Rhee et al. [4] found that lipids, especially serum TAGs of lower carbon numbers and double-bond content, were associated with an increased risk of diabetes, whereas lipids of higher carbon numbers and double-bond content were associated with a decreased risk.

There are data in the lipid profile that point out the possibility NAFLD developing in the HF–RD offspring. The more important results related with this hypothesis mean an increase in liver TAGs and lipogenic index combined with a significant decrease in PCs, and are considered as a hallmark of this pathology [27]. The decrease in PCs may affect the PC–DHA levels that are incorporated in very-low-density lipoproteins (VLDLs) secreted by the liver and transported to peripheral tissues. PCs are required for the assembly/secretion of lipoproteins [28,29]. Formerly, reduced PC levels would be expected to account for at least some of the TAG accumulation in the liver of HF–RD rats due to the reduced secretion of VLDLs from the liver [30].

In conclusion, our lipidomic study indicates that offspring from animals fed an HF–RD diet during pregnancy showed a lipid profile compatible with the early stages of liver steatosis. On the contrary, the supplementation of high-fat diets with slow-digestive carbohydrates during pregnancy had a protective effect on offspring, decreasing the risk of NAFLD. However, no liver histological changes were observed in the different experimental groups (data not shown), probably due to the short period of time (13 weeks after delivery) before the analysis.

The observed changes in the lipidome with a clear effect of the HF–RD diet increasing the TAG content in liver could be explained by changes in the flux from glucose to lipids. Firstly, liver carbohydrate metabolism was studied by measuring the expression of GLUT2 transporter and the PK liver isoenzyme. Our results showed that the expression of GLUT2 was significantly higher in the HF–RD group with respect to the HF–SD and reference groups. GLUT2 is the main glucose and fructose transporter in the liver, which allows large bidirectional fluxes of glucose in and out the cell due to its low affinity and high capacity (high Vmax and Km for glucose) [31]. Therefore, increased expression of GLUT2 in the offspring from the HF–RD group indicates a permanent metabolic adaptation that, by the increase in liver glucose uptake, resembles a situation of hampered insulin response [32]. The increase in GLUT2 did not translated into a higher storage of glycogen in the HF–RD group. This result indicates that, in our experimental setting, dietary treatment during pregnancy did not alter liver glycogen storage and metabolism.

Pyruvate kinase is a glycolytic enzyme that catalyzes the last step, converting phosphoenolpyruvate to pyruvate. Pyruvate kinase exists in a number of isoforms, the liver type (PKLR) being the most abundant in that organ [33]. The combined higher expression of GLUT2 and PKLR in the HF–RD group points to increased glycolytic flux in these animals compared with the reference and HF–SD groups.

The increase in glycolytic flux due to high glycemic index diets is usually combined with an inhibition of liver gluconeogenesis. The decrease of PEPCK in both groups of rats with respect to the reference group supports the inhibition of the gluconeogenic pathway. More interesting, liver PyC was significantly higher in the HF–RD compared with the HF–SD and control groups. This is a remarkable finding since PyC, in addition to its role in gluconeogenesis, is responsible for the anaplerotic replenishment of the Krebs cycle. This replenishment is needed in processes where Krebs cycle intermediaries are used for synthetic purposes, as in lipogenesis from glucose. Therefore, these results are a first insight that in the HF–RD group there was a permanent adaptation to promote lipid synthesis as the ultimate fate of glucose in liver. On the contrary, this channeling was corrected in the HF–SD group.

Up to this point, our liver results, lipid data and higher PyC expression put forward an active conversion of glucose to TAGs in the HF–RD group. Moreover, these animals had higher circulating TAGs and an increase in adipose mass [6]. Liver TAG content was higher in the HF–RD than in the other two groups, in agreement with the results on lipidomics, and corroborating the enhanced conversion of glucose into TAGs in this group. ACC and FAS that catalyze de novo synthesis of fatty acids in liver were significantly higher in the HF–RD than in the HF–SD and reference groups. ACC catalyzes the carboxylation of acetyl-CoA to Malonyl-CoA, the first committed step in long-chain fatty acid synthesis [34]. Later, these fatty acids will be incorporated into TAGs and/or phospholipids. Its expression is regulated at several levels, synthesis of the protein being one of them [35].

In the liver, FAS has long been categorized as a housekeeping protein, producing fat for the storage of energy when nutrients are present in excess. Fatty acids are packed into VLDL particles and delivered to adipose tissue and other extrahepatic tissues through the bloodstream [36]. Therefore, the overexpression of FAS in the HF–RD group, combined with reduced PC levels, explains the higher circulating TAGs observed in these animals [6] together with the increased deposits of TAGs on the liver. On the contrary, the overall analysis of the HF–SD group indicates that the carbohydrate composition of this diet mediates a metabolic state in which a lower use of glucose by glycolysis is combined with a decrease in the synthesis of fatty acids and TAGs.

In addition to regulating the expression of a variety of proteins, dietary carbohydrates also modulate signaling pathways involved in glucidic and lipidic homeostasis as Akt-, AMPK- or mTOR-dependent transduction routes. Insulin stimulates hepatic lipogenesis after feeding through an activation of the PI3-kinase–Akt pathway. Moreover, insulin fuels glycolysis, thus increasing the availability of lipogenic precursors [37,38]. However, high-fat diets are clearly associated with insulin resistance. The HF–SD diet significantly increased the phosphorylation of PKB–Akt, which was measured as a marker of insulin responsiveness in the liver compared with the HF–RD and reference groups. This result could point to better insulin sensitivity in the HF–SD group, which would be in agreement with the data from the postprandial glycemic peak and total glycemic response that were lower in HF–SD rats when compared to the HF–RD [6].

AMPK and mTOR are the main kinases responsible for integrating signals that control growth and development [14,15]. In a favorable energy status, mTOR is active, but when there is a fuel deficiency, activation of AMPK inhibits mTOR signaling [15]. Thus, the availability of nutrients leads to an activation of TOR protein kinase activity, while a depletion of amino acids or glucose points to an attenuation of mTOR phosphorylation [39]. AMPK, a sensor of fuel availability [40], showed similar phosphorylation levels in all groups. On the contrary, the phosphorylation of mTOR was higher in the HF–SD group compared with the other two groups. The regulation of mTOR is controlled by upstream signaling pathways, such as the insulin-dependent signaling pathway, and measured by the phosphorylation status of Akt–PKB and AMPK phosphorylation as a sensor of fuel availability [40]. Therefore, it could be concluded from these results that the higher phosphorylation of mTOR in the HF–SD rats was driven by the activation of the insulin pathway, rather than changes in AMPK activity.

Metabolic flexibility is the capability of a system to regulate fuel (primarily glucose and fatty acids) oxidation in response to nutrient availability. Thus, the possibility to change a substrate catabolism nutritional state depends on the steadiness between oxidation and storage capacities. PDC represents one of the key regulating points in the decision of the cell to use fatty acids or glucose as a fuel. This complex is usually active in tissues in the fed state, but the inhibition of its activity by pyruvate dehydrogenase kinase (PDK) is critical to maintaining energy homeostasis under nutritional conditions [1,41]. PDK isozymes phosphorylate specific serine residues in PDC [42]. Of all the known isozymes, PDK4 is highly expressed in the liver [1,41]. At the same time, PDK4 also controls glycolytic flux, since inactivation of PDC by PDKs blocks acetyl-CoA synthesis from pyruvate, resulting in a shift of pyruvate to the TCA cycle. Since PDK4 expression was similar among groups, it might demonstrate that the effects of the dietary intervention during pregnancy do not translate in a short-term regulation of metabolic flexibility in the liver, where PDK4 is relevant, but maybe involve long-term metabolic regulation of the metabolism at the gene-expression level.

The long-term regulation of metabolism induced by high GI diets is mediated by specific transcription factors that can bind to DNA and mediate the transcription of genes that code for the lipogenic enzymes. A relevant transcription factor is ChREBP, carbohydrate-response element-binding protein, which is activated by glucose-6-phosphate. ChREBP includes in its structure several phosphorylation sites and a glucose-sensing domain. In a low-GI diet or fasting, phosphorylation of ChREBP mediated by cyclic AMP keeps the transcription factor inactive at the cytosol. On the contrary, a high-GI diet increases xylulose-5-phosphate levels, which act as potent activators of the phosphatase A2. In the non-phosphorylated state, ChREBP migrates to the nucleus, and glucose-6-phosphate binds to the glucose-sensing domain. In this state, ChREBP enhances the transcription of the genes involved in the conversion of glucose to fat [16].

Our results indicate that, although the total amount of ChREBP protein (Figure 7) was similar in all experimental groups (Figure 7a), its nuclear concentration was increased in the HF–RD group. This finding could be due to an increase in xylulose-5-phosphate levels in these animals in response to the enhanced glucose uptake mediated by GLUT2. This event would activate phosphatase A2 and nuclear translocation of ChREBP. Furthermore, the increased nuclear levels of ChREBP in the HF–RD group are sufficient to explain the long-term modifications in the levels of key enzymes and transporters leading to a metabolic inflexibility that channels liver glucose to lipid storage, since GLUT2, PKLR, ACC and FAS expression is regulated at the transcription level by this transcription factor.

Transcription of the lipogenic genes is driven in a coordinated way with ChREBP by SREBP1 and SREBP2 transcription factors. These transcription factors have a complex regulation that involves not only a proteolytic activation, but also a regulation by carbohydrates [43]. Our results indicate that there is a synergic regulation through dietary intervention in the liver's nuclear amount of SREBPs. Both transcriptional factors were incremented in the HF–RD group compared with the HF–SD group, and the levels of SRBEPs in the HF–SD group were also significantly lower compared with the reference group.

5. Conclusions

The results presented in this article reinforce the importance of maternal nutrition during the early critical period of development, and suggest that offspring exposed to a maternal HF diet with rapid-digestible carbohydrates have a predisposition to develop NAFLD in later life. In contrast, these negative effects are ameliorated by a maternal HF diet with slow-digestible carbohydrates, probably through an increase in the sensitivity of the insulin signaling pathway, the modulation of key enzymes and transporters and the normalization of the lipid profile. This demonstrates the role carbohydrates play in a maternal diet on the developmental programming of metabolic flexibility in offspring, as well as the relevance of the liver in these adaptive processes.

Supplementary Materials: The following are available online at http://www.mdpi.com/2072-6643/11/4/844/s1, Figure S1: (**a**) Glucokinase activity was measured in liver homogenates as described (Massa, L.; Baltrusch, S.; Okar, D.A.; Lange, A.J.; Lenzen, S.; Tiedge, M., Interaction of 6-phosphofructo-2-kinase/fructose-2, 6-bisphosphatase (PFK-2/FBPase-2) with glucokinase activates glucose phosphorylation and glucose metabolism in insulin-producing cells. Diabetes 2004, 53, 1020–1029). (**b**) Glycogen Phosphorylase b activity was measured spectrophotometrically as described in Board, M.; Hadwen, M.; Johnson, L. N., Effects of novel analogues of D-glucose on glycogen phosphorylase activities in crude extracts of liver and skeletal muscle. Eur. J. Biochem. 1995, 228, 753–761. Results are mean ± SEM (*n* = 6 animals per group). No significant differences were found among experimental groups; Table S1: Experimental diets during rat pregnancy; Table S2: Glucose, insulin and glucagon levels in fasted serum samples from the different experimental groups (*n* = 8); Table S3: Heat map showing relevant changes in liver lipids.

Author Contributions: Conceptualization, M.M. and J.M.L.-P.; Investigation, R.S., M.D.G., A.C. (Ainara Cano), A.C. (Azucena Castro), J.D.V. and E.C.; Writing—original draft, R.S. and M.D.G.; Writing—review & editing, R.S., M.M., M.D.G., A.C. (Ainara Cano), A.C. (Azucena Castro) and J.M.L.-P.

Funding: This research was funded by the European Union's Seventh Framework Programme (FP7/2007–2013): project EarlyNutrition, under grant agreement no. 289346.

Conflicts of Interest: M.M. and J.M.L.-P. have no relevant interests to declare other than their affiliation with Abbott Laboratories. Abbott produces infant formulas and other pediatric and adult nutritional products. The rest of the authors declare no conflict of interest.

References

1. Zhang, S.; Hulver, M.W.; McMillan, R.P.; Cline, M.A.; Gilbert, E.R. The pivotal role of pyruvate dehydrogenase kinases in metabolic flexibility. *Nutr. Metab.* **2014**, *11*, 10. [CrossRef] [PubMed]
2. Nivoit, P.; Morens, C.; Van Assche, F.A.; Jansen, E.; Poston, L.; Remacle, C.; Reusens, B. Established diet-induced obesity in female rats leads to offspring hyperphagia, adiposity and insulin resistance. *Diabetologia* **2009**, *52*, 1133–1142. [CrossRef] [PubMed]
3. Chen, H.; Simar, D.; Morris, M.J. Hypothalamic neuroendocrine circuitry is programmed by maternal obesity: Interaction with postnatal nutritional environment. *PLoS ONE* **2009**, *4*, e6259. [CrossRef] [PubMed]
4. Rhee, E.P.; Cheng, S.; Larson, M.G.; Walford, G.A.; Lewis, G.D.; McCabe, E.; Yang, E.; Farrell, L.; Fox, C.S.; O'Donnell, C.J.; et al. Lipid profiling identifies a triacylglycerol signature of insulin resistance and improves diabetes prediction in humans. *J. Clin. Investig.* **2011**, *121*, 1402–1411. [CrossRef] [PubMed]
5. Maurer, A.D.; Reimer, R.A. Maternal consumption of high-prebiotic fibre or -protein diets during pregnancy and lactation differentially influences satiety hormones and expression of genes involved in glucose and lipid metabolism in offspring in rats. *Br. J. Nutr.* **2011**, *105*, 329–338. [CrossRef]
6. Martin, M.J.; Manzano, M.; Bueno-Vargas, P.; Rueda, R.; Salto, R.; Giron, M.D.; Vilchez, J.D.; Cabrera, E.; Cano, A.; Castro, A.; et al. Feeding a slowly digestible carbohydrate diet during pregnancy of insulin-resistant rats prevents the excess of adipogenesis in their offspring. *J. Nutr. Biochem.* **2018**, *61*, 183–196. [CrossRef]
7. Barr, J.; Caballeria, J.; Martinez-Arranz, I.; Dominguez-Diez, A.; Alonso, C.; Muntane, J.; Perez-Cormenzana, M.; Garcia-Monzon, C.; Mayo, R.; Martin-Duce, A.; et al. Obesity-dependent metabolic signatures associated with nonalcoholic fatty liver disease progression. *J. Proteome Res.* **2012**, *11*, 2521–2532. [CrossRef] [PubMed]
8. Martinez-Una, M.; Varela-Rey, M.; Cano, A.; Fernandez-Ares, L.; Beraza, N.; Aurrekoetxea, I.; Martinez-Arranz, I.; Garcia-Rodriguez, J.L.; Buque, X.; Mestre, D.; et al. Excess S-adenosylmethionine reroutes phosphatidylethanolamine towards phosphatidylcholine and triglyceride synthesis. *Hepatology* **2013**, *58*, 1296–1305. [CrossRef] [PubMed]
9. Martinez-Arranz, I.; Mayo, R.; Perez-Cormenzana, M.; Minchole, I.; Salazar, L.; Alonso, C.; Mato, J.M. Enhancing metabolomics research through data mining. *J. Proteom.* **2015**, *127 Pt B*, 275–288. [CrossRef]
10. Giron, M.D.; Sevillano, N.; Vargas, A.M.; Dominguez, J.; Guinovart, J.J.; Salto, R. The glucose-lowering agent sodium tungstate increases the levels and translocation of GLUT4 in L6 myotubes through a mechanism associated with ERK1/2 and MEF2D. *Diabetologia* **2008**, *51*, 1285–1295. [CrossRef] [PubMed]
11. Chan, T.M.; Exton, J.H. A rapid method for the determination of glycogen content and radioactivity in small quantities of tissue or isolated hepatocytes. *Anal. Biochem.* **1976**, *71*, 96–105. [CrossRef]
12. Hudgins, L.C.; Hellerstein, M.; Seidman, C.; Neese, R.; Diakun, J.; Hirsch, J. Human fatty acid synthesis is stimulated by a eucaloric low fat, high carbohydrate diet. *J. Clin. Investig.* **1996**, *97*, 2081–2091. [CrossRef] [PubMed]
13. Barbier-Torres, L.; Delgado, T.C.; Garcia-Rodriguez, J.L.; Zubiete-Franco, I.; Fernandez-Ramos, D.; Buque, X.; Cano, A.; Gutierrez-de Juan, V.; Fernandez-Dominguez, I.; Lopitz-Otsoa, F.; et al. Stabilization of LKB1 and Akt by neddylation regulates energy metabolism in liver cancer. *Oncotarget* **2015**, *6*, 2509. [CrossRef]
14. Howell, J.J.; Manning, B.D. mTOR couples cellular nutrient sensing to organismal metabolic homeostasis. *Trends Endocrinol. Metab.* **2011**, *22*, 94–102. [CrossRef] [PubMed]
15. Xu, J.; Ji, J.; Yan, X.H. Cross-talk between AMPK and mTOR in regulating energy balance. *Crit. Rev. Food Sci. Nutr.* **2012**, *52*, 373–381. [CrossRef]
16. Uyeda, K.; Repa, J.J. Carbohydrate response element binding protein, ChREBP, a transcription factor coupling hepatic glucose utilization and lipid synthesis. *Cell Metab.* **2006**, *4*, 107–110. [CrossRef]
17. Brumbaugh, D.E.; Friedman, J.E. Developmental origins of nonalcoholic fatty liver disease. *Pediatr. Res.* **2014**, *75*, 140–147. [CrossRef] [PubMed]
18. Popkin, B.M. Global nutrition dynamics: The world is shifting rapidly toward a diet linked with noncommunicable diseases. *Am. J. Clin. Nutr.* **2006**, *84*, 289–298. [CrossRef] [PubMed]
19. Sampey, B.P.; Vanhoose, A.M.; Winfield, H.M.; Freemerman, A.J.; Muehlbauer, M.J.; Fueger, P.T.; Newgard, C.B.; Makowski, L. Cafeteria diet is a robust model of human metabolic syndrome with liver and adipose inflammation: Comparison to high-fat diet. *Obesity* **2011**, *19*, 1109–1117. [CrossRef]

20. Joy, T.; Lahiry, P.; Pollex, R.L.; Hegele, R.A. Genetics of metabolic syndrome. *Curr. Diabetes Rep.* **2008**, *8*, 141–148. [CrossRef]

21. Godfrey, K.M.; Lillycrop, K.A.; Burdge, G.C.; Gluckman, P.D.; Hanson, M.A. Epigenetic mechanisms and the mismatch concept of the developmental origins of health and disease. *Pediatr. Res.* **2007**, *61 Pt 2*, 5R–10R. [CrossRef]

22. Heindel, J.J.; Vandenberg, L.N. Developmental origins of health and disease: A paradigm for understanding disease cause and prevention. *Curr. Opin. Pediatr.* **2015**, *27*, 248–253. [CrossRef] [PubMed]

23. Clapp, J.F., 3rd. Maternal carbohydrate intake and pregnancy outcome. *Proc. Nutr. Soc.* **2002**, *61*, 45–50. [CrossRef] [PubMed]

24. Pereira, T.J.; Fonseca, M.A.; Campbell, K.E.; Moyce, B.L.; Cole, L.K.; Hatch, G.M.; Doucette, C.A.; Klein, J.; Aliani, M.; Dolinsky, V.W. Maternal obesity characterized by gestational diabetes increases the susceptibility of rat offspring to hepatic steatosis via a disrupted liver metabolome. *J. Physiol.* **2015**, *593*, 3181–3197. [CrossRef] [PubMed]

25. Oben, J.A.; Mouralidarane, A.; Samuelsson, A.M.; Matthews, P.J.; Morgan, M.L.; McKee, C.; Soeda, J.; Fernandez-Twinn, D.S.; Martin-Gronert, M.S.; Ozanne, S.E.; et al. Maternal obesity during pregnancy and lactation programs the development of offspring non-alcoholic fatty liver disease in mice. *J. Hepatol.* **2010**, *52*, 913–920. [CrossRef] [PubMed]

26. Petry, C.J.; Koulman, A.; Lu, L.; Jenkins, B.; Furse, S.; Prentice, P.; Matthews, L.; Hughes, I.A.; Acerini, C.L.; Ong, K.K.; et al. Associations between the maternal circulating lipid profile in pregnancy and fetal imprinted gene alleles: A cohort study. *Reprod. Biol. Endocrinol.* **2018**, *16*, 82. [CrossRef] [PubMed]

27. Puri, P.; Baillie, R.A.; Wiest, M.M.; Mirshahi, F.; Choudhury, J.; Cheung, O.; Sargeant, C.; Contos, M.J.; Sanyal, A.J. A lipidomic analysis of nonalcoholic fatty liver disease. *Hepatology* **2007**, *46*, 1081–1090. [CrossRef]

28. West, A.A.; Yan, J.; Jiang, X.; Perry, C.A.; Innis, S.M.; Caudill, M.A. Choline intake influences phosphatidylcholine DHA enrichment in nonpregnant women but not in pregnant women in the third trimester. *Am. J. Clin. Nutr.* **2013**, *97*, 718–727. [CrossRef]

29. Sherriff, J.L.; O'Sullivan, T.A.; Properzi, C.; Oddo, J.L.; Adams, L.A. Choline, Its Potential Role in Nonalcoholic Fatty Liver Disease, and the Case for Human and Bacterial Genes. *Adv. Nutr.* **2016**, *7*, 5–13. [CrossRef]

30. Cano, A.; Alonso, C. Deciphering non-alcoholic fatty liver disease through metabolomics. *Biochem. Soc. Trans.* **2014**, *42*, 1447–1452. [CrossRef]

31. Leturque, A.; Brot-Laroche, E.; Le Gall, M. GLUT2 mutations, translocation, and receptor function in diet sugar managing. *Am. J. Physiol. Endocrinol. Metab.* **2009**, *296*, E985–E992. [CrossRef]

32. Marks, J.; Carvou, N.J.; Debnam, E.S.; Srai, S.K.; Unwin, R.J. Diabetes increases facilitative glucose uptake and GLUT2 expression at the rat proximal tubule brush border membrane. *J. Physiol.* **2003**, *553 Pt 1*, 137–145. [CrossRef]

33. Han, S.M.; Namkoong, C.; Jang, P.G.; Park, I.S.; Hong, S.W.; Katakami, H.; Chun, S.; Kim, S.W.; Park, J.Y.; Lee, K.U.; et al. Hypothalamic AMP-activated protein kinase mediates counter-regulatory responses to hypoglycaemia in rats. *Diabetologia* **2005**, *48*, 2170–2178. [CrossRef]

34. Tong, L. Acetyl-coenzyme A carboxylase: Crucial metabolic enzyme and attractive target for drug discovery. *Cell. Mol. Life Sci.* **2005**, *62*, 1784–1803. [CrossRef]

35. Brownsey, R.W.; Boone, A.N.; Elliott, J.E.; Kulpa, J.E.; Lee, W.M. Regulation of acetyl-CoA carboxylase. *Biochem. Soc. Trans.* **2006**, *34 Pt 2*, 223–227. [CrossRef]

36. Eissing, L.; Scherer, T.; Todter, K.; Knippschild, U.; Greve, J.W.; Buurman, W.A.; Pinnschmidt, H.O.; Rensen, S.S.; Wolf, A.M.; Bartelt, A.; et al. De novo lipogenesis in human fat and liver is linked to ChREBP-beta and metabolic health. *Nat. Commun.* **2013**, *4*, 1528. [CrossRef]

37. Jensen-Urstad, A.P.; Semenkovich, C.F. Fatty acid synthase and liver triglyceride metabolism: Housekeeper or messenger? *Biochim. Biophys. Acta* **2012**, *1821*, 747–753. [CrossRef]

38. Rui, L. Energy metabolism in the liver. *Compr. Physiol.* **2014**, *4*, 177–197. [CrossRef]

39. Sengupta, S.; Peterson, T.R.; Sabatini, D.M. Regulation of the mTOR complex 1 pathway by nutrients, growth factors, and stress. *Mol. Cell* **2010**, *40*, 310–322. [CrossRef]

40. Huang, J.; Manning, B.D. The TSC1-TSC2 complex: A molecular switchboard controlling cell growth. *Biochem. J.* **2008**, *412*, 179–190. [CrossRef]

41. Parillo, M.; Licenziati, M.R.; Vacca, M.; de Marco, D.; Iannuzzi, A. Metabolic changes after a hypocaloric, low-glycemic-index diet in obese children. *J. Endocrinol. Investig.* **2012**, *35*, 629–633. [CrossRef]

Nutrients **2019**, *11*, 844

42. Sugden, M.C.; Holness, M.J. Mechanisms underlying regulation of the expression and activities of the mammalian pyruvate dehydrogenase kinases. *Arch. Physiol. Biochem.* **2006**, *112*, 139–149. [CrossRef] [PubMed]

43. Osborne, T.F. Sterol regulatory element-binding proteins (SREBPs): Key regulators of nutritional homeostasis and insulin action. *J. Biol. Chem.* **2000**, *275*, 32379–32382. [CrossRef] [PubMed]

nutrients

MDPI

Article

Sialic Acid and Sialylated Oligosaccharide Supplementation during Lactation Improves Learning and Memory in Rats

Elena Oliveros [1,2,*], Enrique Vázquez [1], Alejandro Barranco [1], María Ramírez [1], Agnes Gruart [3], Jose María Delgado-García [3], Rachael Buck [4], Ricardo Rueda [1] and María J. Martín [1]

[1] R&D Abbott Nutrition, 18004 Granada, Spain; enrique.vazquez@abbott.com (E.V.);
    alejandro.barranco@abbott.com (A.B.); maria.ramirez@abbott.com (M.R.); ricardo.rueda@abbott.com (R.R.);
    maria.j.martin@abbott.com (M.J.M.)
[2] Doctoral programme in Biomedicine, School of Health Sciences, University of Granada,
    18071 Granada, Spain
[3] Division of Neurosciences, Pablo de Olavide University, 41013 Seville, Spain; agrumas@upo.es (A.G.);
    jmdelgar@upo.es (J.M.D.-G.)
[4] R&D Abbott Nutrition, Columbus, OH 43219, USA; rachael.buck@abbott.com
*   Correspondence: elena.oliveros@abbott.com; Tel.: +34-958-249-822

Received: 3 September 2018; Accepted: 11 October 2018; Published: 16 October 2018

Abstract: Sialic acids (Sia) are postulated to improve cognitive abilities. This study evaluated Sia effects on rat behavior when administered in a free form as N-acetylneuraminic acid (Neu5Ac) or conjugated as 6'-sialyllactose (6'-SL). Rat milk contains Sia, which peaks at Postnatal Day 9 and drops to a minimum by Day 15. To bypass this Sia peak, a cohort of foster mothers was used to raise the experimental pups. A group of pups received a daily oral supplementation of Neu5Ac to mimic the amount naturally present in rat milk, and another group received the same molar amount of Sia as 6'-SL. The control group received water. After weaning, rats were submitted to behavioral evaluation. One year later, behavior was re-evaluated, and in vivo long-term potentiation (LTP) was performed. Brain samples were collected and analyzed at both ages. Adult rats who received Sia performed significantly better in the behavioral assessment and showed an enhanced LTP compared to controls. Within Sia groups, 6'-SL rats showed better scores in some cognitive outcomes compared to Neu5Ac rats. At weaning, an effect on polysialylated-neural cell adhesion molecule (PSA-NCAM) levels in the frontal cortex was only observed in 6'-SL fed rats. Providing Sia during lactation, especially as 6'-SL, improves memory and LTP in adult rats.

Keywords: milk oligosaccharides; infant formula; 6'-sialyllactose; cognitive development; sialic acid

---

1. Introduction

Breast feeding is associated with multiple health benefits. Infants fed human milk have lower rates of intestinal and autoimmune diseases and food allergies and show higher intellectual quotient (IQ) scores compared to formula-fed infants [1]. The unique composition of human milk plays a key role in optimal development of newborns, and sialic acids (Sia) are considered as one of the components responsible for the multiple benefits. They are present in most mammalian milks, mainly conjugated to proteins and oligosaccharides, but also as free monosaccharides. Within this last group of monosaccharides, there are some key core structures such as N-acetylneuraminic acid (Neu5Ac), N-glycolylneuraminic acid (Neu5Gc) and deaminated neuraminic acid (Kdn), which encompass the so-called Sia molecules [2]. Neu5Ac is the only form of Sia synthesized by humans [2] and is present throughout the body, including brain [3].

Nutrients **2018**, *10*, 1519

Most of the Sia in human milk, approximately 73% [4], are bound to oligosaccharides, generating sialyloligosaccharides, which represent a significant fraction of human milk oligosaccharides (HMOs). Oligosaccharides in human milk are at a higher concentration than in other mammalian milks, ranging from 13.3–23 g/L in colostrum to 3.5–14 g/L in mature milk [4,5], with structural diversity and over 200 different molecules identified so far [6]. Conversely, bovine milk, which is the basis of most infant formula, has an oligosaccharide content over 100-fold [7] lower than human milk. Although goat milk-based infant formula is now available, its use is not as extensive as bovine milk-based infant formulas. Interestingly, goat milk oligosaccharides have significant structural similarities to human milk oligosaccharides [8].

Milk oligosaccharides from any mammal consist of acidic oligosaccharides, predominantly sialylated ones, and neutral oligosaccharides, predominantly fucosylated compounds. There is a significant difference in the distribution of these compounds between human and bovine milk [9]. Sialylated structures represent 70% of bovine milk oligosaccharides compared with 10–20% in human milk [8], while fucosylated glycans are the most abundant form in human milk [10]. Within sialylated compounds, a recent study showed 6′-sialyllactose (6′-SL) is the major representative during the first two months, when it starts to sharply decrease so that 3′-sialyllactose becomes the predominant sialyl-oligosaccharide of human milk beyond the fourth month [11].

Milk oligosaccharides have been linked to different biological functions in the newborn, such as anti-infective activity [12,13], immune system and gastrointestinal development [14,15], modulation of inflammation [16], bifidogenic activity [17], modulation of gut motility and activation of enteric neurons [18] and even enhancement of central nervous system (CNS) functions [19]. Sia has been described as an essential nutrient in the development of infant brain [20], since its highest concentration in milk during early lactation concurs with a rapid increase of brain gangliosides [21]. Furthermore, the content of gangliosides and protein-bound sialic acid is higher in the brain of breast-fed infants in comparison to formula-fed infants [22]. Sia is a key component of brain gangliosides and polysialic acid (polySia) attached to the oligosaccharide chains on neural cell adhesion molecule (NCAM); both brain gangliosides and polysialylated-NCAM (PSA-NCAM) are crucial to ensure synaptic connections and memory formation or neuronal outgrowth [3,20].

HMOs and, by extension, sialyloligosaccharides have been reported as important factors for optimal development and maturation of the immune system [23]. HMOs are also prebiotics favoring the growth of healthy gut microbiota and as pathogen decoys, inhibiting their adhesion to the intestinal mucosa [24–26].

Several studies performed both in rodents [27] and pigs [28] have shown a relationship between Sia and cognitive function. The mechanism of action for the role of Sia on cognitive development has not been proven to date. However, recent research has shown the effects of HMOs on the gut-brain axis (GBA) in rodents [29,30], which compliments their role as prebiotics. The gut-brain axis consists of a bidirectional interaction between the central and the enteric nervous system [31]. The influences of gut microbiota on human health and disease is an important topic in medicine that is believed to occur via gut-brain microbe communication [32]. A recent study demonstrated associations between gut microbiota and cognition in infants [33].

The first two years of life are a critical window for brain development, not only for rapid proliferation and growth of neural cells, but also for increases in synapse connections [34]. Thus, adequate early life nutrition is crucial to ensure optimal cognitive development in infants and children.

The present study was aimed at determining if early Sia supplementation, as a free or conjugated form, plays a role during neural development in rats and whether effects could be detected later in life. For that purpose, rat pups were orally supplemented with Neu5Ac or 6′-SL during the lactation period and were evaluated using behavioral tests and electrophysiological measurements in both young and adult rats.

2. Materials and Methods

2.1. Animals

Pregnant Sprague-Dawley rats (*n* = 47) in the second week of gestation were purchased from Charles River (Charles River, France) and kept under controlled environmental conditions of temperature (22 °C ± 2), humidity (55% ± 10) and lighting (12 h light/dark cycles). Rats were fed a standard diet ad libitum. Animal experimental protocols were approved by the Ethics Committee of the Estación Experimental del Zaidín-CSIC (Consejo Superior de Investigaciones Científicas, Granada, Spain. Approval number: CBA EEZ-2011/20), and the experiment was performed in accordance with the Spanish and European regulations for the care and use of experimental animals for research.

2.2. Experimental Design

Rodent milk contains sialylated oligosaccharides, in particular 3'-sialyllactose (3'-SL) and 6'-SL [35]. In rat milk, total Sia levels peak at Postnatal Day 9 (PN9), sharply drop until Day 15 and then slightly decrease until the end of lactation [36]. To bypass this peak of Sia, two cohorts of pregnant rats were used with a time lag on the delivery day of 13 days, as shown in the experimental design scheme (see Scheme 1). Animals were maintained on a chow diet for gestation, lactation and growth (2018 Teklad Global 18% Protein Diet) until delivery. The first cohort (*n* = 20), hereafter called foster mothers, was kept with its respective offspring until PN16. Experimental pups were born from the second cohort of mothers (*n* = 27) and remained with them until PN3. On that day, 3-day-old pups were removed from their dams, sexed and randomly distributed into 10 pups (5 males and 5 females) groups with balanced weights and litters. Subsequently, they substituted for the 16 day-old pups in foster mother cages. These dams willingly accepted the new litters and took care of them for the rest of the lactation period. Afterwards, foster mothers and their litters were then distributed into 3 experimental groups of 4 dams each.

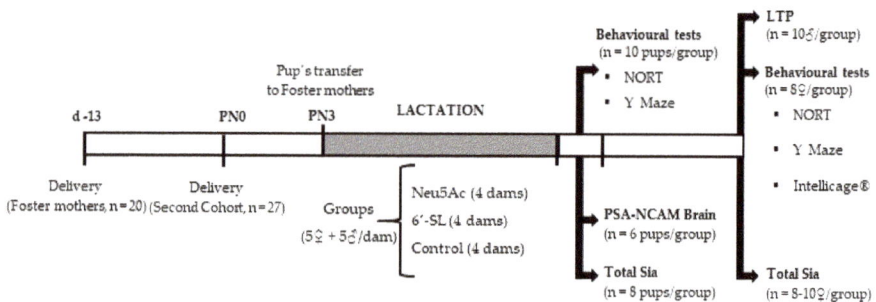

Scheme 1. Experimental design scheme. PN, postnatal day; Neu5Ac, N-acetylneuraminic acid; 6'-SL, 6'-sialyllactose; NORT, Novel Object Recognition Test; PSA-NCAM, polysialylated-neural cell adhesion molecule; LTP, long-term potentiation; Sia, sialic acids; ♀, female; ♂, male.

Since the content of sialic acid in rat milk reaches the lowest values from PN16, all pups raised by foster mothers received less Sia from milk than normally raised rats during early development. To compensate for the deficiency of Sia in milk, one group of pups received an oral supplementation of Neu5Ac from PN3 to weaning, and another group received the same molar amount of Sia given as 6'-SL. The control pups were given water and, therefore, received approximately 5-times less Sia than normally raised rats. In contrast, the 2 experimental groups received the same Sia levels naturally present in rat milk, but from two different sources. Therefore, the model assessed adequate Sia intake during lactation and the influence of the form in which it was provided.

Pups remained undisturbed with the foster rats except for a brief time while they received supplementation (8 times/day; max. 5 min/time). After weaning, all mothers and 2 pups per litter were sacrificed by an intraperitoneal overdose of anesthetic. Brains obtained from the pups were separated into hemispheres. One hemisphere was analyzed to determine Sia content at weaning by high performance liquid chromatography (HPLC), and the other one was used to determine NCAM and PSA-NCAM expression in the frontal cortex by Western blotting. Another set of weaned rats was submitted to classical behavioral tests such as NORT (Novel Object Recognition Test) and Y maze tests. Pups were then maintained for 1 year on a standard chow diet. At that age, males were used for LTP measurements; meanwhile, females were subjected again to psychological tests to evaluate long-term effects on learning and memory. Rats were then sacrificed, and brains were analyzed by HPLC to determine Sia content at adulthood.

2.3. Sialic Acid Doses

Rat milk contains both forms of free Sia, Neu5Ac and Neu5Gc; since Neu5Ac is the most abundant in humans [3] and is present throughout the body including brain, this was the form chosen as a free Sia source to be used in the supplementation.

Rat milk concentration of sialic acid varies through the lactation period having a Sia peak ($\approx$8 mg/mL) at PN9 and dropping to lower levels by Postnatal Day 15 [35] (Figure 1a).

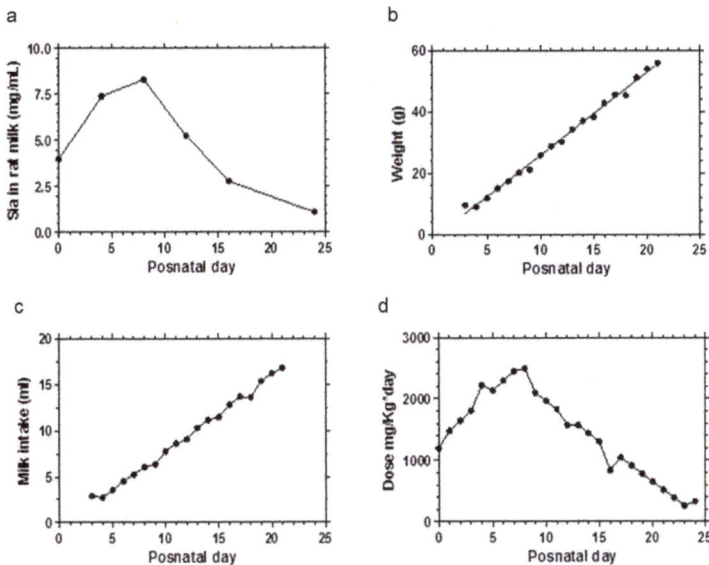

Figure 1. (**a**) Sia concentration (mg/mL) in rat milk throughout the lactation period per the literature [36]. (**b**) Experimental data of rat pups' weight evolution from birth until weaning. (**c**) Estimated data of milk volume intake (mL) in rat pups from birth until weaning per the literature [37]. (**d**) Sialic acid dose (mg/kg/day) that rat pups were receiving from rat milk considering the data shown in graphs (**a–c**).

To determine the effects of Sia, a model was designed in which one of the groups received less Sia than that from normal lactation, while the other groups received the same levels of Sia as naturally-nursed pups, either in the free form Neu5Ac or as 6'-SL. The daily dose of Sia was calculated based on Sia content of rat milk, pup body weights and milk intake during lactation according to data found in the literature [36,37] and to internal data from previous experiments (Figure 1b). Figure 1d shows the theoretical Sia dose (mg/kg weight/day) that rat pups would normally receive from rat milk. Equimolar stock solutions with Neu5Ac and 6'-SL were prepared. Variable volumes of these stock solutions were given to the pups per day to adjust the Sia dose, as shown in Figure 1d.

2.4. Analytical Determination of Sialic Acid Content by HPLC

Right hemispheres were analyzed to determine the total amount of sialic acid. After homogenization of the brain, 35 mg were weighed and re-suspended at 0.2 mg/μL in deionized water. Sia were released by mild hydrolysis in 0.2N H_2SO_4 at 80 °C for 1 h. After filtering and centrifuging during 30 min at 10,000 rpm, supernatants were used for derivatization with 5-(difluoromethoxy)-2-mercapto-1H-benzimidazole (DMB) as described by Hara et al. [38] with some modifications. Thus, 50 μL of samples and 50 μL of DMB reagent (8 mM DMB, 1.5 M acetic acid, 14 mM sodium hydrosulfite, 0.8 M 2-mercaptoethanol (Sigma Aldrich, Saint Louis, MO, USA)) were kept for 2.5 h at 50 °C in the dark. A volume of 10 μL of the derivatized solution was injected on an Alliance 2695 HPLC system equipped with a 474-fluorescence detector from Waters (Mildford, MA, USA) as described by Martin et al. [39]. A LiChrosorb RP-18 column (5 μm, 250 mm × 4.6 mm) with a LiChrosorb RP-18 guard cartridge (5 μm), both from Supelco (Bellefonte, PA, USA), was used. DMB derivatives of Sia were isocratically eluted using 7% (*v/v*) methanol, 8% (*v/v*) acetonitrile in water for 40 min at a flow rate of 0.9 mL min^{-1}. All injections were performed at room temperature. The eluent was monitored for fluorescence at 373 nm (excitation wavelength) and 448 nm (emission wavelength). The gain was fixed at 1, and the attenuation at 64 for the 474-fluorescence detector. A set of standards (25–250 ng Neu5Ac) was injected with every set of samples. Data were integrated by the Millennium 4.0 software (Waters, Milford, MA, USA) coupled to the HPLC system. Standard regression and sample quantitation were calculated by Microsoft Excel 2011 and Graph Pad (Prism 4, San Diego, CA, USA).

2.5. Western Blotting

For Western analysis, ≈15–20 mg of frontal cortex samples of pups at weaning (PN22) were homogenized in cold suspension buffer (PBS, 0.1% TritonX100 and protease inhibitor cocktail from Sigma Aldrich, Saint Louis, MO, USA), centrifuged at 10,000 *g* for 15 min, and the supernatant collected. The protein content was determined using the Bradford assay (Bio-Rad, Hercules, CA, USA). Ten micrograms of protein, diluted in phosphate-buffered saline (PBS), were loaded into Criterion XT 4–20% Bis-Tris gels (Bio-Rad). Separation was carried out using MOPS buffer (0.05 M 3-Morpholinopropane-1-sulfonic acid (MOPS), 0.05 M Tris base, 0.003 M SDS, 0.8 mM EDTA) and run at 200 V for 45 min. Separated proteins were transferred onto nitrocellulose membranes for 3.5 h using a Trans-Blot electrophoresis transfer cell. Six samples were run per group. After blocking non-specific binding sites for 1 h with 3% bovine serum albumin (BSA, Sigma Aldrich, Saint Louis, MO, USA) in TBS-TritonX100 (0.025% TritonX100, 20 mM Tris base, 150 mM NaCl, pH 7.6), blots were incubated overnight at 4 °C with one of the following monoclonal antibodies: anti-NCAM (Santa Cruz, CA, USA) at 1:2500; anti-PSA-NCAM (Millipore/Merck, Darmstadt, Germany) at 1:1000. All the antibodies were diluted in TBS containing 1% BSA. After three quick washes in TBS-TritonX100, NCAM blots were incubated for 2 h at room temperature with anti-IgG HRP at 1:5000 (Sigma Aldrich), while PSA-NCAM were incubated with anti-IgG HRP (Bethyl laboratories, Montgomery, TX, USA) at 1:5000 for 2 h. β-actin was used as a control for quantitation and detected with monoclonal antibody (Sigma Aldrich, 1:5000). After five washes in TBS-TritonX100, membranes were developed using Pierce Supersignal West-Pico substrate (Thermo Fisher Scientific, Waltham, MA, USA) and quantified using a Chemidoc XRS system from Bio-Rad (Hercules, CA, USA).

2.6. Long-Term Potentiation Measurement

One-year old male offspring were assigned to this analysis. Animals were anesthetized and implanted with stimulating and recording electrodes in the hippocampus to measure LTP in vivo, as previously described [40]. Based on prior research [41,42], animals were implanted with stimulating electrodes at the Schaffer collateral-commissural pathway of the dorsal hippocampus (3.5 mm lateral and 3.2 mm posterior to Bregma) as per stereotaxic coordinates. Four recording electrodes were also implanted targeting the ipsilateral stratum radiatum beneath the CA1 area (2.5 mm lateral and 3.6 mm posterior to Bregma). Stimulation occurred, and the field excitatory post-synaptic potential (fEPSP) was recorded. After the high-frequency stimulation (HFS) protocols, fEPSPs were recorded for 30 min. Additional 15-min recording sessions were carried out during the following days.

2.7. Classical Behavioral Tests

A video tracking system, Sony Camera SSC-G213A (Sony Electronics Inc., Park Ridge, NJ, USA), was used to record all animal trials. Rat performances were then analyzed offline using the analytical software Viewer (Biobserve GmbH, Bonn, Germany) by a technician that was blinded to treatment. An alcohol dilution was used to clean the surface of Y maze and NORT structures after each animal performance to prevent odor cues.

2.7.1. Novel Object Recognition Test

NORT is based on the natural tendency of rodents to explore novel stimuli presented to them; thus, when novel and familiar stimuli are present at the same time, the novel one will be naturally explored for longer [43]. Therefore, this test was run in a black opened plastic chamber (40 × 40 cm), known as the arena. Three objects differing in material, form and color were chosen. After 3 days of habituation to the arena, an acquisition phase took place, and the animals were faced with two different objects (familiar objects) for 10 min. One day later, animals were submitted to a retention test in which a new object (novel object) substituted one of the familiar objects, allowing the animals to explore them for 5 min. Animals were considered as exploring the objects when they approached their whiskers at approximately 1–2 cm or licked them; exploration was never considered when they sat on the objects [44]. The key parameter evaluated in this task was time spent by the rats exploring each object, novel or familiar, during the retention phase.

2.7.2. Y Maze with Blocked Arm Test

The Y maze with blocked arm test measures the ability of rodents to explore new environments and assesses exploratory behavior and memory [45]. The experimental design used was that proposed by Dellu et al. [46] with some modifications. A Y-shape stainless-steeled structure (20 × 10.5 × 50 cm) was used. There are two consecutive phases in the paradigm: an acquisition trial and a retention test. During the acquisition trial, one of the three arms of the maze was blocked (novel arm), and rats could explore the other two arms for 15 min. Four hours later, animals went through the test session in which all arms were open to be explored for 5 min. Novelty was represented by the arm that was not accessible in the acquisition phase. When introduced in the maze after a few hours, animal normal behavior should be to explore first the previously hidden arm; thus, to evaluate animal memory skills, it was analyzed whether animals went first into the novel arm.

2.8. IntelliCage® Protocol

The IntelliCage® (NewBehavior AG, Zurich, Switzerland) is a computer-based, fully-automated testing apparatus used to analyze the spontaneous and learning behavior of rodents. This system consists of a cage that presents 4 operant conditioning corners, which can locate one rat at a time. Each corner is equipped with 2 motorized doors, which block or allow access to water bottles placed on both sides of the corner. When a rat tries to access through whichever of the 2 doors (nose poke action), the interruption of a light-beam sensor at either door triggers one or the other doors to open and allow access to a water bottle. Radio frequency identification (RFID) transponders are implanted under the rat skin, allowing for individual recognition; thus, rat entries into the corners are detected through RFID antennas located there. Using this technology, conditioning protocols can be implemented to evaluate the behavioral activities of the rats. Animals were implanted with the RFID-transponders, and 1 day later, they were placed into the IntelliCage® ($n = 8$ rats/group) and maintained there during

2 weeks for habituation and testing. The habituation process was counted with different stages: 1 day of free exploration with all doors opened; 5 days in which doors were opened only upon a visit to the corner; in the following 2 days, doors opened after a nose poke in the right place; in the last 4 days, one nose poke opened any door only during two drinking sessions of 90 min per night, while the doors remained closed during the rest of the night and day.

Thereafter, a paradigm to measure cognitive outcomes was tested by restricting water access to a specific corner and time periods. During the two-day test, access to water was restricted to one corner (correct corner) only during the two drinking sessions. If the rat nose poked any other corner, water was not available. Each rat was randomly assigned a correct corner. Every visit to this assigned corner was counted as correct, while incorrect visits were those to any other corner. Thus, only rats with adequate cognitive skills could learn which is the correct corner in this place learning paradigm.

2.9. Statistical Analysis

Data are presented as the mean $\pm$ SEM. Graph displays and statistical analysis were done using SPSS (SPSS Inc., Chicago, IL, USA) and GraphPad (GraphPad Inc., La Jolla, CA, USA) software. Comparisons between groups were performed by Student's *t*-test and one-way analysis of variance (ANOVA) in NORT and Western blotting. Statistical differences between groups were determined with a two-way repeated measure of the analysis of variance (ANOVA) in LTP measurements. The Y maze test and IntelliCage® paradigm were analyzed through a contingency table analysis using Fisher's test. The significance level was established at $p < 0.05$ for all tests.

3. Results

3.1. Sialic Acid Content in Brain

The content of Neu5Ac in the right brain hemisphere of rats at weaning (PN22) and at 1 year of age is shown in Table 1. Brain concentration of sialic acid (μg NeuAc/mg brain) did not show significant differences among groups at any age.

Table 1. Concentration of Sia (μg Neu5Ac/mg) in brain in weaned and 1-year old rats supplemented with different Sia sources or water during the lactation period. Data are the mean $\pm$ SEM.

	Sia Concentration (μg Neu5Ac/mg Brain)			
	Neu5Ac Group	6′-SL Group	Control Group	One-Way ANOVA *p*-Value
Pups rats (*n* = 8/group)	1.399 $\pm$ 0.067	1.405 $\pm$ 0.124	1.422 $\pm$ 0.058	*p* = 0.8336
1-year old rats (*n* = 8–10/group)	1.018 $\pm$ 0.106	1.067 $\pm$ 0.025	0.9745 $\pm$ 0.096	*p* = 0.2143

Sia, sialic acids; Neu5Ac, N-acetylneuraminic acid; 6′-SL, 6′-sialyllactose.

3.2. Western Blot

Frontal cortex samples of weaned pups (PN22) were analyzed to determine NCAM and PSA-NCAM expression by Western blotting (6 pups/group). Actin was used as a control for quantitation. The control group was used as the reference group and ratios were calculated between normalized values and the density mean for the control group.

As shown in Figure 2, no differences among groups related to NCAM expression ($p = 0.8819$) were found. However, rats fed 6′-SL in the lactation period expressed significantly more PSA-NCAM in the frontal cortex when compared to rats supplemented with Neu5Ac ($p = 0.012$) or control animals ($p = 0.041$).

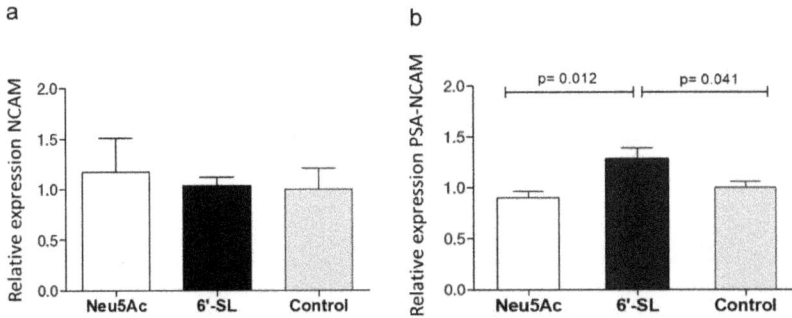

Figure 2. Relative protein expression of NCAM (**a**) and PSA-NCAM (**b**) in frontal cortex samples of rat pups of the three experimental groups at Postnatal Day 22. One-way ANOVA and Student's *t*-test were used for statistical analysis. Data are expressed as the mean ± SEM (*n* = 6 rats/group), *p* < 0.05 denotes significant differences. NCAM, neural cell adhesion molecule; PSA-NCAM, polysialylated-neural cell adhesion molecule; Neu5Ac, N-acetylneuraminic acid; 6′-SL, 6′-sialyllactose.

3.3. In Vivo LTP

At one-year of age, male rats were submitted to in vivo LTP (10 rats/group). A significant response in LTP was evoked in all the groups following the HFS session (Figure 3). LTP responses were significantly improved in male rats supplemented with 6′-SL during lactation compared to controls. Differences were detectable several days after the stimulation. Neu5Ac group also reached a more intense LTP than the control group, although the difference was not statistically significant.

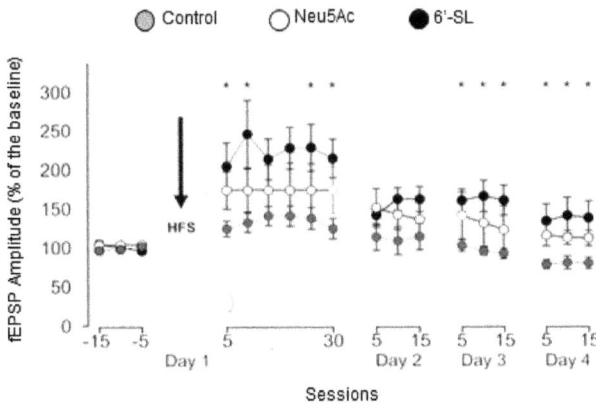

Figure 3. Experimentally-evoked LTP in the groups of animals at one year of age. Data collected from the three groups of animals before and after high-frequency stimulation (HFS) session (indicated by the downward arrow). The HFS was presented after 15 min of baseline recordings. LTP evolution was followed for up to four days. Statistical differences between groups were determined with a two-way repeated measure of the analysis of variance (ANOVA). The three groups of animals presented a significant (*p* ≤ 0.05) LTP in relation with baseline values. In addition, field excitatory post-synaptic potential (fEPSP) values evoked after the HFS session were significantly larger (*, *p* ≤ 0.05) for the 6′-SL group in comparison with values collected from controls. Data are represented as the mean ± SEM (*n* = 10 animals/group). Neu5Ac, N-acetylneuraminic acid; 6′-SL, 6′-sialyllactose.

3.4. Classical Behavioral Tests

A behavioral assessment with different paradigms such as NORT and the Y maze test was performed at weaning and when animals were one year old. Results obtained at weaning ($n = 10$ pups/group) did not show any difference in performance of the three groups tested (data not shown). Conversely, data found at adult age ($n = 8$ rats/group) were more conclusive.

3.4.1. NORT

Regarding NORT, the key parameter to study was the time spent exploring the novel object versus the familiar one in the retention phase. Thus, animals with good cognitive abilities tended to explore the novel object for longer. As shown in Figure 4a, rats belonging to the groups that received Sia supplementation, both Neu5Ac and 6'-SL, spent a significantly longer time exploring the novel object than the familiar one. A significant difference in the exploration time of both objects was not found in the control group.

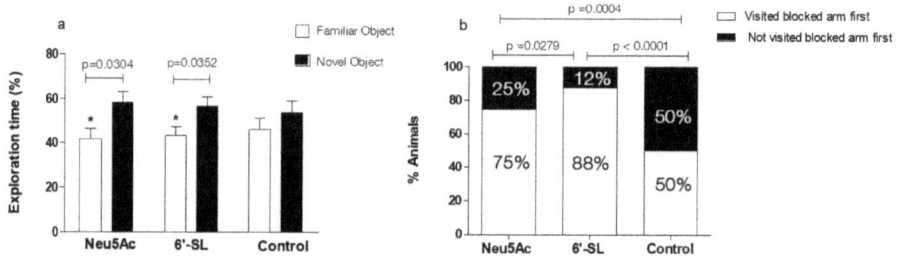

Figure 4. (**a**) Percentage of time spent by the animals exploring a familiar object against a novel object in NORT at one year of age. One-way ANOVA and Student's *t*-test were applied for statistical analysis; **, p < 0.05* denotes significant differences. (**b**) Percentage of animals that chose the novel arm first in the Y maze during the retention test versus those that did not. Fisher's test was applied for statistical analysis; *p < 0.05* denotes significant differences ($n = 8$ rats/group). Neu5Ac, N-acetylneuraminic acid; 6'-SL, 6'-sialyllactose.

3.4.2. Y Maze with Blocked Arm Test

As for the Y maze task, one-year old rats that received Neu5Ac or 6'-SL during lactation performed significantly better than the control group by clearly identifying the blocked arm. During the retention test, rats with superior cognitive abilities explored the arm first, since it represents novelty. Figure 4b shows the percentage of animals from each group that visited the previously blocked arm first versus the percentage of animals that chose one of the other two arms as their first option. Rats receiving Sia achieved the best score, i.e., 88% of 6'-SL animals visited the novel arm first and 75% of Neu5Ac rats. By contrast, the control group visited the novel arm with the same probability as the other two combined. From a statistical standpoint, the 6'-SL group score was significantly higher, not only when compared to the control group ($p < 0.0001$), but also compared to the Neu5Ac group ($p = 0.0279$). There were also significant differences between Neu5Ac rats and control animals ($p = 0.0004$).

3.5. IntelliCage® Protocol

In the place learning paradigm performed with the IntelliCage® system ($n = 8$ rats/group), results were aligned to the behavior observed in the classical tests. As previously explained, access to water was restricted to the "correct corner" during the two drinking sessions of two consecutive nights. The percentages of visits to the correct corner and visits to incorrect corners were analyzed. Neu5Ac and 6'-SL groups obtained 41% and 39% correct visits, respectively, while the control group reached 25% of visits to the correct corner (Figure 5). Thereby, groups supplemented with Neu5Ac or 6′-SL exhibited a significantly better performance compared to the control group.

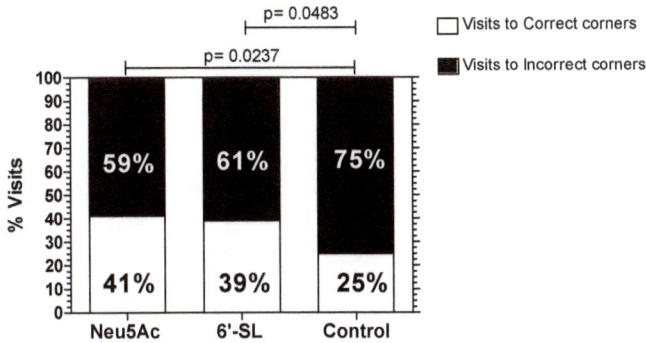

Figure 5. Percentage of total visits to the correct corner versus incorrect corner in the IntelliCage® paradigm for each group of animals during the four drinking sessions of the two days of place learning. Fisher's test was applied for statistical analysis; $p < 0.05$ denotes significant differences; $n = 8$ rats/group; one year of age. Neu5Ac, N-acetylneuraminic acid; 6′-SL, 6′-sialyllactose.

4. Discussion

In the current study, the impact of two forms of sialic acid, either free form or conjugated given during lactation, on cognitive skills later in life was assessed. We evaluated cognitive functions with behavioral and electrophysiological measurements, demonstrating that these sources of Sia given at early stages after birth maintain cognitive function in adulthood when compared to a group that received a lower amount of Sia, and that the provision of Sia as 6′-SL may confer some advantages over the use of the free form.

Pioneering studies in the 1980s showed that exogenous Sia could be incorporated into brain gangliosides and glycoproteins when injected intraperitoneally in rat pups [27]. Wang et al. investigated the metabolic fate of intravenously-administrated ^{14}C-Neu5Ac in piglets, concluding that an exogenous source of sialic acid could cross the blood-brain barrier and be incorporated into various tissues [47]. Carlson and House also demonstrated that Sia given by intraperitoneal injection or orally significantly impacted the concentration of brain gangliosides and glycoproteins [48]. Another study confirmed the impact of orally-administered sialic acid on cortical ganglioside concentration after feeding rat pups a solid diet prepared with different Sia doses from PN17–31. Cortical gangliosides Sia concentration was significantly higher in rats that received a Sia-supplemented diet provided by a protein-bound source of Sia (casein glycomacropeptide) [49].

Considerable research has been carried out to date to elucidate the effects of supplementation with Sia early in life on brain composition and behavior. In a study by Morgan and Winick [27], rat pups were intraperitoneally injected with Neu5Ac or glucose (control group), and an improvement of performance was observed in the Neu5Ac group. A piglet model was used by Wang et al. [28] to evaluate the effect of a sow milk replacer supplemented with several doses of casein glycomacropeptide for 35 days after weaning. Neu5Ac brain concentration, expression of two learning associated genes and learning and memory abilities were positively impacted by the Sia supplementation. However, a research team has recently reported two studies in which supplementation with sialyllactose from PN2–PN22 or PN32 in piglets showed no significant effect on recognition memory [50], but did show a sialic acid increase in hippocampus, prefrontal cortex and corpus callosum [51].

The novelty of the present work is three-fold: (1) the use of a unique foster rat mother model, whereby rat pups were breastfed during the entire lactation period, but received milk with a lower amount of Sia from PN3 until end of lactation; in addition, the animal intervention was performed at an early, developmental stage, which is challenging due to the vulnerability of pups; (2) the timeframe in which Sia content in brain and cognitive outcomes in rats were measured; previous analyses were carried out after weaning, when the Sia supplementation had just finished; herein, brain composition analyses and behavior assessments were performed at the end of Sia supplementation, but also in adult animals that had spent almost one year without receiving additional Sia; and (3) the use of 6′-SL as a source of Sia in comparison to the free form.

Improved cognitive skills reported here were not attributable to an increased Sia content in a brain hemisphere. In fact, Sia content in brain hemispheres samples was the same for the three experimental groups at weaning and in adulthood. According to Wang et al., the ganglioside and glycoprotein sialic acid concentrations in the brain frontal cortex are higher in breastfed infants when compared to formula-fed infants [22]. In a study performed with piglets, it was observed that supplementation with an exogenous source of Sia increased protein-bound sialic acid concentrations in the frontal cortex [28]. Jacobi et al. also demonstrated in piglets that a formula supplemented with dietary 6′-SL and 3′-SL increased the sialic acid bound to gangliosides in several areas of the brain [52]. As described above, Mudd and co-workers have recently published a study in which piglets were fed several diets containing different doses of sialyllactose from PN2–PN32 and showed that dietary sialyllactose increased conjugated Neu5Ac in the prefrontal cortex, among other brain structures [51]. In our study, Neu5Ac determination by HPLC was done in the whole hemisphere instead of in separate structures. Thus, significant differences among groups could have been missed because Neu5Ac may be accumulated in certain brain structures such as frontal cortex. Furthermore, it could be hypothesized that Neu5Ac content in the brain of the different groups was different during an early phase of development, promoting higher rates of axonal growth and enhanced connections between neurons that would support the behavioral differences found in adulthood.

To further explore this, we analyzed NCAM and PSA-NCAM expression levels in the frontal cortex of early weaned pups (PN22). NCAM is a widely-expressed protein involved in the stabilization and modulation of CNS [3]. Polysialic acid (PSA) is a linear homopolymer of α2-8-linked Neu5Ac. PSA is added to NCAM by a regulated post-translational process and varies through development [20]. Polysialylation supports the maintenance of an immature phenotype, allowing the neurites to grow and sprout to connect the complex circuitry of the brain while its graded downregulation enables fine-tuning of NCAM-dependent cell-cell interactions, stabilizing the newly-formed structures. Following the findings of the studies previously cited [28,52], our results in weaned rats showed that PSA-NCAM expression in the frontal cortex was significantly higher in the 6′-SL group when compared to Neu5Ac and the control groups; this result suggests that conjugated Sia such as 6′-SL might be taken up preferentially by the brain, enhancing the sialylation process of proteins highly involved in brain development. The maintenance of PSA-NCAM levels suggests neuroplasticity may be higher in the 6′-SL group.

With regards to cognitive evaluation, several behavioral tests were implemented consisting of two classical tests including the Y Maze with blocked arm and NORT, as well as a novel cognitive paradigm performed in the IntelliCage® system. Y maze and NORT have been used in previous animal studies to evaluate the effects of nutrition on cognition [53,54] and were conducted at both ages in the present work. IntelliCage® paradigm was only performed in adulthood. Data from weaned rats did not show differences among groups in any of the classical tests. In fact, performing cognitive tests at early ages of life entails certain methodological challenges. Leussis and Bolivar [55] pointed out that pre-weaning and weaning rodents normally habituate slower to novel environments, showing low exploratory activity when compared to older animals. It has been suggested that younger animals would perform better in rewarded tasks. Longer periods of exposure or habituation have also been proposed to increase the success of these tests. Based on our experience, the intense physical activity and the lack of focus in very young rodents are the main hurdles found when carrying out cognitive evaluation of weanling animals.

Conversely, significant differences were found in the assessment of cognitive skills in adulthood. In NORT, groups receiving Sia during lactation could distinguish better between novel and familiar objects than the control group. Impairment of the perirhinal cortex, the main brain structure involved in object recognition memory [44,56] has been suggested [57] as the reason why animals are not able to get a good score on memory tasks like NORT. Results obtained from the IntelliCage® paradigm confirmed that animals receiving Sia at early stages performed significantly better than the control group in adulthood. As for the Y maze test, Sia-supplemented animals performed better than the control rats, since they were able to recognize the novel arm easily. An interesting finding to highlight was that, within groups supplemented with Sia, 6′-SL rats had better scores than animals that received Neu5Ac during lactation, suggesting 6′-SL may be a better source of Sia.

An added value of the present study is that animal cognitive assessments were also supported by electrophysiological data based on in vivo LTP measurements. LTP is the basic mechanism for certain types of hippocampal learning [40]. The best results were observed in Sia-supplemented animals, mainly those of the 6′-SL group, while the control group performance was significantly lower than that of 6′-SL rats. These data suggest early supplementation with Sia might enhance processes related to neuronal plasticity.

Remarkably, control animals under-performed in all tests compared to Sia groups. In NORT, there were no differences in exploration time for both objects. In the Y maze, no preference for visiting the blocked arm first was shown. In the IntelliCage®, these animals scored poorly for correct visits, and a low increase in fEPSP amplitude was found in LTP. These results highlighted a certain degree of deficiency in control animals that could be explained by the lower amount of sialic acid and/or other nutrients such as minerals [58] that they received during the lactation period. The data indicate that the exogenous Sia provided by milk is necessary for the establishment of a neuronal network that ensures long-lasting cognitive skills. Lower levels of Sia during this critical timeframe may lead to premature aging of certain brain functions, while an exogenous supply of Sia was shown to be effective in maintaining cognitive skills for a longer duration. This phenomenon agrees with our previous research in which early supplementation during lactation with the HMO 2′-fucosyllactose (2′-FL) induced cognitive benefits in adult animals [42]. This new perspective highlights the concept of early nutrition [59] and the role of sialic acid in brain development.

Although the mechanism of action through which HMOs might exert this positive effect on cognitive outcomes is not yet known, there are several hypotheses. The improved cognitive performance observed in the Sia-supplemented groups could be due to an increase in the polysialylation of NCAM [60] or to an increase of sialylated compounds in brain due to the additional supply of Sia. There is a consistent body of evidence highlighting the potential biological importance of sialylated compounds in brain. Gangliosides are known to be key molecules in many CNS functions such as synaptogenesis, neurite and axonal growth and neural transmission [61–63].

Alternatively, 2′-FL orally given to adult mice also improves cognition and is dependent on gut-brain crosstalk through the vagus nerve [29]. The gut-brain axis (GBA) is a complex bidirectional network that communicates between the brain and gastrointestinal tract and modulates the respective functions [31]. This gut-brain crosstalk is driven by neural pathways and immune and endocrine mechanisms. It is known that one of the main factors able to trigger or modulate GBA is the intestinal microbiota [64] by producing different metabolites, such as short chain fatty acids or neurotransmitters. Several studies have reported how HMOs promote intestinal bacteria such as *Bifidobacterium* and *Lactobacillus* [65,66] based on their prebiotic role. Tarr et al. reported that dietary supplementation with both 6′-SL or 3′-SL effectively altered colonic microbiota communities, as well as diminished stressor-induced alterations in colonic mucosa structure, anxiety-like behavior and immature neuron cell numbers regardless of immune or endocrine functionality [30]. Despite this evidence, the role of 6′-SL as a GBA regulator and the mechanism underlying the reported effects are yet to be explored. Thus, the reported benefits on CNS function could be related either to its prebiotic capability, with its uptake by the brain during the critical neurodevelopment period, inducing long-lasting biochemical changes that ultimately enhance long-term cognitive skills or via GBA regulation.

With regards to the Sia sources used in this study, our results show that free Sia was less effective than 6′-SL, possibly because free Sia is more rapidly eliminated in the urine and, therefore, less bio-available. Nöhle and Schauer [67] suggested that metabolism of sialic acids bound to oligosaccharides might take more time than free Sia, since the former would require hydrolysis. In addition, oral treatment with 6′-SL in a mouse model of symptomatic *GNE* myopathy led to greater restoration of sialylation in muscle and improvement in muscle size and function as compared to the group treated with free Sia [68]. Based on these data, 6′-SL may be a better source of bio-available Sia.

5. Conclusions

In summary, our findings confirm the beneficial effects of supplementation of Neu5Ac or 6′-SL during lactation on long-term cognitive development in rats. Both compounds had positive outcomes; however, the results showed improved scores in behavior and electrophysiological analysis in those animals that received 6′-SL compared to Sia. Further research is needed to delineate the mechanisms of action underlying cognitive benefits and the potential role of intestinal microbiota.

6. Patents

There is one patent resulting from the work reported in this manuscript: the method of achieving memory and learning improvement by the administration of sialic acid. Publication number: WO/2015/085077. International Application Number: PCT/US2014/068582.

Author Contributions: Formal analysis, E.O., A.B., A.G. and J.M.D.-G. Investigation, E.O., A.B., A.G. and J.M.D.-G. Methodology, E.V., M.R. and M.J.M. Supervision, M.J.M. Writing, original draft, E.O. Writing, review and editing, E.V., M.R., R.B., R.R. and M.J.M. The results described in this manuscript are part of the doctoral thesis project of E.O. within the doctoral programme in Biomedicine of the University of Granada.

Funding: This research received no external funding.

Acknowledgments: The authors would like to thank the technical staff for their assistance.

Conflicts of Interest: E.O., M.R., E.V., A.B., R.B., R.R. and M.J.M. are employees of Abbott Laboratories, a company that manufactures infant formulas. A.G. and J.M.D.-G., have no conflicts of interest to declare.

References

1. Anderson, J.W.; Johnstone, B.M.; Remley, D.T. Breast-feeding and cognitive development: A meta-analysis. *Am. J. Clin. Nutr.* **1999**, *70*, 525–535. [CrossRef] [PubMed]
2. Varki, A.; Schauer, R. Sialic acids. In *Essentials of Glycobiology*, 2nd ed.; Varki, A., Cummings, R.D., Esko, J.D., Freeze, H.H., Stanley, P., Bertozzi, C.R., Hart, G.W., Etzler, M.E., Eds.; Cold Spring Harbor Laboratory Press: Cold Spring Harbor, NY, USA, 2009.
3. Wang, B. Sialic acid is an essential nutrient for brain development and cognition. *Annu. Rev. Nutr.* **2009**, *29*, 177–222. [CrossRef] [PubMed]
4. Ten Bruggencate, S.J.; Bovee-Oudenhoven, I.M.; Feitsma, A.L.; van Hoffen, E.; Schoterman, M.H. Functional role and mechanisms of sialyllactose and other sialylated milk oligosaccharides. *Nutr. Rev.* **2014**, *72*, 377–389. [CrossRef] [PubMed]
5. Coppa, G.V.; Pierani, P.; Zampini, L.; Carloni, I.; Carlucci, A.; Gabrielli, O. Oligosaccharides in human milk during different phases of lactation. *Acta Paediatr. Suppl.* **1999**, *88*, 89–94. [CrossRef] [PubMed]
6. Ninonuevo, M.R.; Park, Y.; Yin, H.; Zhang, J.; Ward, R.E.; Clowers, B.H.; German, J.B.; Freeman, S.L.; Killeen, K.; Grimm, R.; et al. A strategy for annotating the human milk glycome. *J. Agric. Food Chem.* **2006**, *54*, 7471–7480. [CrossRef] [PubMed]
7. Jantscher-Krenn, E.; Bode, L. Human milk oligosaccharides and their potential benefits for the breast-fed neonate. *Minerva Pediatr.* **2012**, *64*, 83–99. [PubMed]
8. Kiskini, A.; Difilippo, E. Oligosaccharides in goat milk: Structure, health effects and isolation. *Cell. Mol. Biol.* **2013**, *59*, 25–30. [PubMed]

9.   Martin-Sosa, S.; Martin, M.J.; Garcia-Pardo, L.A.; Hueso, P. Sialyloligosaccharides in human and bovine milk and in infant formulas: Variations with the progression of lactation. *J. Dairy Sci.* **2003**, *86*, 52–59. [CrossRef]

10. Peterson, R.; Cheah, W.Y.; Grinyer, J.; Packer, N. Glycoconjugates in human milk: Protecting infants from disease. *Glycobiology* **2013**, *23*, 1425–1438. [CrossRef] [PubMed]

11. Austin, S.; De Castro, C.A.; Bénet, T.; Hou, Y.; Sun, H.; Thakkar, S.K.; Vinyes-Pares, G.; Zhang, Y.; Wang, P. Temporal Change of the Content of 10 Oligosaccharides in the Milk of Chinese Urban Mothers. *Nutrients* **2016**, *8*, 346. [CrossRef] [PubMed]

12. Weichert, S.; Jennewein, S.; Hufner, E.; Weiss, C.; Borkowski, J.; Putze, J.; Schroten, H. Bioengineered 2′-fucosyllactose and 3-fucosyllactose inhibit the adhesion of pseudomonas aeruginosa and enteric pathogens to human intestinal and respiratory cell lines. *Nutr. Res.* **2013**, *33*, 831–838. [CrossRef] [PubMed]

13. Weichert, S.; Koromyslova, A.; Singh, B.K.; Hansman, S.; Jennewein, S.; Schroten, H.; Hansman, G.S. Structural basis for norovirus inhibition by human milk oligosaccharides. *J. Virol.* **2016**, *90*, 4843–4848. [CrossRef] [PubMed]

14. Goehring, K.C.; Marriage, B.J.; Oliver, J.S.; Wilder, J.A.; Barrett, E.G.; Buck, R.H. Similar to those who are breastfed, infants fed a formula containing 2′-fucosyllactose have lower inflammatory cytokines in a randomized controlled trial. *J. Nutr.* **2016**, *146*, 2559–2566. [CrossRef] [PubMed]

15. Donovan, S.M.; Comstock, S.S. Human milk oligosaccharides influence neonatal mucosal and systemic immunity. *Ann. Nutr. Metab.* **2016**, *69*, 42–51. [CrossRef] [PubMed]

16. He, Y.; Liu, S.; Kling, D.E.; Leone, S.; Lawlor, N.T.; Huang, Y.; Feinberg, S.B.; Hill, D.R.; Newburg, D.S. The human milk oligosaccharide 2′-fucosyllactose modulates cd14 expression in human enterocytes, thereby attenuating lps-induced inflammation. *Gut* **2016**, *65*, 33–46. [CrossRef] [PubMed]

17. Jost, T.; Lacroix, C.; Braegger, C.; Chassard, C. Impact of human milk bacteria and oligosaccharides on neonatal gut microbiota establishment and gut health. *Nutr. Rev.* **2015**, *73*, 426–437. [CrossRef] [PubMed]

18. Bienenstock, J.; Buck, R.H.; Linke, H.; Forsythe, P.; Stanisz, A.M.; Kunze, W.A. Fucosylated but not sialylated milk oligosaccharides diminish colon motor contractions. *PLoS ONE* **2013**, *8*, e76236. [CrossRef] [PubMed]

19. Nakano, T.; Sugawara, M.; Kawakami, H. Sialic acid in human milk: Composition and functions. *Acta Paediatr. Taiwan* **2001**, *42*, 11–17. [PubMed]

20. Wang, B. Molecular mechanism underlying sialic acid as an essential nutrient for brain development and cognition. *Adv. Nutr. Int. Rev. J.* **2012**, *3*, 465S–472S. [CrossRef] [PubMed]

21. Asakuma, S.; Akahori, M.; Kimura, K.; Watanabe, Y.; Nakamura, T.; Tsunemi, M.; Arai, I.; Sanai, Y.; Urashima, T. Sialyl oligosaccharides of human colostrum: Changes in concentration during the first three days of lactation. *Biosci. Biotechnol. Biochem.* **2007**, *71*, 1447–1451. [CrossRef] [PubMed]

22. Wang, B.; McVeagh, P.; Petocz, P.; Brand-Miller, J. Brain ganglioside and glycoprotein sialic acid in breastfed compared with formula-fed infants. *Am. J. Clin. Nutr.* **2003**, *78*, 1024–1029. [CrossRef] [PubMed]

23. Chichlowski, M.; German, J.B.; Lebrilla, C.; Mills, D.A. The influence of milk oligosaccharides on microbiota of infants: Opportunities for formulas. *Annu. Rev. Food Sci. Technol.* **2011**, *2*, 331–351. [CrossRef] [PubMed]

24. German, J.B.; Freeman, S.L.; Lebrilla, C.B.; Mills, D.A. Human milk oligosaccharides: Evolution, structures and bioselectivity as substrates for intestinal bacteria. *Nestle Nutr. Workshop Ser. Pediatr. Program.* **2008**, *62*, 205–218. [PubMed]

25. Jantscher-Krenn, E.; Marx, C.; Bode, L. Human milk oligosaccharides are differentially metabolised in neonatal rats. *Br. J. Nutr.* **2013**, *110*, 640–650. [CrossRef] [PubMed]

26. Bode, L. Human milk oligosaccharides: Every baby needs a sugar mama. *Glycobiology* **2012**, *22*, 1147–1162. [CrossRef] [PubMed]

27. Morgan, B.L.; Winick, M. Effects of administration of n-acetylneuraminic acid (nana) on brain nana content and behavior. *J. Nutr.* **1980**, *110*, 416–424. [CrossRef] [PubMed]

28. Wang, B.; Yu, B.; Karim, M.; Hu, H.; Sun, Y.; McGreevy, P.; Petocz, P.; Held, S.; Brand-Miller, J. Dietary sialic acid supplementation improves learning and memory in piglets. *Am. J. Clin. Nutr.* **2007**, *85*, 561–569. [CrossRef] [PubMed]

29. Vazquez, E.; Barranco, A.; Ramirez, M.; Gruart, A.; Delgado-Garcia, J.M.; Jimenez, M.L.; Buck, R.; Rueda, R. Dietary 2′-fucosyllactose enhances operant conditioning and long-term potentiation via gut-brain communication through the vagus nerve in rodents. *PLoS ONE* **2016**, *11*. [CrossRef] [PubMed]

30. Tarr, A.J.; Galley, J.D.; Fisher, S.E.; Chichlowski, M.; Berg, B.M.; Bailey, M.T. The prebiotics 3'sialyllactose and 6'sialyllactose diminish stressor-induced anxiety-like behavior and colonic microbiota alterations: Evidence for effects on the gut-brain axis. *Brain Behav. Immun.* **2015**, *50*, 166–177. [CrossRef] [PubMed]

31. Carabotti, M.; Scirocco, A.; Maselli, M.A.; Severi, C. The gut-brain axis: Interactions between enteric microbiota, central and enteric nervous systems. *Ann. Gastroenterol.* **2015**, *28*, 203–209. [PubMed]

32. Dinan, T.G.; Cryan, J.F. Gut instincts: Microbiota as a key regulator of brain development, ageing and neurodegeneration. *J. Physiol.* **2017**, *595*, 489–503. [CrossRef] [PubMed]

33. Carlson, A.L.; Xia, K.; Azcarate-Peril, M.A.; Goldman, B.D.; Ahn, M.; Styner, M.A.; Thompson, A.L.; Geng, X.; Gilmore, J.H.; Knickmeyer, R.C. Infant gut microbiome associated with cognitive development. *Biol. Psychiatry* **2018**, *83*, 148–159. [CrossRef] [PubMed]

34. Benton, D.; ILSI Europe a.i.s.b.l. The influence of children's diet on their cognition and behavior. *Eur. J. Nutr.* **2008**, *47* (Suppl. 3), 25–37. [CrossRef]

35. Sprenger, N.; Duncan, P.I. Sialic acid utilization. *Adv. Nutr.* **2012**, *3*, 392S–397S. [CrossRef] [PubMed]

36. Dickson, J.J.; Messer, M. Intestinal neuraminidase activity of suckling rats and other mammals. Relationship to the sialic acid content of milk. *Biochem. J.* **1978**, *170*, 407–413. [CrossRef] [PubMed]

37. Dominguez, H.D.; Thomas, J.D. Artificial Rearing. In *Alcohol: Methods and Protocols*; Nagy, L.E., Ed.; Humana Press: Totowa, NJ, USA, 2008; Volume 447, pp. 85–100.

38. Hara, S.; Takemori, Y.; Yamaguchi, M.; Nakamura, M.; Ohkura, Y. Fluorometric high-performance liquid chromatography of N-acetyl- and N-glycolylneuraminic acids and its application to their microdetermination in human and animal sera, glycoproteins, and glycolipids. *Anal. Biochem.* **1987**, *164*, 138–145. [CrossRef]

39. Martin, M.J.; Vazquez, E.; Rueda, R. Application of a sensitive fluorometric HPLC assay to determine the sialic acid content of infant formulas. *Anal. Bioanal. Chem.* **2007**, *387*, 2943–2949. [CrossRef] [PubMed]

40. Gruart, A.; Munoz, M.D.; Delgado-Garcia, J.M. Involvement of the ca3-ca1 synapse in the acquisition of associative learning in behaving mice. *J. Neurosci.* **2006**, *26*, 1077–1087. [CrossRef] [PubMed]

41. Vazquez, E.; Barranco, A.; Ramirez, M.; Gruart, A.; Delgado-García, J.M.; Martinez-Lara, E.; Blanco, S.; Martin, M.J.; Castanys-Muñoz, E.; Buck, R.H.; et al. Effects of a human milk oligosaccharide, 2'-fucosyllactose, on hippocampal long term potentiation and learning capabilities in rodents. *JNB* **2015**, *26*, 455–465. [PubMed]

42. Oliveros, E.; Ramirez, M.; Vazquez, E.; Barranco, A.; Gruart, A.; Delgado-Garcia, J.M.; Buck, R.; Rueda, R.; Martin, M.J. Oral supplementation of 2'-fucosyllactose during lactation improves memory and learning in rats. *J. Nutr. Biochem.* **2016**, *31*, 20–27. [CrossRef] [PubMed]

43. Antunes, M.; Biala, G. The novel object recognition memory: Neurobiology, test procedure, and its modifications. *Cogn. Process.* **2012**, *13*, 93–110. [CrossRef] [PubMed]

44. Ennaceur, A.; Delacour, J. A new one-trial test for neurobiological studies of memory in rats. 1: Behavioral data. *Behav. Brain Res.* **1988**, *31*, 47–59. [CrossRef]

45. Wolf, A.; Bauer, B.; Abner, E.L.; Ashkenazy-Frolinger, T.; Hartz, A.M. A comprehensive behavioral test battery to assess learning and memory in 129s6/tg2576 mice. *PLoS ONE* **2016**, *11*. [CrossRef] [PubMed]

46. Dellu, F.; Mayo, W.; Cherkaoui, J.; Le Moal, M.; Simon, H. A two-trial memory task with automated recording: Study in young and aged rats. *Brain Res.* **1992**, *588*, 132–139. [CrossRef]

47. Wang, B.; Downing, J.A.; Petocz, P.; Brand-Miller, J.; Bryden, W.L. Metabolic fate of intravenously administered n-acetylneuraminic acid-6-14c in newborn piglets. *Asia Pac. J. Clin. Nutr.* **2007**, *16*, 110–115. [PubMed]

48. Carlson, S.E.; House, S.G. Oral and intraperitoneal administration of n-acetylneuraminic acid: Effect on rat cerebral and cerebellar n-acetylneuraminic acid. *J. Nutr.* **1986**, *116*, 881–886. [CrossRef] [PubMed]

49. Scholtz, S.A.; Gottipati, B.S.; Gajewski, B.J.; Carlson, S.E. Dietary sialic acid and cholesterol influence cortical composition in developing rats. *J. Nutr.* **2013**, *143*, 132–135. [CrossRef] [PubMed]

50. Fleming, S.A.; Chichlowski, M.; Berg, B.M.; Donovan, S.M.; Dilger, R.N. Dietary sialyllactose does not influence measures of recognition memory or diurnal activity in the young pig. *Nutrients* **2018**, *10*, 395. [CrossRef] [PubMed]

51. Mudd, A.T.; Fleming, S.A.; Labhart, B.; Chichlowski, M.; Berg, B.M.; Donovan, S.M.; Dilger, R.N. Dietary sialyllactose influences sialic acid concentrations in the prefrontal cortex and magnetic resonance imaging measures in corpus callosum of young pigs. *Nutrients* **2017**, *12*, 1297. [CrossRef] [PubMed]

52. Jacobi, S.K.; Yatsunenko, T.; Li, D.; Dasgupta, S.; Yu, R.K.; Berg, B.M.; Chichlowski, M.; Odle, J. Dietary isomers of sialyllactose increase ganglioside sialic acid concentrations in the corpus callosum and cerebellum and modulate the colonic microbiota of formula-fed piglets. *J. Nutr.* **2016**, *146*, 200–208. [CrossRef] [PubMed]

53. Cutuli, D.; De Bartolo, P.; Caporali, P.; Laricchiuta, D.; Foti, F.; Ronci, M.; Rossi, C.; Neri, C.; Spalletta, G.; Caltagirone, C.; et al. N-3 polyunsaturated fatty acids supplementation enhances hippocampal functionality in aged mice. *Front. Aging Neurosci.* **2014**, *6*, 220. [CrossRef] [PubMed]

54. Hiratsuka, S.; Honma, H.; Saitoh, Y.; Yasuda, Y.; Yokogoshi, H. Effects of dietary sialic acid in n-3 fatty acid-deficient dams during pregnancy and lactation on the learning abilities of their pups after weaning. *J. Nutr. Sci. Vitaminol. (Tokyo)* **2013**, *59*, 136–143. [CrossRef] [PubMed]

55. Leussis, M.P.; Bolivar, V.J. Habituation in rodents: A review of behavior, neurobiology, and genetics. *Neurosci. Biobehav. Rev.* **2006**, *30*, 1045–1064. [CrossRef] [PubMed]

56. Aggleton, J.P.; Albasser, M.M.; Aggleton, D.J.; Poirier, G.L.; Pearce, J.M. Lesions of the rat perirhinal cortex spare the acquisition of a complex configural visual discrimination yet impair object recognition. *Behav. Neurosci.* **2010**, *124*, 55–68. [CrossRef] [PubMed]

57. Albasser, M.M.; Davies, M.; Futter, J.E.; Aggleton, J.P. Magnitude of the object recognition deficit associated with perirhinal cortex damage in rats: Effects of varying the lesion extent and the duration of the sample period. *Behav. Neurosci.* **2009**, *123*, 115–124. [CrossRef] [PubMed]

58. Keen, C.L.; Lönnerdal, B.; Clegg, M.; Hurley, L.S. Developmental changes in composition of rat milk: Trace elements, minerals, protein, carbohydrate and fat. *J. Nutr.* **1981**, *111*, 226–236. [CrossRef] [PubMed]

59. Anjos, T.; Altmae, S.; Emmett, P.; Tiemeier, H.; Closa-Monasterolo, R.; Luque, V.; Wiseman, S.; Perez-Garcia, M.; Lattka, E.; Demmelmair, H.; et al. Nutrition and neurodevelopment in children: Focus on nutrimenthe project. *Eur. J. Nutr.* **2013**, *52*, 1825–1842. [CrossRef] [PubMed]

60. Becker, C.G.; Artola, A.; Gerardy-Schahn, R.; Becker, T.; Welzl, H.; Schachner, M. The polysialic acid modification of the neural cell adhesion molecule is involved in spatial learning and hippocampal long-term potentiation. *J. Neurosci. Res.* **1996**, *45*, 143–152. [CrossRef]

61. Schnaar, R.L.; Gerardy-Schahn, R.; Hildebrandt, H. Sialic acids in the brain: Gangliosides and polysialic acid in nervous system development, stability, disease, and regeneration. *Physiol. Rev.* **2014**, *94*, 461–518. [CrossRef] [PubMed]

62. Hwang, H.M.; Wang, J.T.; Chiu, T.H. Effects of exogenous gm1 ganglioside on ltp in rat hippocampal slices perfused with different concentrations of calcium. *Neurosci. Lett.* **1992**, *141*, 227–230. [CrossRef]

63. Ledeen, R.W.; Wu, G.; Lu, Z.H.; Kozireski-Chuback, D.; Fang, Y. The role of gm1 and other gangliosides in neuronal differentiation. Overview and new finding. *Ann. N. Y. Acad. Sci.* **1998**, *845*, 161–175. [CrossRef] [PubMed]

64. Burokas, A.; Moloney, R.D.; Dinan, T.G.; Cryan, J.F. Microbiota regulation of the mammalian gut-brain axis. *Adv. Appl. Microbiol.* **2015**, *91*, 1–62. [PubMed]

65. Coppa, G.V.; Zampini, L.; Galeazzi, T.; Facinelli, B.; Ferrante, L.; Capretti, R.; Orazio, G. Human milk oligosaccharides inhibit the adhesion to caco-2 cells of diarrheal pathogens: *Escherichia coli*, vibrio cholerae, and salmonella fyris. *Pediatr. Res.* **2006**, *59*, 377–382. [CrossRef] [PubMed]

66. Bindels, L.B.; Delzenne, N.M.; Cani, P.D.; Walter, J. Towards a more comprehensive concept for prebiotics. *Nat. Rev. Gastroenterol. Hepatol.* **2015**, *12*, 303–310. [CrossRef] [PubMed]

67. Nöhle, U.; Schauer, R. Metabolism of sialic acids from exogenously administered sialyllactose and mucin in mouse and rat. *Hoppe Seylers Z. Physiol. Chem.* **1984**, *365*, 1457–1467. [CrossRef] [PubMed]

68. Yonekawa, T.; Malicdan, M.C.; Cho, A.; Hayashi, Y.K.; Nonaka, I.; Mine, T.; Yamamoto, T.; Nishino, I.; Noguchi, S. Sialyllactose ameliorates myopathic phenotypes in symptomatic gne myopathy model mice. *Brain* **2014**, *137*, 2670–2679. [CrossRef] [PubMed]

nutrients

MDPI

Review
Energy Intake Requirements in Pregnancy

Jasper Most [1], Sheila Dervis [2], Francois Haman [2], Kristi B Adamo [2] and Leanne M Redman [1,*]

[1]   Pennington Biomedical Research Center, Baton Rouge, LA 70808, USA
[2]   School of Human Kinetics, University of Ottawa, Ottawa, ON K1N 6N5, Canada
*   Correspondence: leanne.redman@pbrc.edu; Tel.: +1-225-763-0947

Received: 1 March 2019; Accepted: 8 April 2019; Published: 6 August 2019

Abstract: Energy intake requirements in pregnancy match the demands of resting metabolism, physical activity, and tissue growth. Energy balance in pregnancy is, therefore, defined as energy intake equal to energy expenditure plus energy storage. A detailed understanding of these components and their changes throughout gestation can inform energy intake recommendations for minimizing the risk of poor pregnancy outcomes. Energy expenditure is the sum of resting and physical activity-related expenditure. Resting metabolic rate increases during pregnancy as a result of increased body mass, pregnancy-associated physiological changes, i.e., cardiac output, and the growing fetus. Physical activity is extremely variable between women and may change over the course of pregnancy. The requirement for energy storage depends on maternal pregravid body size. For optimal pregnancy outcomes, women with low body weight require more fat mass accumulation than women with obesity, who do not require to accumulate fat mass at all. Given the high energy density of fat mass, these differences affect energy intake requirements for a healthy pregnancy greatly. In contrast, the energy stored in fetal and placental tissues is comparable between all women and have small impact on energy requirements. Different prediction equations have been developed to quantify energy intake requirements and we provide a brief review of the strengths and weaknesses and discuss their application for healthy management of weight gain in pregnant women.

Keywords: pregnancy; energy expenditure; energy intake; physical activity; metabolic rate; fetal development

1. Introduction

Pregnancy is a determining period of future health for women and children. For the mother, poor pregnancy outcomes including excess gestational weight gain, gestational diabetes, hypertension and preeclampsia, or having a cesarean section increase the risk for future obesity, type 2 diabetes and cardiovascular diseases [1,2]. An infant born to a mother who experienced poor pregnancy outcomes is at increased risk for preterm birth, macrosomia and growth restriction, catch-up growth, adiposity and insulin resistance [3–5]. This increased risk for metabolic disease is not only limited to infancy or childhood, but is carried into adulthood [6,7].

One of the key factors for achieving optimal pregnancy outcomes is energy balance or the relationship between energy intake, energy expenditure and energy storage in maternal and fetal tissues. The energy intake requirement for a pregnant woman reflects the amount of energy needed to support maternal and fetal metabolism (energy expenditure) and fetal growth and accumulation of energy depots during pregnancy (energy storage). Individual energy requirements depend on numerous factors such as maternal pregravid body size, physical activity, and the physiological demands of pregnancy in each trimester.

Between 1980 and 2010, various groups studied the energy requirements of pregnant women (Table 1). However, these studies come with limitations which are seldom considered when translating the findings into energy requirement recommendations. First, except for one study [8], the energy costs of weight gain are estimated based on all women studied. As such, women with excess weight gain likely cause an overestimation of the energy intake requirements because they deposited more energy than necessary, and their larger body size caused larger energy expenditure. Second, only one study estimated energy costs for recommended weight gain [8], but classification of both maternal pregravid body mass index (BMI) and recommendations for weight gain in pregnancy have since been revised. Third, women with obesity are significantly underrepresented. In 2014, the prevalence of obesity in women of childbearing age was 37% [9], whereas only three women with obesity have been enrolled in energy requirement studies in total [10]. Similarly, the prevalence of obesity is even larger among women of ethnic minorities (African American 57%, Hispanics 43%), yet the vast majority of women studied were Caucasian. These are important considerations for the assessment of energy requirements because both obesity and race have well-established effects on energy expenditure. The observation that offspring of African-American mothers are already prone to obesity suggests that energy metabolism may not only be differently regulated between races, but that this differential regulation is transmitted from one generation to the next, independent of environmental factors or lifestyle. Lastly, the main determinant of energy expenditure is body weight and thus, due to weight gain, energy expenditure increases during pregnancy. An accurate estimation of the energy costs that are a result of the increase in body weight as compared to the increased metabolic rate per body mass, or due to the fetus requires appropriate adjustment for the increase in weight gain. Herein, we review the energy intake requirements specific to the components of energy expenditure, including resting metabolic rate, fetal development, and physical activity. We also provide an estimate of the energy required to gain the appropriate amount of fat mass that relates to the recommended weight gain.

Table 1. Subject Characteristics of Cohorts in Energy Requirement Studies of Pregnancy.

First Author	Measurement Time Points	Cohort Size	Ethnicity	Age	BMI	Excess GWG
	Weeks Gestation	N	White, AA, Other			
Butte, UW [10]	0, 22, 36	17	15, 0, 2	31 ± 4	18.9 ± 0.8	18%
Butte, NW [10]	0, 22, 36	34	24, 4, 5	30 ± 3	22.1 ± 1.5	35%
Butte, OV/OB [10]	0, 22, 36	12	9, 2, 1	31 ± 5	28.8 ± 2.6	100%
Forsum[1] [11]	0, 17, 30	22		29 ± 4	22.3 ± 3.1	
Forsum[2] [11]	0, 36	19		28 ± 4	22.1 ± 3.4	
Goldberg [12]	0, 6, 12, 18, 24, 30, 36	12	12, 0, 0	29 ± 3	23.0 ± 3.3	
Kopp [13]	0, 9, 25, 35	10		29 ± 5	23.1 ± 2.1	10%
Lof [14]	0, 14, 34	23		30 ± 4	24.2 ± 4.8	
Most [15]	15, 36	54	28, 22, 4	28 ± 5	35.8 ± 5.0	67%

Age and BMI (body mass index) are presented as mean ± SD at enrollment. AA: African-American, Others: Hispanic, Asian, Biracial, 'Other', Excess GWG: gestational weight gain exceeding the guidelines at the time (1990), UW: underweight, NW: normal weight, OV: overweight, OB: obesity, Forsum[1] and Forsum[2] indicate the two different cohorts in the study: both measured before pregnancy, but at different times during gestation; Butte reported data per BMI class: 'UW' indicates underweight: BMI ≤ 19.8 kg/m², 'NW' indicates normal weight: BMI 19.8–26 kg/m², and 'OV/OB' indicates overweight/obesity: BMI ≥ 26 kg/m².

To assist women and healthcare providers in achieving healthy rates of gestational weight gain, prediction models have been developed to estimate energy intake requirements. In the final Section 5 of this review, we discuss the accuracy, strengths and limitations of these prediction models and propose future directions for studies that are needed to better understand the high prevalence of excess gestational weight gain among pregnant women.

2. Overview of Energy Intake Requirements of Healthy Pregnancy

2.1. Definition of Energy Requirements

Energy intake requirements are defined as dietary intake that is essential for an individual to sustain optimal health outcomes [16]. In a non-pregnant population, this is generally weight maintenance. In that case, the goal is to achieve long-term energy balance, by maintaining an energy intake that approximates energy expenditure. Total daily energy expenditure (TDEE) is the sum of energy expended to support the resting metabolic rate (RMR), diet-induced thermogenesis and physical activity. RMR accounts for 60–80% of TDEE and is dependent on body mass, sex, age and race. Diet-induced thermogenesis or the energy expended in response to the digestion and processing of food accounts for 10% of TDEE. Lastly, activity-related energy expenditure, the most variable component, includes energy expended during activity associated with daily living and structured bouts of exercise. The quality of the diet and type of activity may affect the determinants of energy balance, but these interactions are not discussed.

2.2. Energy Requirements in Pregnancy

Energy intake requirements in pregnancy are defined as dietary intake necessary to support optimal development of maternal tissues as well as to support fetal growth and development [1]. Therefore, requirements encompass energy intake that not only balances maternal and fetal energy expenditure but also provides additional energy for fetal growth, as well as the growth of maternal tissues such as fat mass, breast tissue, uterus and placenta. Thus, energy intake requirements in pregnancy are not aimed at weight maintenance, but for appropriate rates of weight gain, which in turn, minimizes the risks of adverse outcomes in the mother and her offspring.

For women of normal weight (body mass index 18.5 to 24.9 kg/m^2), the recommended amount of weight gain during pregnancy is 11.5 to 16 kg, according to the 2009 Institute of Medicine guidelines [1]. Within this range of weight gain, the risk for non-elective cesarean delivery, postpartum weight retention, preterm birth, small- or large-for-gestational-age birth, and childhood obesity is minimized. The weight gain recommendations are inversely proportional to maternal body mass index at conception and, therefore, allow for more weight gain for women classified as underweight (12.5–18 kg) and less weight gain for women who are classified as overweight (7–11.5 kg) and with obesity (5–9 kg). For women with obesity, the weight gain recommendations have recently been challenged. Acknowledging limitations of available data [17–19], the Institute of Medicine was unable to suggest weight gain recommendations by obesity class [1]. Since then epidemiological studies propose that optimal pregnancy outcomes in women with severe obesity (body mass index ≥ 35 kg/m^2) are achieved if weight gain is restricted to less than 5 kg or weight is maintained across pregnancy [20–26].

3. Energy Requirements for Expenditure

3.1. Total Energy Expenditure

3.1.1. Appropriate Adjustment of Energy Expenditure Data

As in non-pregnant populations, there is a linear association between energy expenditure and body size in pregnant women such that energy expenditure is higher in heavier women, see Figure 1 [10,27,28]. To be able to assess metabolic or behavioral differences or adaptations in energy expenditure between women, energy expenditure needs to be compared independent of body weight and this requires appropriate statistical adjustment of the raw (or absolute) data [27]. The consensus among energy metabolism experts is that the relationship between energy expenditure and body weight (or body composition) is best characterized by a linear regression with an intercept greater than 0 (EE = a + b × BW or EE = a + b × FFM + c × FM), and not by a simple ratio (EE = b × BW). A ratio assumes that the intercept of the relationship between energy expenditure and weight (or body composition) is equal to 0, see Figure 1A. Using a linear regression rather than a ratio to describe the relationship between body weight and energy expenditure reduces the coefficients 'b' for body weight (regression: 11.97, ratio: 21.56), improves the fit of the regression line (regression R^2 = 0.66, ratio R^2 = 0.22), and eliminates systematic bias for the estimation of individual variability, see Figure 1B,C. Using a ratio, women with lower body weight will be falsely characterized as having high energy expenditure ('Residual'), and vice versa, women with more body weight will be falsely characterized as having low energy expenditure.

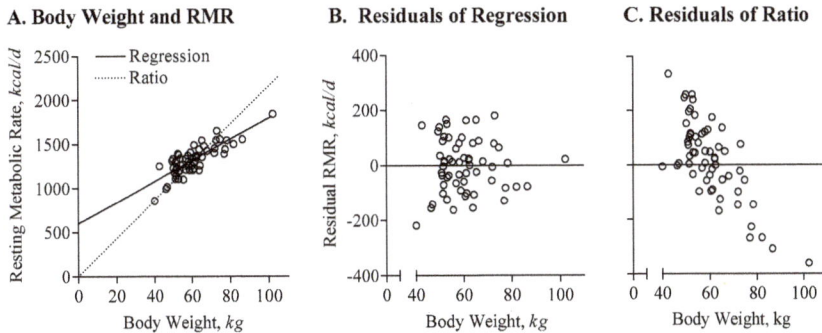

Figure 1. Appropriate adjustment of energy-expenditure data. (**A**) The association between body weight and resting metabolic rate is characterized by linear regression (solid line, RMR[*kcal/d*] = 598 + 11.97 × BW[*kg*], R^2 = 0.66) and ratio (pointed line, RMR[*kcal/d*] = 21.56 × BW[*kg*], R^2 = 0.22); of note, logarithmic or polynomial approximations do not improve the prediction of RMR (for both, R^2 = 0.67). (**B**) Residual resting metabolic rate (RMR), calculated as measured RMR minus predicted RMR; based on the regression equation in (**A**); the residuals show no structural over- or underestimate based on body weight. (**C**) Residual RMR, based on the ratio equation; women with increased body weight have significant lower residual RMR, and would be considered as having a low adjusted metabolic rate.

3.1.2. Measurement by Doubly Labeled Water

Ideally, TDEE is objectively measured in free-living conditions by doubly labeled water [29]. The principle behind doubly labeled water is to indirectly assess carbon dioxide production from the differential dilution rates of deuterium (^{2}H) and oxygen-18 isotope (^{18}O) after a prescribed dose of the isotope-cocktail is ingested. After ingestion of the water, deuterium dilutes into the body water pool and is excreted from the body via urine, sweat and water vapor in breath. In contrast, the oxygen isotope is excreted not only through the loss of water but also as carbon dioxide. Thus, the difference in dilution rates of ^{2}H and ^{18}O is proportional to carbon dioxide production. Then, estimation of TDEE from carbon dioxide production requires an assumption or quantification of the oxygen consumption, which can be inferred by measurement or estimation of the respiratory quotient (EE[*kcal/d*] = VCO$_2$ [*L/d*] × (1.1 + 3.9/RQ)) [29]. In most cases, the respiratory quotient or the ratio of carbon dioxide production to oxygen consumption will not be measured throughout the doubly labeled water period, and thus, requires assumptions. The best estimate for the respiratory quotient, which describes the macronutrient composition of the energy expenditure, is the food quotient, which describes the macronutrient composition of the energy intake. In most (pregnant and non-pregnant) studies, a respiratory quotient equivalent to a standard western-style diet of 0.86 is used (35–40% energy from fat) [30]. In 12 h overnight measurements of the energy expenditure, we observed no significant deviation from this commonly used value (unpublished).

The use of doubly labeled water to estimate TDEE is limited to a period of 7–14 days. Measurements of energy expenditure throughout pregnancy, therefore, require repeated, independent assessments of TDEE at the beginning and end of the assessment period of interest. The mean energy expenditure throughout pregnancy can then be calculated by the trapezoid method [31–35]. A limitation of this approach is the assumption of linearity between TDEE measurements. More frequent measurements of body weight and activity throughout this period, which is more feasible than TDEE, increase specificity, e.g., towards trimester-specific requirements.

3.1.3. First Trimester

Groups have performed cross-sectional comparisons of TDEE between pregnant and non-pregnant women [36–39]; however, these are outside of the scope of the current review. In Figure 2, we show published, time-series data of TDEE throughout pregnancy. On average, TDEE increases by approximately 15 kcal per day per week or by 420 kcal per day from pre-pregnancy to delivery. Prospective measurements of TDEE prior to pregnancy and in the first trimester have been reported in four studies of non-obese women [10–12,14]. One study included women with obesity ($n = 3$), but they were combined with overweight women [10]; another did not report changes in body weight during pregnancy [13]. On average, TDEE does not change during the first trimester, at an average increase in body weight of 3.2 kg [10–12,14]. Based on reported average changes in body mass and composition in these studies, we estimated the changes in TDEE after adjustment for changes in body mass. Using the linear regression equation of TDEE proposed by Gilmore and Butte et al. [31] that estimates TDEE from fat-free mass, fat mass and age ($\text{TDEE}_{pred}[kcal/d] = 1528 + 29.4 \times \text{FFM}[kg] + 13.6 \times \text{FM}[kg] - 21 \times \text{age}[years]$), body-weight-adjusted TDEE declines by 55 kcal/d from pre-pregnancy until 13 weeks of gestation across all studies. The components of TDEE that contribute to this decline could be of metabolic and/or behavioral origin, which we will discuss in the following sections.

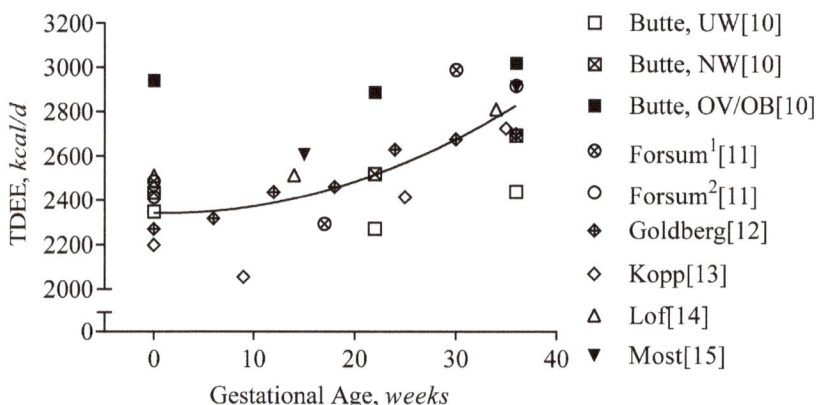

Figure 2. Total Daily Energy Expenditure during Gestation. Total daily energy expenditure (TDEE) is presented per study and gestational age (GA). The regression line represents the average increase during gestation ($\text{TDEE}[kcal/d] = 2343 - 1 \times \text{GA}[weeks] + 0.4 \times \text{GA}^2[weeks]$, $R^2 = 0.62$). Forsum[1] and Forsum[2] indicate the two different cohorts in the study: both measured before pregnancy, but at different times during gestation; Butte reported the changes in TDEE per BMI class: 'UW' indicates underweight: BMI $\leq$ 19.8 kg/m^2, 'NW' indicates normal weight: BMI 19.8–26 kg/m^2, and 'OV/OB' indicates overweight/obesity: BMI $\geq$ 26 kg/m^2.

3.1.4. Second and Third Trimester

Only half a dozen studies have obtained prospective measurements of TDEE during the second and third trimester of pregnancy using doubly labeled water [10–15]. On average, TDEE increased by 18 kcal/d/week or ~420 kcal/d from week 13 to 36, at an average increase in body weight of 10.3 kg [10–15]. After adjusting for the change in body weight, the increase in TDEE was 170 kcal/d from early to late pregnancy (13 to 36 weeks). Using individual data from 112 women [10,15], we estimated that after adjustment for body weight change, the increase in TDEE was only ~75 kcal/d; 110 kcal/d in non-obese women and 45 kcal/d in women with obesity.

Based on these findings, energy expenditures from the second to the third trimester, are linearly related to weight gain with an estimated 45–170 kcal/d increase. Aside from metabolic and/or behavioral adaptations to pregnancy by the mother, the additional increase in TDEE also reflects the contribution of the fetus, since fetal metabolic rate has a disproportionately high energy expenditure to mass (100 kcal/d/kg body weight vs. ~25 kcal/d/kg in adults) [40,41]. The metabolic contribution of the placenta to TDEE is poorly understood but is likely an additional contributing factor. Therefore, to understand the individual determinants of energy requirements in pregnancy, it is important to assess the components of TDEE separately.

3.2. Basal Metabolism

3.2.1. Measurement by Indirect Calorimetry

Differentiation of the metabolic and behavioral adaptations to pregnancy requires the assessment of individual components of TDEE, i.e., resting metabolic rate (RMR), diet-induced thermogenesis and activity-related energy expenditure. The largest component of TDEE is RMR, which accounts for 60–80% of daily energy requirements [10–12,14,28]. In majority of studies, RMR is measured by indirect calorimetry using a bedside ventilated hood system. In contrast to using doubly labeled water, oxygen consumption and carbon dioxide production are directly measured. The measurement of RMR occurs during fasting conditions over 30–60 min and extrapolated to 24 h. The RMR measurements have a high degree of accuracy, estimated at 3% [42], by test-retest reliability studies.

3.2.2. First Trimester

As with TDEE, only a few studies have assessed RMR before pregnancy and in early pregnancy. In Figure 3, we present data from studies with prospective assessments of RMR and TDEE. On average, RMR increased by 60 kcal/d during the first trimester (pre-pregnancy until week 13 [10–12,14] weeks gestation), and by 20 kcal/d after adjustment for the increase in body weight. A comparable increase in RMR (+30 kcal/d) was observed using individual data obtained by Butte and colleagues [10]. Thus, aside from changes in energy expenditure due to body weight, metabolic rate increases to a small extent, likely due to increased cardiac output [43] and/or hormonal changes, e.g., progesterone, human chorionic gonadotropin [44].

3.2.3. Second and Third Trimester

During the second and third trimester, RMR increases by 390 kcal/d (or 17 kcal/d/week). Considering the change in body weight, the increase in RMR across studies was predicted at 170 kcal/d [31]. Thus >50% of the increase in RMR (215 kcal/d) is not explained by body weight but by the increased metabolic cost of pregnancy [10–12,14] that includes increased cardiac output [43], increased work of breathing [45], fetal activity, metabolic rate of fetal tissues [40].

3.3. Physical Activity/Exercise

Physical activity throughout pregnancy is crucial for healthy gestation, despite the increased physiological demands placed on the body [46–48]. Accordingly, many specialized guidelines and recommendations have been established [49–53] to promote women to accrue sufficient activity throughout gestation. These guidelines recommend at least 150 min of moderate physical activity every week. In support of these physical activity guidelines, meta-analyses [46–48] report that habitual exercise reduces the risks of excessive gestational weight gain by 32%, gestational diabetes by 38%, hypertension by 39%, moderate depression by 67% and of macrosomia by 39%. Despite the concomitant benefits of following physical activity guidelines during pregnancy, presently, less than 50% of women adhere to the guidelines [54–58].

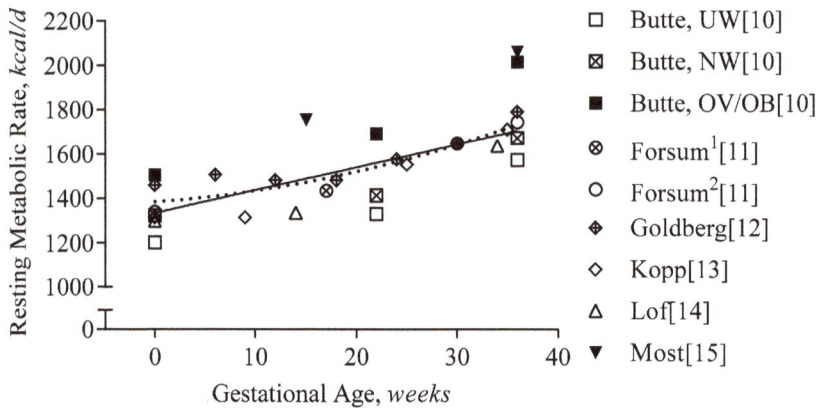

Figure 3. Resting Metabolic Rate during Gestation. Resting metabolic rate is presented per study and gestational age (GA). The regression line represents the average increase in the resting metabolic rate during gestation (RMR[*kcal/d*] = 1334 + 10.3 × GA[*weeks*], R^2 = 0.55, using non-linear regression (pointed line) did not improve the prediction, R^2 = 0.57). Forsum[1] and Forsum[2] indicate the two different cohorts in the study: both measured before pregnancy, but at different times during gestation; Butte reported the changes in RMR per BMI class: 'UW' indicates underweight: BMI ≤ 19.8 kg/m^2, 'NW' indicates normal weight: BMI 19.8–26 kg/m^2, and 'OV/OB' indicates overweight/obesity: BMI ≥ 26 kg/m^2.

3.3.1. Measurement of Physical Activity

Physical activity can be quantified through measurement of the energy expenditure related to physical activity or through measurement of body movement. Using available calorimetry data, the easiest manner of quantifying activity-related energy expenditure is by subtracting metabolic rate and diet-induced thermogenesis (10% of TDEE) from total daily energy expenditure (activity-related EE = TDEE − 0.1x TDEE − RMR). Because body size not only affects RMR but also activity-related energy expenditure, the most widely used adjustment is determined by assessing the physical activity level (PAL), calculated as TDEE divided by RMR (PAL = TDEE/RMR). Of note, PAL may underestimate physical activity levels due to the increased RMR during pregnancy, but validation studies have yet to be performed. To assess the inter-individual variability within a cohort or over time, activity-related energy expenditure can also be assessed as the residual of the linear regression, TDEE = a + b × RMR; this value is positive for subjects with higher physical activity than average and is negative for subjects with lower physical activity than average independent of metabolic body size. Because residual activity-related energy expenditure is adjusted for the RMR, this value is directly proportional to the amount of physical activity. Calorimetric assessments of activity-related energy expenditure are limited by their relatively high costs and are, therefore, limited to small cohorts.

Devices that objectively measure physical activity such as accelerometers (i.e., Actigraph, Actical, RT3, FitBit) are widely used in scientific studies, and by the public to overcome the limitations associated with calorimetric measurements. Accelerometry assesses the magnitude and duration of movement and gravitational forces of activity to yield outcome variables such as time-stamped activity counts, duration and intensity. An advantageous feature that some accelerometers possess (e.g., Actical) is the inclusion of a pedometer or step counter, which are shown to be valuable objective tools for assessing physical activity interventions in pregnant women [59–61]. Accelerometers can be combined with heart rate monitors to obtain a more objective measure of intensity, relative to an individual's fitness level [62,63]. To obtain a calorimetric estimate of physical activity (or total energy expenditure), accelerometry-variables are imputed into proprietary equations that predict, but do not measure, energy expenditure from body weight, height, and age of an individual.

Lastly, the easiest, cheapest and most established method for the evaluation of the level of physical activity include physical activity questionnaires, physical activity recall records and diaries. Specifically, during pregnancy, the 'gold standard' Physical Activity Questionnaire (PPAQ) is a widely accepted tool to assess the level of physical activity [64]. This self-administered questionnaire requires that participants report time spent in 32 different categories of activities including household/caregiving (13 activities), occupational (5 activities), sports/exercise (8 activities), transportation (3 activities), and in-activities (3 activities) [64]. Limitations of self-report methods include their low accuracy and reliability [65–67], as demonstrated by a significant overestimation of physical activity compared to accelerometers [65,68–70].

3.3.2. First Trimester

In Figure 4, we provide the physical activity level data of the calorimetry studies of pregnancy. During the first trimester, physical activity declined minimally in energy-requirement studies (−60 kcal/d, −2%, −0.08 PAL). Early in pregnancy, physical-activity-related energy expenditure accounted for 26–39% of energy expenditure in non-obese women [10–12,14] and for 23% in women with obesity [28].

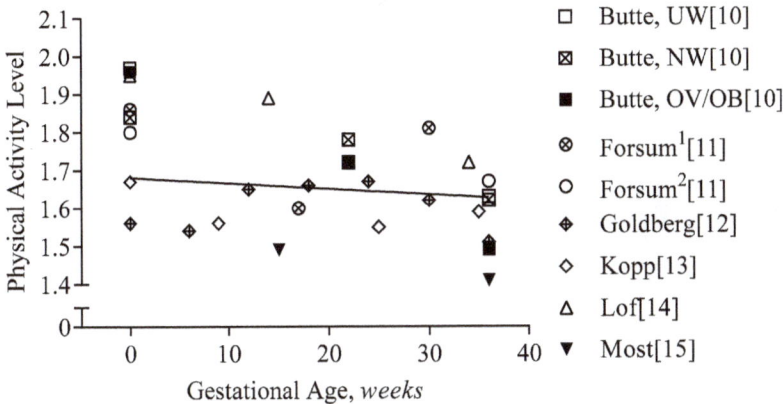

Figure 4. Physical activity level during gestation. Physical activity level is presented per study and gestational age (GA). The regression line represents the average decline in the physical activity level during gestation (PAL = 1.68 − 0.0015 × GA[*weeks*], R^2 = 0.02). Forsum[1] and Forsum[2] indicate the two different cohorts in the study: both measured before pregnancy, but at different times during gestation; Butte reported the changes in PAL per BMI class: 'UW' indicates underweight: BMI ≤ 19.8 kg/m[2], 'NW' indicates normal weight: BMI 19.8–26 kg/m[2], and 'OV/OB' indicates overweight/obesity: BMI ≥ 26 kg/m[2].

3.3.3. Second and Third Trimester

During the second and third trimesters, physical-activity-related energy expenditure decreased further by 3% (PAL, −0.12) [10–12,14]. Importantly, there is a significant variation between the studies; some report an increase in PAL of 0.37 [11], and others a decrease of −0.38 in women with overweight and obesity [10]; in a cohort of women who were exclusively obese (*n* = 54), PAL declined by 0.12 [15].

The variability in changes in activity is confirmed by studies using accelerometry. Some data show a decrease in physical activity level [55,71], while others reported no change [55,72,73]. The discrepancy in the literature may also relate to inconsistent methods of measurement, i.e., self-reported questionnaires vs. different various objective measures. Among a wide variety of demographic variables that may account for differences in physically active and inactive pregnant women, education and socio-economic status are consistently reported as predictors [54,74]. Therefore, while health care professionals should be encouraging all women to adopt an active lifestyle adherent to the recommendations of ≥150 min per week, they need to be cognizant of a woman's demographic and socio-economic status and ensure their guidance is relevant and feasible.

4. Energy Cost for Weight Gain

4.1. Determination of Appropriate Weight Gain and Composition

In addition to the energy intake requirements needed to balance energy expenditure, total energy intake requirements also encompass energy deposition which during pregnancy includes maternal fat accumulation and fetal growth. The amount and composition of healthy weight gain are highly variable between women, particularly by BMI class [8]. Large epidemiological studies have assessed the range of gestational weight gain for each BMI class that are associated with the lowest risk for adverse pregnancy outcomes [17–19]. Based on these studies, the recommendations for appropriate weight gain are inversely related to pregravid BMI [1]. Translating these weight gain goals into energy intake requirements involves an understanding of weight gain composition because the components, i.e., fat mass and fat-free mass, have different contributions to the total weight gain and have different energy densities.

The differences in recommended weight gain have been attributed to the variance in fat mass accumulation, but not fat-free mass [8,15,75,76]. Thus, recommendations for higher weight gain, e.g., in underweight or normal weight as compared to overweight and obese women, imply that more fat mass is accumulated while the variability in fat-free mass is smaller, and from the perspective of energy content, less significant.

In most pregnancy studies, the density of fat-free mass is assumed to be 771 kcal/kg and the density of fat mass 9500 kcal/kg [10,31,34,35]. Thus, the determination of energy intake requirements for pregnancy is developed through the energy balance equation as:

$$EIR \: [kcal/d] = TDEE + (771[kcal/kg] \times dFFM[kg] + 9500 \: [kcal/kg] \times dFM[kg])$$

in which EIR is the estimated energy intake requirement, TDEE is the average total energy expenditure during pregnancy, and dFM and dFFM are the changes in fat and fat-free mass during pregnancy.

4.2. Measurement of Body Composition in Pregnancy

The measurement of changes in body composition during pregnancy is challenging (time-constraints) and costly, thus limited to small cohorts. Available methods include air displacement plethysmography, isotope dilution, skinfold thickness measurement, bioimpedance, and magnetic resonance imaging, which we have recently reviewed for their applications in pregnancy [77]. While all methods can be used in pregnancy, the trade-off between cost, time and precision determines the best choice. The most important consideration for the use of all methods in pregnancy is the adjustment for an increase in fat-free mass hydration during pregnancy. Based on classic data by van Raaij et al. [78], we have developed a prediction equation to estimate fat-free mass hydration dependent on gestational age that allows standardization of methods across studies:

$$Hydration \: FFM \: [L/kg] = 0.724 + 0.00008484 \times GA[weeks] + 0.00001435 \times GA[weeks]^2$$

where FFM is fat-free mass and GA is gestational age; $R^2 = 0.998$, $p < 0.001$, and

$$Density \: FFM \: [kg/L] = 1.1 - 0.00002988 \times GA[weeks] - 0.00000731 \times GA[weeks]^2$$

where FFM is fat-free mass and GA is gestational age; $R^2 = 0.999$, $p < 0.001$.

4.3. First Trimester

Independent of BMI class, the weight gain recommendations during the first trimester are 0.5 to 2.0 kg for all women [1]. To date, no study has assessed the composition of early pregnancy weight gain, specifically for women that adhered to this recommendation. Using a conservative estimate for the energy density of overall weight gain (7400 kcal/kg) [79], the energy requirements for weight gain in early pregnancy are ~3,700–14,800 kcal, or 40–165 kcal/d over 13 weeks.

4.4. Second and Third Trimester

The accumulation of fat-free mass during the third and second trimester is ~7–8 kg and comparable between women of different BMI classes [8,10,77]. Assuming an energy density of 771 kcal/kg fat-free mass, the energy required to support this fat-free mass accumulation is 5000–6000 kcal or 30–35 kcal per day, while studies report ~4,000–12,000 kcal [80,81] or 20–60 kcal/d. Considering the low energy density of fat-free mass, the variability in fat-free mass accumulation does not strongly impact energy requirements. In contrast, the variability in fat mass accumulation largely affects the energy requirements for weight gain in pregnancy.

Several studies have assessed the energy requirements for weight and fat mass gain during the second and third trimesters, yet few studies have determined these requirements specific to women who gained the recommended amount of weight. Based on Lederman et al. [8], pregnant women with a normal weight and recommended weight gain (BMI 19.8–26 kg/m^2 and 11.5–16 kg, respectively) accumulated ~4 kg of fat mass between 14 and 37 weeks gestation. The energy intake requirements to support this energy deposition in fat is ~240 kcal/d over 22.7 weeks (i.e., 4 kg × 9500 kcal × kg^{-1}/159 d = 240 kcal/d). For underweight or overweight women, gains in fat mass were 6.0 and 2.8 kg, and the respective energy intake requirements were 360 kcal/d and 165 kcal/d. For women with obesity, we have recently shown that recommended weight gain is achieved with an average 2.5 kg loss of fat mass (15 to 36 weeks of gestation) [15]. Therefore, instead of women with obesity depositing energy, 160 kcal/d of energy are mobilized from fat mass in the second and third trimester, and can, therefore, be subtracted from energy intake requirements. Given that 160 kcal/d is mobilized from fat tissue, while the accumulation of fat-free mass only requires 20–60 kcal/d, the factor of energy storage in the energy balance equation is negative and thus energy intake in pregnant women with obesity should be lower than energy expenditure.

5. Models that Estimate Energy Intake Requirements

Experimental reports evaluating energy intake requirements have been an essential tool to educate women and clinicians on the optimal energy intake during pregnancy. The Institute of Medicine, using available data at the time published 'Equations to Estimate Energy Requirement for Pregnant Women' (p. 316) [1]. The equation is based on the requirements of non-pregnant women (EIR = 354 − 6.91 × age [*years*] + PA × 9.36 × BW [*kg*] + 726 × height [*m*], in which EIR are estimated energy requirements, PA is physical activity (1.00 for PAL < 1.4, 1.12 for 1.4 ≤ PAL ≤ 1.59, 1.27 for 1.6 ≤ PAL ≤ 1.89, and 1.45 for 1.9 ≤ PAL ≤ 2.5), and BW is body weight. Importantly, by including physical activity as a variable, this model accounts for individual differences in the level and changes in activity. Energy costs for weight gain are assumed to be constant (180 kcal/d).

A dynamic model for the estimation of energy requirements for optimal weight gain has been developed to better acknowledge the different contributions of fat and fat-free mass in different BMI classes [34]. This model based on the most comprehensive studies of energy requirements in women without obesity [10,12,13], estimates the energy requirements to achieve the recommended rate for weight gain. The TDEE in this model is estimated using maternal age, body weight, and height, but not physical activity.

Importantly, both the Institute of Medicine and the model by Thomas and colleagues, have included all available data to develop an energy requirement prediction, and, therefore, have included data of women with excess and inadequate weight gain. Future models can, thus, be improved by restricting models to those women with appropriate weight gain only.

5.1. First Trimester

For the recommended weight gain of 0.5–2 kg in the first trimester [1], the Institute of Medicine and American College of Obstetricians and Gynecologists recommended women maintain pregravid energy intake throughout the first trimester because the energy costs for weight gain are considered minimal [1,82].

The energy intake requirement model suggests that all women should consume an additional 100–200 kcal/d to support first-trimester weight gain, assuming that physical activity does not decline during the first trimester [10,34]. For women with obesity, we have recently shown that the model significantly overestimates energy requirements by 400 kcal/d due to unaccounted lower levels of physical activity as compared to the non-obese cohorts, on which the model is based [28].

Similar to the Institute of Medicine, we estimate the energy requirements to be 50–150 kcal/d. An average decline in physical activity (PAL, −0.08 or −5%, or 60 kcal/d) may compensate those energy costs, but this compensation may differ greatly between individuals.

5.2. Second and Third Trimester

The Institute of Medicine estimates the energy requirements during the second and third trimester to be 340 kcal/d and 452 kcal/d, respectively [83]. The estimate is calculated as the sum of the increased energy expenditure over pregnancy (8 kcal/d/week) and the costs of energy deposition (180 kcal/d). The increase in energy expenditure is consistent with our estimates, whereas the proposed energy cost for deposition neglects BMI-specific weight gain recommendations. In 2016, the American College of Obstetricians and Gynecologists have, thus, adjusted their recommendations by acknowledging that 'if you were overweight or obese, you may need fewer extra calories' [82], without specific guidance.

The energy intake requirements model by Thomas et al. [34] estimated surplus energy requirements to be only 400–600 kcal/d for underweight and normal-weight women, and 220–350 kcal per day for overweight or obese pregnant women [34]. Because our estimates are based on the same data, they agree for underweight, normal-weight and overweight women. However, the model was developed from women without obesity or with obesity coupled with excess gestational weight gain, and, therefore, for women with obesity, the model overestimates fat mass gains and thus energy requirements [15].

During the second and third trimester, we estimate that the energy requirements for expenditure increase by 15 kcal/d per week, half of which is explained by the increase in body weight and half by increased metabolic rate largely due to cardiac output and fetal metabolism. The energy requirements for recommended weight gain differ by BMI class and range between 360 kcal/d for underweight women, to −165 kcal/d for women with obesity, with the variation explained by fat mass accumulation [8,77] at a constant fat-free mass accumulation equivalent of ~50 kcal/d.

6. Limitations and Practical Considerations

6.1. First Trimester

In clinical practice and research settings, it is challenging to quantify changes occurring in the first trimester. Such an assessment requires a measurement prior to pregnancy and, ideally, close to conception. In most studies, pre-pregnancy weights are estimated or abstracted from charts, which lacks the control required to define energy requirements accurately. Moreover, access to weight data only precludes interpretation of changes in body composition. Given the importance of the changes during the first trimester, innovative study designs need to be developed to overcome this barrier.

6.2. Variability in Energy Expenditure

Based on findings by Ravussin et al. in the late 1980s [84], studies have investigated whether variability in the RMR would increase the risk of increased fat accumulation. Indeed, multiple studies [27,85–87] support this hypothesis, showing that women with low RMR, that is residual RMR adjusted for body weight or fat-free mass, gained more weight than women with higher adjusted RMR. However, these studies adjusted energy expenditure using ratios, which prohibits them from reliable conclusions. Nevertheless, in a re-analysis of the data generated by Butte (1994), we confirmed the hypothesis of excess weight gain in women with lower adjusted RMR, who were not obese [27]; however, we did not observe this in women with obesity [15]. The origin of these differences in RMR remains to be determined since body mass and composition explain only ~70% of its variation [28].

One of the determinants of energy expenditure is race. As in non-pregnant cohorts [88], RMR in pregnant African-American women is lower than in Caucasian women (−80 kcal/d or −5%) [33]. This difference does not change over time and, therefore, adjusting energy intake recommendations early in pregnancy is likely sufficient to consider the lower energy requirements of African-American women during pregnancy.

We have reported that TDEE adjusted for body composition is ~600 kcal/d lower in women with obesity [28] as compared to women without obesity [10,27,31]. Obesity status itself, after adjustment for body weight, does not seem to affect RMR [28]. The difference in TDEE likely relates to differences in physical activity, the timing of energy expenditure measurements and ethnicity of the cohorts. Indeed, women with obesity were less active than women without obesity (PAL, 1.5 vs. 1.8) [28]; energy expenditure in the cohort with obesity was measured ~8 weeks earlier than in the cohort without obesity (8 weeks × 15 kcal/week = 120 kcal/d) [10,31,77]; and 50% of women in the obese cohort were African-American women, who have significantly lower energy expenditure (−8%) [33].

6.3. Communication of Recommendations

The communication of energy intake recommendations must be improved. Studies show that only one in four women feels appropriately informed on weight gain or energy intake goals suitable for pregnancy [89]. Aside from the notion that some form of communication has to be initiated, qualitative studies are needed to inform how such information should be best communicated to patients. For example, based on our recent study in women with obesity, the same energy intake target can be expressed differently; e.g., 2700 kcal/d, 17 kcal/d/kg body weight, the diet should be equivalent to early pregnancy energy expenditure, or eat ~5% less during pregnancy. Possible ways of communicating these results to patients (pregnant women) that may be received and perceived differently are 'try and follow the energy intake requirement estimates using this equation/model' [1,34], 'try to maintain your usual diet throughout pregnancy', or 'try to eat 5% fewer calories than you are expending'. Importantly, such recommendations need to consider potential declines in physical activity. More work is needed to establish best practices for knowledge translation and dissemination of this information. How to best inform and guide women to make appropriate decisions and choices while pregnant is an active field of research.

7. Summary

In summary, energy intake requirements during the first trimester are minimally different from requirements before pregnancy (Table 2). Therefore, the Institute of Medicine model adopted from non-pregnancy provides valid estimates of energy requirements to match energy expenditure. During the second and third trimester, the development of the fetus and pregnancy-associated changes in the mother occur, i.e., increased blood volume, cardiac output, increase energy requirements by ~390 kcal/d, with small variations between women of differing body sizes. We estimate that 50% of this increase is proportional to the increase in body mass, and 50% is due to the increased metabolic costs of pregnancy and metabolic rate of the fetus. However, the energy requirements for weight gain differ significantly.

We estimate that to achieve the recommendations for healthy gestational weight gain [83], underweight women store 360 kcal/d as fat tissue, normal-weight women store ~240 kcal/d, overweight women store 165 kcal/d, and women with obesity do not store energy at all, but instead mobilize 160 kcal/d from adipose tissue energy stores.

Table 2. Summary of changes in body mass, composition and energy expenditure, proportional and independent of body weight changes.

	Tri-Mester	Body Weight	Fat-Free Mass	Fat Mass	TDEE	TDEE, Pred	TDEE, Res	RMR	RMR, Pred	RMR, Res
Non-Obesity										
Forsum1	1	2.1	0.2	1.9	−146	26	−172	73	25	48
Goldberg	1	1.7	0.2	1.5	181	22	159	26	21	5
Butte, UW	1	4.6	2.1	2.5	−45	91	−136	76	67	9
Butte, NW	1	3.4	1.7	1.7	51	68	−18	53	50	3
Butte, OV/OB	1	5.0	1.9	3.1	−31	93	−124	111	71	41
Lof	1	2.3	0.9	1.4	2	41	−39	33	32	1
Average	1	3.2	1.2	2.0	2	57	−55	62	44	18
Non-Obesity										
Forsum1	2 and 3	11.5	10.4	1.1	1226	312	914	381	199	181
Goldberg	2 and 3	8.0	5.9	2.0	252	193	59	298	129	168
Butte, UW	2 and 3	8.7	7.1	1.6	274	221	54	399	145	254
Butte, NW	2 and 3	11.7	8.9	2.8	284	290	−5	427	192	235
Butte, OV/OB	2 and 3	13.1	9.7	3.5	219	323	−104	531	215	316
Lof	2 and 3	10.8	7.5	3.3	341	256	85	349	173	176
Average		10.6	8.2	2.4	433	266	167	397	176	222
Obesity										
Most	2 and 3	8.5	8.2	0.3	335	237	98	335	150	185

Changes in kilograms and kcal per day. TDEE: Total daily energy expenditure, Pred: predicted, changes proportional to changes in body weight, Res: Residual, changes independent of changes in body weight, RMR: resting metabolic rate. Forsum1 indicate the two different cohorts in the study: both measured before pregnancy, but at different times during gestation; Butte reported the changes in PAL per BMI class: 'UW' indicates underweight: BMI $\leq$ 19.8 kg/m^2, 'NW' indicates normal weight: BMI 19.8–26 kg/m^2, and 'OV/OB' indicates overweight/obesity: BMI $\geq$ 26 kg/m^2.

Current methods to assess energy intake requirements are limited in their prediction of energy expenditure and energy storage. Specifically, body composition and physical activity data during the first trimester are scarce. Given the importance of the first trimester for placental and fetal development [90], more studies are required to better inform patients and clinicians how to achieve healthier pregnancies. Particularly in women with obesity, energy expenditure (i.e., physical activity) and energy deposition are overestimated and thus would allow women to consume an amount of energy that would result in excess gestational weight gain. The level of physical activity is the most variable component of energy expenditure and requires consideration on an individual level. Commercially available accelerometers or built-in devices in mobile phones may assist women to maintain or increase their level of physical activity during pregnancy; currently, less than 50% of women meet the recommendations by leading institutions dedicated to promoting a healthy pregnancy. Adherence to recommendations related to energy intake and energy expenditure (physical activity) are key factors that will assist women in maintaining an appropriate energy balance thereby supporting the needs of pregnancy and the growing fetus.

Author Contributions: Conceptualization, J.M, K.B.A., L.M.R.; methodology, J.M., S.D.; software, J.M.; validation, K.B.A., F.H., L.M.R.; formal analysis, J.M.; investigation, J.M., S.D., F.H., K.B.A., L.M.R.; K.B.A., L.M.R.; data curation, J.M.; writing—original draft preparation, J.M., S.D.; writing—review and editing, L.M.R.; visualization, J.M.; supervision, K.B.A., L.M.R; project administration, K.B.A, L.M.R.; funding acquisition, K.B.A., L.M.R.

Funding: This research was funded by NIH, grant number R01DK099175 (L.M.R.), and by CIHR, grant number MOP 142298 (K.B.A.).

Nutrients **2019**, *11*, 1812

Acknowledgments: We would like to acknowledge Drs Nancy Butte and Anne Gilmore for their work on energy requirements in women without obesity, and for providing us with individual data of this study.

Conflicts of Interest: The authors declare no conflict of interest. The funders had no role in the design of the study; in the collection, analyses, or interpretation of data; in the writing of the manuscript, or in the decision to publish the results.

References

1. Institute of Medicine (US) and National Research Council (US) Committee to Reexamine IOM Pregnancy Weight Guidelines. Determining optimal weight gain. In *Weight Gain during Pregnancy: Reexamining the Guidelines*; Rasmussen, K.M., Yaktine, A.L., Eds.; National Academies Press: Washington, DC, USA, 2009.
2. Kominiarek, M.A.; Peaceman, A.M. Gestational weight gain. *Am. J. Obstet. Gynecol.* **2017**, *217*, 642–651. [CrossRef] [PubMed]
3. Laitinen, J.; Jaaskelainen, A.; Hartikainen, A.L.; Sovio, U.; Vaarasmaki, M.; Pouta, A.; Kaakinen, M.; Jarvelin, M.R. Maternal weight gain during the first half of pregnancy and offspring obesity at 16 years: A prospective cohort study. *BJOG* **2012**, *119*, 716–723. [CrossRef] [PubMed]
4. Catalano, P.M.; Farrell, K.; Thomas, A.; Huston-Presley, L.; Mencin, P.; de Mouzon, S.H.; Amini, S.B. Perinatal risk factors for childhood obesity and metabolic dysregulation. *Am. J. Clin. Nutr.* **2009**, *90*, 1303–1313. [CrossRef] [PubMed]
5. Sridhar, S.B.; Darbinian, J.; Ehrlich, S.F.; Markman, M.A.; Gunderson, E.P.; Ferrara, A.; Hedderson, M.M. Maternal gestational weight gain and offspring risk for childhood overweight or obesity. *Am. J. Obstet. Gynecol.* **2014**, *211*, 259.e1–259.e8. [CrossRef] [PubMed]
6. Reynolds, R.M.; Allan, K.M.; Raja, E.A.; Bhattacharya, S.; McNeill, G.; Hannaford, P.C.; Sarwar, N.; Lee, A.J.; Bhattacharya, S.; Norman, J.E. Maternal obesity during pregnancy and premature mortality from cardiovascular event in adult offspring: Follow-up of 1 323 275 person years. *BMJ* **2013**, *347*, f4539. [CrossRef] [PubMed]
7. Mamun, A.A.; Kinarivala, M.; O'Callaghan, M.J.; Williams, G.M.; Najman, J.M.; Callaway, L.K. Associations of excess weight gain during pregnancy with long-term maternal overweight and obesity: Evidence from 21 y postpartum follow-up. *Am. J. Clin. Nutr.* **2010**, *91*, 1336–1341. [CrossRef] [PubMed]
8. Lederman, S.A.; Paxton, A.; Heymsfield, S.B.; Wang, J.; Thornton, J.; Pierson, R.N., Jr. Body fat and water changes during pregnancy in women with different body weight and weight gain. *Obstet. Gynecol.* **1997**, *90*, 483–488. [CrossRef]
9. Flegal, K.M.; Kruszon-Moran, D.; Carroll, M.D.; Fryar, C.D.; Ogden, C.L. Trends in obesity among adults in the united states, 2005 to 2014. *JAMA* **2016**, *315*, 2284–2291. [CrossRef]
10. Butte, N.F.; Wong, W.W.; Treuth, M.S.; Ellis, K.J.; O'Brian Smith, E. Energy requirements during pregnancy based on total energy expenditure and energy deposition. *Am. J. Clin. Nutr.* **2004**, *79*, 1078–1087. [CrossRef]
11. Forsum, E.; Kabir, N.; Sadurskis, A.; Westerterp, K. Total energy expenditure of healthy swedish women during pregnancy and lactation. *Am. J. Clin. Nutr.* **1992**, *56*, 334–342. [CrossRef]
12. Goldberg, G.R.; Prentice, A.M.; Coward, W.A.; Davies, H.L.; Murgatroyd, P.R.; Wensing, C.; Black, A.E.; Harding, M.; Sawyer, M. Longitudinal assessment of energy expenditure in pregnancy by the doubly labeled water method. *Am. J. Clin. Nutr.* **1993**, *57*, 494–505. [CrossRef] [PubMed]
13. Kopp-Hoolihan, L.E.; van Loan, M.D.; Wong, W.W.; King, J.C. Longitudinal assessment of energy balance in well-nourished, pregnant women. *Am. J. Clin. Nutr.* **1999**, *69*, 697–704. [CrossRef] [PubMed]
14. Lof, M.; Forsum, E. Activity pattern and energy expenditure due to physical activity before and during pregnancy in healthy swedish women. *Br. J. Nutr.* **2006**, *95*, 296–302. [CrossRef] [PubMed]
15. Most, J.; St Amant, M.; Hsia, D.; Altazan, A.; Thomas, D.; Gilmore, A.; Vallo, P.; Beyl, R.; Ravussin, E.; Redman, L. Evidence-based recommendations for energy intake in pregnant women with obesity. *J. Clin. Invest.* **2019**, *130*. [CrossRef] [PubMed]
16. Institute of Medicine. Energy. In *Dietary Reference Intake for Energy, Carbohydrate, Fiber, Fat, Fatty Acids, Cholesterol, Protein, and Amino Acids*; The National Academies Press: Washington, DC, USA, 2005; pp. 107–264.
17. Bracero, L.A.; Byrne, D.W. Optimal maternal weight gain during singleton pregnancy. *Gynecol. Obstet. Investig.* **1998**, *46*, 9–16. [CrossRef] [PubMed]

18. Cedergren, M.I. Optimal gestational weight gain for body mass index categories. *Obstet. Gynecol.* **2007**, *110*, 759–764. [CrossRef] [PubMed]

19. Nohr, E.A.; Vaeth, M.; Baker, J.L.; Sorensen, T.; Olsen, J.; Rasmussen, K.M. Combined associations of prepregnancy body mass index and gestational weight gain with the outcome of pregnancy. *Am. J. Clin. Nutr.* **2008**, *87*, 1750–1759. [CrossRef]

20. Blomberg, M. Maternal and neonatal outcomes among obese women with weight gain below the new institute of medicine recommendations. *Obstet. Gynecol.* **2011**, *117*, 1065–1070. [CrossRef]

21. Bodnar, L.M.; Pugh, S.J.; Lash, T.L.; Hutcheon, J.A.; Himes, K.P.; Parisi, S.M.; Abrams, B. Low gestational weight gain and risk of adverse perinatal outcomes in obese and severely obese women. *Epidemiology* **2016**, *27*, 894–902. [CrossRef]

22. Bodnar, L.M.; Siminerio, L.L.; Himes, K.P.; Hutcheon, J.A.; Lash, T.L.; Parisi, S.M.; Abrams, B. Maternal obesity and gestational weight gain are risk factors for infant death. *Obesity* **2016**, *24*, 490–498. [CrossRef]

23. Faucher, M.A.; Barger, M.K. Gestational weight gain in obese women by class of obesity and select maternal/newborn outcomes: A systematic review. *Women Birth* **2015**, *28*, e70–e79. [CrossRef] [PubMed]

24. Robillard, P.Y.; Dekker, G.; Boukerrou, M.; Le Moullec, N.; Hulsey, T.C. Relationship between pre-pregnancy maternal bmi and optimal weight gain in singleton pregnancies. *Heliyon* **2018**, *4*, e00615. [CrossRef] [PubMed]

25. Durie, D.E.; Thornburg, L.L.; Glantz, J.C. Effect of second-trimester and third-trimester rate of gestational weight gain on maternal and neonatal outcomes. *Obstet. Gynecol.* **2011**, *118*, 569–575. [CrossRef] [PubMed]

26. Santos, S.; Eekhout, I.; Voerman, E.; Gaillard, R.; Barros, H.; Charles, M.A.; Chatzi, L.; Chevrier, C.; Chrousos, G.P.; Corpeleijn, E.; et al. Gestational weight gain charts for different body mass index groups for women in europe, north america, and oceania. *BMC Med.* **2018**, *16*, 201. [CrossRef] [PubMed]

27. Most, J.; Redman, L.M. Does energy expenditure influence body fat accumulation in pregnancy? *Am. J. Obstet. Gynecol.* **2019**, *220*, 119–120. [CrossRef] [PubMed]

28. Most, J.; Vallo, P.M.; Gilmore, L.A.; St Amant, M.; Hsia, D.S.; Altazan, A.D.; Beyl, R.A.; Ravussin, E.; Redman, L.M. Energy expenditure in pregnant women with obesity does not support energy intake recommendations. *Obesity* **2018**, *26*, 992–999. [CrossRef] [PubMed]

29. International Atomic Energy Agency. *Assessment of Body Composition and Total Energy Expenditure in Humans Using Stable Isotope Techniques*; IAEA: Vienna, Austria, 2009.

30. Black, A.E.; Prentice, A.M.; Coward, W.A. Use of food quotients to predict respiratory quotients for the doubly-labelled water method of measuring energy expenditure. *Hum. Nutr. Clin. Nutr.* **1986**, *40*, 381–391. [PubMed]

31. Gilmore, L.A.; Butte, N.F.; Ravussin, E.; Han, H.; Burton, J.H.; Redman, L.M. Energy intake and energy expenditure for determining excess weight gain in pregnant women. *Obstet. Gynecol.* **2016**, *127*, 884–892. [CrossRef] [PubMed]

32. Gilmore, L.A.; Ravussin, E.; Bray, G.A.; Han, H.; Redman, L.M. An objective estimate of energy intake during weight gain using the intake-balance method. *Am. J. Clin. Nutr.* **2014**, *100*, 806–812. [CrossRef] [PubMed]

33. Most, J.; Gilmore, L.A.; Altazan, A.D.; St Amant, M.; Beyl, R.A.; Ravussin, E.; Redman, L.M. Propensity for adverse pregnancy outcomes in african-american women may be explained by low energy expenditure in early pregnancy. *Am. J. Clin. Nutr.* **2018**, *107*, 957–964. [CrossRef]

34. Thomas, D.M.; Navarro-Barrientos, J.E.; Rivera, D.E.; Heymsfield, S.B.; Bredlau, C.; Redman, L.M.; Martin, C.K.; Lederman, S.A.; Linda, M.C.; Butte, N.F. Dynamic energy-balance model predicting gestational weight gain. *Am. J. Clin. Nutr.* **2012**, *95*, 115–122. [CrossRef] [PubMed]

35. Most, J.; Broskey, N.T.; Altazan, A.D.; Beyl, R.A.; St Amant, M.; Hsia, D.S.; Ravussin, E.; Redman, L.M. Is energy balance in pregnancy involved in the etiology of gestational diabetes in women with obesity? *Cell Metab.* **2019**, *29*, 231–233. [CrossRef] [PubMed]

36. Heini, A.; Schutz, Y.; Diaz, E.; Prentice, A.M.; Whitehead, R.G.; Jequier, E. Free-living energy expenditure measured by two independent techniques in pregnant and nonpregnant gambian women. *Am. J. Physiol.* **1991**, *261*, E9–E17. [CrossRef] [PubMed]

37. Singh, J.; Prentice, A.M.; Diaz, E.; Coward, W.A.; Ashford, J.; Sawyer, M.; Whitehead, R.G. Energy expenditure of gambian women during peak agricultural activity measured by the doubly-labelled water method. *Br. J. Nutr.* **1989**, *62*, 315–329. [CrossRef] [PubMed]

38. Lof, M. Physical activity pattern and activity energy expenditure in healthy pregnant and non-pregnant swedish women. *Eur. J. Clin. Nutr.* **2011**, *65*, 1295–1301. [CrossRef] [PubMed]

39. Gradmark, A.; Pomeroy, J.; Renstrom, F.; Steiginga, S.; Persson, M.; Wright, A.; Bluck, L.; Domellof, M.; Kahn, S.E.; Mogren, I.; et al. Physical activity, sedentary behaviors, and estimated insulin sensitivity and secretion in pregnant and non-pregnant women. *BMC Pregnancy Childbirth* **2011**, *11*, 44. [CrossRef] [PubMed]

40. Diem, K. Metabolisme basal. In *Tables Scientifiques*; Geigy, J.R.S.A., Ed.; Departement Pharmaceutique: Basel, Switzerland, 1963; Volume 6, pp. 637–643.

41. Sparks, J.W.; Girard, J.R.; Battaglia, F.C. An estimate of the caloric requirements of the human fetus. *Neonatology* **1980**, *38*, 113–119. [CrossRef] [PubMed]

42. Lam, Y.Y.; Ravussin, E. Analysis of energy metabolism in humans: A review of methodologies. *Mol. Metab.* **2016**, *5*, 1057–1071. [CrossRef] [PubMed]

43. Meah, V.L.; Cockcroft, J.R.; Backx, K.; Shave, R.; Stohr, E.J. Cardiac output and related haemodynamics during pregnancy: A series of meta-analyses. *Heart* **2016**, *102*, 518–526. [CrossRef] [PubMed]

44. Napso, T.; Yong, H.E.J.; Lopez-Tello, J.; Sferruzzi-Perri, A.N. The role of placental hormones in mediating maternal adaptations to support pregnancy and lactation. *Front. Physiol.* **2018**, *9*, 1091. [CrossRef]

45. Mottola, M.F. Physical activity and maternal obesity: Cardiovascular adaptations, exercise recommendations, and pregnancy outcomes. *Nutr. Rev.* **2013**, *71* (Suppl. 1), S31–S36. [CrossRef]

46. Ferraro, Z.M.; Contador, F.; Tawfiq, A.; Adamo, K.B.; Gaudet, L. Gestational weight gain and medical outcomes of pregnancy. *Obstet. Med.* **2015**, *8*, 133–137. [CrossRef] [PubMed]

47. Davenport, M.H.; Sobierajski, F.; Mottola, M.F.; Skow, R.J.; Meah, V.L.; Poitras, V.J.; Gray, C.E.; Jaramillo Garcia, A.; Barrowman, N.; Riske, L.; et al. Glucose responses to acute and chronic exercise during pregnancy: A systematic review and meta-analysis. *Br. J. Sports Med.* **2018**, *52*, 1357–1366. [CrossRef] [PubMed]

48. Ruchat, S.M.; Mottola, M.F.; Skow, R.J.; Nagpal, T.S.; Meah, V.L.; James, M.; Riske, L.; Sobierajski, F.; Kathol, A.J.; Marchand, A.A.; et al. Effectiveness of exercise interventions in the prevention of excessive gestational weight gain and postpartum weight retention: A systematic review and meta-analysis. *Br. J. Sports Med.* **2018**, *52*, 1347–1356. [CrossRef] [PubMed]

49. Pate, R.R.; Pratt, M.; Blair, S.N.; Haskell, W.L.; Macera, C.A.; Bouchard, C.; Buchner, D.; Ettinger, W.; Heath, G.W.; King, A.C.; et al. Physical activity and public health. A recommendation from the centers for disease control and prevention and the american college of sports medicine. *JAMA* **1995**, *273*, 402–407. [CrossRef] [PubMed]

50. Ferguson, B. Acsm's guidelines for exercise testing and prescription 9th ed. 2014. *J. Can. Chiropr. Assoc.* **2014**, *58*, 328.

51. Lauer, E.E.; Jackson, A.W.; Martin, S.B.; Morrow, J.R., Jr. Meeting usdhhs physical activity guidelines and health outcomes. *Int. J. Exerc. Sci.* **2017**, *10*, 121–127. [PubMed]

52. Davenport, M.H.; Ruchat, S.M.; Mottola, M.F.; Davies, G.A.; Poitras, V.J.; Gray, C.E.; Garcia, A.J.; Barrowman, N.; Adamo, K.B.; Duggan, M.; et al. 2019 canadian guideline for physical activity throughout pregnancy: Methodology. *J. Obstet. Gynaecol. Can.* **2018**, *40*, 1468–1483. [CrossRef] [PubMed]

53. Mottola, M.F.; Davenport, M.H.; Ruchat, S.-M.; Davies, G.A.; Poitras, V.J.; Gray, C.E.; Jaramillo Garcia, A.; Barrowman, N.; Adamo, K.B.; Duggan, M.; et al. 2019 canadian guideline for physical activity throughout pregnancy. *Br. J. Sports Med.* **2018**, *52*, 1339. [CrossRef] [PubMed]

54. Ribeiro Santos, P.C.; Abreu, S.; Moreira, C.; Lopes, D.; Santos, R.; Alves, O.; Silva, P.; Montenegro, N.; Mota, J. Impact of compliance with different guidelines on physical activity during pregnancy and perceived barriers to leisure physical activity. *J. Sports Sci.* **2014**, *32*, 1398–1408. [CrossRef]

55. Hesketh, K.R.; Evenson, K.R. Prevalence of us pregnant women meeting 2015 acog physical activity guidelines. *Am. J. Prev. Med.* **2016**, *51*, e87–e89. [CrossRef] [PubMed]

56. Di Fabio, D.R.; Blomme, C.K.; Smith, K.M.; Welk, G.J.; Campbell, C.G. Adherence to physical activity guidelines in mid-pregnancy does not reduce sedentary time: An observational study. *Int. J. Behav. Nutr. Phys. Act.* **2015**, *12*, 27. [CrossRef] [PubMed]

57. Hayman, M.; Short, C.; Reaburn, P. An investigation into the exercise behaviours of regionally based australian pregnant women. *J. Sci. Med. Sport* **2016**, *19*, 664–668. [CrossRef] [PubMed]

58. Smith, K.M.; Campbell, C.G. Physical activity during pregnancy: Impact of applying different physical activity guidelines. *J. Pregnancy* **2013**, *2013*, 165617. [CrossRef] [PubMed]

59. Renault, K.M.; Norgaard, K.; Nilas, L.; Carlsen, E.M.; Cortes, D.; Pryds, O.; Secher, N.J. The treatment of obese pregnant women (top) study: A randomized controlled trial of the effect of physical activity intervention assessed by pedometer with or without dietary intervention in obese pregnant women. *Am. J. Obstet. Gynecol.* **2014**, *210*, 134.e1–134.e9. [CrossRef] [PubMed]

60. Medek, H.; Halldorsson, T.; Gunnarsdottir, I.; Geirsson, R.T. Physical activity of relatively high intensity in mid-pregnancy predicts lower glucose tolerance levels. *Acta Obstet. Gynecol. Scand.* **2016**, *95*, 1055–1062. [CrossRef] [PubMed]

61. Renault, K.M.; Carlsen, E.M.; Haedersdal, S.; Nilas, L.; Secher, N.J.; Eugen-Olsen, J.; Cortes, D.; Olsen, S.F.; Halldorsson, T.I.; Norgaard, K. Impact of lifestyle intervention for obese women during pregnancy on maternal metabolic and inflammatory markers. *Int. J. Obes.* **2017**, *41*, 598–605. [CrossRef]

62. Kraal, J.J.; Sartor, F.; Papini, G.; Stut, W.; Peek, N.; Kemps, H.M.; Bonomi, A.G. Energy expenditure estimation in beta-blocker-medicated cardiac patients by combining heart rate and body movement data. *Eur. J. Prev. Cardiol.* **2016**, *23*, 1734–1742. [CrossRef]

63. Kim, D.; Cho, J.; Oh, H.; Chee, Y.; Kim, I. The estimation method of physical activity energy expenditure considering heart rate variability. In Proceedings of the 2009 36th Annual Computers in Cardiology Conference (CinC), Park City, UT, USA, 13–16 September 2009; pp. 413–416.

64. Chasan-Taber, L.; Schmidt, M.D.; Roberts, D.E.; Hosmer, D.; Markenson, G.; Freedson, P.S. Development and validation of a pregnancy physical activity questionnaire. *Med. Sci. Sports Exerc.* **2004**, *36*, 1750–1760. [CrossRef]

65. Ara, I.; Aparicio-Ugarriza, R.; Morales-Barco, D.; Nascimento de Souza, W.; Mata, E.; Gonzalez-Gross, M. Physical activity assessment in the general population; validated self-report methods. *Nutr. Hosp.* **2015**, *31* (Suppl. 3), 211–218.

66. Bell, R.; Tennant, P.W.; McParlin, C.; Pearce, M.S.; Adamson, A.J.; Rankin, J.; Robson, S.C. Measuring physical activity in pregnancy: A comparison of accelerometry and self-completion questionnaires in overweight and obese women. *Eur. J. Obstet. Gynecol. Reprod. Biol.* **2013**, *170*, 90–95. [CrossRef] [PubMed]

67. Harrison, C.L.; Thompson, R.G.; Teede, H.J.; Lombard, C.B. Measuring physical activity during pregnancy. *Int. J. Behav. Nutr. Phys. Act.* **2011**, *8*, 19. [CrossRef] [PubMed]

68. Brett, K.E.; Wilson, S.; Ferraro, Z.M.; Adamo, K.B. Self-report pregnancy physical activity questionnaire overestimates physical activity. *Can. J. Public Health* **2015**, *106*, e297–e302. [CrossRef] [PubMed]

69. Lemmens, P.M.C.; Sartor, F.; Cox, L.G.E.; den Boer, S.V.; Westerink, J. Evaluation of an activity monitor for use in pregnancy to help reduce excessive gestational weight gain. *BMC Pregnancy Childbirth* **2018**, *18*, 312. [CrossRef] [PubMed]

70. Conway, M.R.; Marshall, M.R.; Schlaff, R.A.; Pfeiffer, K.A.; Pivarnik, J.M. Physical activity device reliability and validity during pregnancy and postpartum. *Med. Sci. Sports Exerc.* **2018**, *50*, 617–623. [CrossRef] [PubMed]

71. Mudd, L.M.; Evenson, K.R. Review of impacts of physical activity on maternal metabolic health during pregnancy. *Curr. Diabetes Rep.* **2015**, *15*, 6. [CrossRef] [PubMed]

72. Acog Committee Opinion. Exercise during pregnancy and the postpartum period. *Int. J. Gynaecol. Obstet.* **2002**, *77*, 79–81. [CrossRef]

73. Horns, P.N.; Ratcliffe, L.P.; Leggett, J.C.; Swanson, M.S. Pregnancy outcomes among active and sedentary primiparous women. *J. Obstet. Gynecol. Neonatal Nurs.* **1996**, *25*, 49–54. [CrossRef]

74. Gaston, A.; Cramp, A. Exercise during pregnancy: A review of patterns and determinants. *J. Sci. Med. Sport* **2011**, *14*, 299–305. [CrossRef]

75. Butte, N.F.; Ellis, K.J.; Wong, W.W.; Hopkinson, J.M.; Smith, E.O. Composition of gestational weight gain impacts maternal fat retention and infant birth weight. *Am. J. Obstet. Gynecol.* **2003**, *189*, 1423–1432. [CrossRef]

76. Berggren, E.K.; Groh-Wargo, S.; Presley, L.; Hauguel-de Mouzon, S.; Catalano, P.M. Maternal fat, but not lean, mass is increased among overweight/obese women with excess gestational weight gain. *Am. J. Obstet. Gynecol.* **2016**, *214*, e741–e745. [CrossRef] [PubMed]

77. Most, J.; Marlatt, K.L.; Altazan, A.D.; Redman, L.M. Advances in assessing body composition during pregnancy. *Eur. J. Clin. Nutr.* **2018**, *72*, 645–656. [CrossRef] [PubMed]

78. van Raaij, J.M.; Peek, M.E.; Vermaat-Miedema, S.H.; Schonk, C.M.; Hautvast, J.G. New equations for estimating body fat mass in pregnancy from body density or total body water. *Am. J. Clin. Nutr.* **1988**, *48*, 24–29. [CrossRef] [PubMed]

79. Tataranni, P.A.; Harper, I.T.; Snitker, S.; Del Parigi, A.; Vozarova, B.; Bunt, J.; Bogardus, C.; Ravussin, E. Body weight gain in free-living pima indians: Effect of energy intake vs. expenditure. *Int. J. Obes. Relat. Metab. Disord.* **2003**, *27*, 1578–1583. [CrossRef] [PubMed]

80. Butte, N.F.; King, J.C. Energy requirements during pregnancy and lactation. *Public Health Nutr.* **2005**, *8*, 1010–1027. [CrossRef] [PubMed]

81. Durnin, J.V. Energy requirements of pregnancy. *Diabetes* **1991**, *40* (Suppl. 2), 152–156. [CrossRef]

82. American College of Obstetricians and Gynecologists. Nutrition in pregnancy. In *Your Pregnancy and Childbirth: Month to Month*, 6th ed.; ACOG: Washington, DC, USA, 2016; pp. 313–327.

83. IOM (Institute of Medicine); NRC (National Research Council). *Weight Gain during Pregnancy: Reexamining the Guidelines*; The National Academies Press: Washington, DC, USA, 2009.

84. Ravussin, E.; Lillioja, S.; Knowler, W.C.; Christin, L.; Freymond, D.; Abbott, W.G.; Boyce, V.; Howard, B.V.; Bogardus, C. Reduced rate of energy expenditure as a risk factor for body-weight gain. *N. Engl. J. Med.* **1988**, *318*, 467–472. [CrossRef] [PubMed]

85. Berggren, E.K.; O'Tierney-Ginn, P.; Lewis, S.; Presley, L.; De-Mouzon, S.H.; Catalano, P.M. Variations in resting energy expenditure: Impact on gestational weight gain. *Am. J. Obstet. Gynecol.* **2017**, *217*, 445.e1–445.e6. [CrossRef] [PubMed]

86. Lof, M.; Olausson, H.; Bostrom, K.; Janerot-Sjoberg, B.; Sohlstrom, A.; Forsum, E. Changes in basal metabolic rate during pregnancy in relation to changes in body weight and composition, cardiac output, insulin-like growth factor I, and thyroid hormones and in relation to fetal growth. *Am. J. Clin. Nutr.* **2005**, *81*, 678–685. [CrossRef]

87. Meng, Y.; Groth, S.W.; Stewart, P.; Smith, J.A. An exploration of the determinants of gestational weight gain in african american women: Genetic factors and energy expenditure. *Biol. Res. Nurs.* **2018**, *20*, 118–125. [CrossRef]

88. Katzmarzyk, P.T.; Most, J.; Redman, L.M.; Rood, J.; Ravussin, E. Energy expenditure and substrate oxidation in white and african american young adults without obesity. *Eur. J. Clin. Nutr.* **2018**, *72*, 920–922. [CrossRef] [PubMed]

89. Deputy, N.P.; Sharma, A.J.; Kim, S.Y.; Olson, C.K. Achieving appropriate gestational weight gain: The role of healthcare provider advice. *J. Womens Health* **2018**, *27*, 552–560. [CrossRef] [PubMed]

90. Stephenson, J.; Heslehurst, N.; Hall, J.; Schoenaker, D.; Hutchinson, J.; Cade, J.E.; Poston, L.; Barrett, G.; Crozier, S.R.; Barker, M.; et al. Before the beginning: Nutrition and lifestyle in the preconception period and its importance for future health. *Lancet* **2018**, *391*, 1830–1841. [CrossRef]

nutrients

MDPI

Article

Uncontrolled Eating during Pregnancy Predicts Fetal Growth: The Healthy Mom Zone Trial

Jennifer S. Savage [1,2,*], Emily E. Hohman [1], Katherine M. McNitt [1,2], Abigail M. Pauley [3], Krista S. Leonard [3], Tricia Turner [4], Jaimey M. Pauli [5,6], Alison D. Gernand [2], Daniel E. Rivera [7] and Danielle Symons Downs [3,5]

[1] Center for Childhood Obesity Research, The Pennsylvania State University, University Park, State College, PA 16802, USA; eeh12@psu.edu (E.E.H.); kmm6054@psu.edu (K.M.M.)
[2] Department of Nutritional Sciences, The Pennsylvania State University, University Park, State College, PA 16802, USA; adg14@psu.edu
[3] Exercise Psychology Laboratory, Department of Kinesiology, The Pennsylvania State University, University Park, State College, PA 16802, USA; amp34@psu.edu (A.M.P.); kbl5167@psu.edu (K.S.L.); dsd11@psu.edu (D.S.D.)
[4] Diagnostic Medical Sonography, South Hills School of Business and Technology, State College, PA 16801, USA; tslturner3@gmail.com
[5] Department of Obstetrics and Gynecology, Penn State College of Medicine, Hershey, PA 17033, USA; jpauli@pennstatehealth.psu.edu
[6] Department of Maternal & Fetal Medicine, Penn State College of Medicine, Hershey, PA 17033, USA
[7] Control Systems Engineering Laboratory, School for Engineering of Matter, Transport, and Energy, Arizona State University, Tempe, AZ 85287, USA; daniel.rivera@asu.edu
* Correspondence: jfs195@psu.edu; Tel.: +814-865-0514; Fax: +814-863-0057

Received: 28 February 2019; Accepted: 18 April 2019; Published: 21 April 2019

Abstract: Excess maternal weight gain during pregnancy elevates infants' risk for macrosomia and early-onset obesity. Eating behavior is also related to weight gain, but the relationship to fetal growth is unclear. We examined whether Healthy Mom Zone, an individually tailored, adaptive gestational weight gain intervention, and maternal eating behaviors affected fetal growth in pregnant women ($n = 27$) with a BMI > 24. At study enrollment (6–13 weeks gestation) and monthly thereafter, the Three-Factor Eating Questionnaire was completed. Ultrasounds were obtained monthly from 14–34 weeks gestation. Data were analyzed using multilevel modeling. Higher baseline levels of uncontrolled eating predicted faster rates of fetal growth in late gestation. Cognitive restraint was not associated with fetal growth, but moderated the effect of uncontrolled eating on fetal growth. Emotional eating was not associated with fetal growth. Among women with higher baseline levels of uncontrolled eating, fetuses of women in the control group grew faster and were larger in later gestation than those in the intervention group (study group × baseline uncontrolled eating × gestational week interaction, $p = 0.03$). This is one of the first intervention studies to use an individually tailored, adaptive design to manage weight gain in pregnancy to demonstrate potential effects on fetal growth. Results also suggest that it may be important to develop intervention content and strategies specific to pregnant women with high vs. low levels of disinhibited eating.

Keywords: pregnancy; gestational weight gain intervention; eating behavior; restraint; disinhibition; uncontrolled and emotional eating; fetal growth; overweight and obesity; generalized linear models

1. Introduction

Over 50% of women in the United States enter pregnancy already having overweight or obesity; the majority of these women (60%) gain more weight in pregnancy than is recommended by the Institute of Medicine (IOM) [1]. Data also indicate that 23% of US 2–5-year-olds have overweight [2] and 9%–14%

have obesity [3,4], underscoring the need for research on early obesity prevention. The prenatal period may be an opportune time to intervene and break the intergenerational cycle of obesity by reducing fetal exposure to an "obesogenic" intrauterine environment [5,6] through promoting maternal energy balance (EB) (i.e., a nutrient-rich, low-energy dense diet and physical activity) [7]. Infant birth weight is commonly used as a surrogate marker for intrauterine growth and as an indicator of the conditions experienced in utero [8]. Maternal weight gain during pregnancy is necessary for healthy fetal development, but high gestational weight gain elevates infant risk for macrosomia and early onset obesity [1]. Further, infant birth weight is a positive, independent predictor of obesity during later childhood [9–11], and high birth weight is associated with increased risk for adult-onset obesity and diabetes [12].

Behavioral interventions have effectively managed weight among normal weight women, but similar interventions have had mixed effects among women with overweight or obesity [13]. Together, these data indicate the need for effective interventions that promote healthy gestational weight gain and are able to influence the etiology of obesity for offspring at a critical time in development. Based on past research and our own pilot data [7,14–17], we developed Healthy Mom Zone [18], an individually tailored and "intensively adaptive" intervention to effectively manage gestational weight gain and promote optimal maternal and child health in women with overweight or obesity. This novel intervention provides each woman with the support needed to manage her gestational weight gain, while adapting the dose and intensity of the intervention to her unique needs across pregnancy. Healthy Mom Zone is also innovative, applying principles/methods from systems science and engineering [19] including dynamical modeling, simulation, and controller design to optimize the intervention. Intensive longitudinal data (e.g., weight, energy intake, eating behavior, physical activity, and psychosocial factors) were collected and used in making decisions about adapting the intervention.

Throughout pregnancy, maternal nutritional status and dietary intake are key factors influencing the health of both the child and mother [20]. However, misreporting of dietary intake is common, making it difficult to identify how dietary intake patterns, macronutrient intakes, or even energy density are associated with gestational weight gain and fetal programming [21]. Among pregnant women with overweight or obesity, as many as one-third to one-half under-report dietary intake, with rates higher in late compared to early pregnancy [22], a finding supported by our own Healthy Mom Zone data [23].

Moreover, an individual's eating behavior and attitudes influence food choices, impact the regulation of food intake, and ultimately affect weight. Given the recall bias of self-reported dietary intake, alternative modifiable factors such as eating behavior (e.g., restrained eating, uncontrolled eating, emotional eating) that are relatively easy to measure may be amenable to intervention during pregnancy. Dietary restraint and disinhibited eating (emotional and uncontrolled), as measured by the Eating Inventory (EI) (also referred to as the Three Factor Eating Questionnaire [24,25]), are psychological constructs that assess behavioral control and attitudes toward food and eating. Cognitive dietary restraint is defined as a tendency to consciously restrict or control food intake, regardless of physiological signs of hunger and satiety. Uncontrolled eating is defined as a tendency to overeat relative to physiologic need and feeling of lack of control, whereas emotional eating is the tendency to overeat during depressed and melancholic states [26].

Eating behaviors may be a "proxy" for dietary intake in lieu of accurate, self-reported dietary intake [27]. In the general population, disinhibited eating is positively associated with food intake and weight [28–32], but the association between dietary restraint and emotional eating with these outcomes remains unclear [29–31,33–37]. In our current obesogenic environment, cognitive controls of eating may be necessary to prevent overeating and moderate weight gain [38]. For example, a randomized weight loss study reported that cognitive restrained eating was associated with lower energy intakes [39]. In pregnant women, restrained eating and emotional eating have been positively associated with poorer diet quality and greater overall weight gain [40–43]. However, cross-sectional studies [30,31,44] and longitudinal data [45] in nonpregnant individuals have revealed that it is not the independent effects of restraint and disinhibition but their interaction that predicts food intake and ultimately body weight, with restraint moderating the impact of uncontrolled eating. No study, to our knowledge, has examined the long-term effects of the interaction between restrained and disinhibited eating (uncontrolled or emotional) in pregnant women with overweight/obesity on fetal growth.

The primary aim of this study was to determine the effect of an individually tailored, adaptive intervention designed to help pregnant women with overweight and obesity to manage their gestational weight gain within the IOM guidelines on fetal growth, compared to a standard of care control. A secondary aim was to explore the interactive effects of maternal cognitive restraint, with uncontrolled and emotional eating during pregnancy, on fetal growth. Based on previous data [27,32,36,40,41,45], we expect that: (1) fetuses of women receiving the treatment would gain weight slower on average compared to control fetuses; (2) higher levels of emotional and uncontrolled eating will predict faster fetal weight gain; and (3) cognitive restraint will attenuate the positive relationship between emotional and uncontrolled eating and fetal weight gain. Findings from this study may inform counseling during pregnancy.

2. Methods

2.1. Study Design and Participants

Data for this analysis were collected as part of an ancillary study to measure fetal growth in women participating in Healthy Mom Zone, an optimization trial within the multiphase optimization strategy (MOST) framework [46], of an adaptive intervention to manage gestational weight gain among pregnant women with overweight or obesity [18]. This study was approved by the Pennsylvania State University Institutional Review Board (study ID #00003752, initial approval date 12/1/2015), and participants provided written consent for their participation. Women were eligible if they were 6–13 weeks pregnant with a single fetus, English-speaking, non-smoking, had a prepregnancy BMI > 24, and were free of significant pregnancy complications or medical conditions. The study was powered on the primary outcome of gestational weight gain; a sample size of 24 participants (12 per group) was determined to yield 80% power to detect a standardized effect size for gestational weight gain of 1.2 using a two-sided test with a significance level of $p = 0.05$. We aimed to recruit 30 participants, accounting for up to 25% drop out [18]. Participants ($n = 31$) were randomized to the intervention or a standard care control group. From this initial group, three women had a miscarriage at <12 weeks gestation, and one woman dropped out of the study at 22 weeks gestation, representing 87.1% sample retention (9.7% miscarriage rate and 3.2% drop rate). For the purpose of analysis, women with a prepregnancy BMI 24.1–29.9 were classified as having overweight (OW) and those with BMI ≥ 30 were classified as having obesity (OB). Fetal growth data were collected on a total of 27 women (intervention: 9 OW/4 OB; control: 7 OW/7 OB). One woman (control group, OW) did not complete the baseline survey measure of eating behavior, resulting in a sample of $n = 26$ women for those analyses.

2.2. Intervention

Healthy Mom Zone is an individually tailored, adaptive intervention to optimize gestational weight gain among pregnant women with overweight and obesity. This optimization trial was built within the multiphase optimization strategy (MOST) framework [46] with the aim of developing an intervention to efficiently and effectively manage weight gain over pregnancy. Details of the Healthy Mom Zone intervention are published elsewhere [18], but in short, principles from the theory of planned behavior (TPB) [47] and self-regulation [48] provide the conceptual framework for the intervention. The key intervention components include education, goal-setting/action plans, self-monitoring, and active learning. Participants in the intervention group met weekly with a study dietitian and were given customized calorie and physical activity goals. Plots of each woman's weight, physical activity, and dietary intake via real-time data collection procedures were plotted to generate a report illustrating gestational weight gain in the context of the IOM guidelines. This report was discussed by the dietitian.

In addition, emails were sent discussing maternal behaviors that impact baby health and growth. Lastly, gestational weight gain was evaluated weekly and decision rules were used in 3–4 week cycles to decide when and how the intervention may be adapted for an individual. When a participant's weight gain trajectory exceeded the recommended rate, they received an intervention step-up consisting of additional healthy eating active learning sessions (e.g., cooking demonstrations, grocery store tours) and onsite physical activity sessions with a fitness instructor. All women in the study received standard prenatal care (e.g., regular visits, prenatal counseling) with their personal healthcare provider.

2.3. Measures

At study enrollment, height and weight were measured by clinical research center staff. Participants completed a baseline background questionnaire that assessed prepregnancy weight, height, age, years of education, combined family income, and general health. At study enrollment and every four weeks thereafter, women completed the three-factor eating questionnaire (TFEQ-21) [24] via online survey through REDCap [49]. This 21-item scale, with a four point response scale ranging from (1) definitely true to (4) definitely false, measures three subscales: cognitive restraint (e.g., "I consciously hold back on how much I eat at meals to keep from gaining weight."), uncontrolled eating (e.g., "Sometimes when I start eating, I just can't seem to stop."), and emotional eating (e.g., "I start to eat when I feel anxious."). Scores for each subscale were calculated by averaging respective items.

Standard ultrasounds were obtained at approximately 14, 18, 22, 26, 30, and 34 weeks gestation by a sonographer using the same ultrasound machine (Philips iU22 MATRIX, Philips Healthcare, Andover, MA, USA). Standard fetal biometry measures including biparietal diameter (BPD), head circumference (HC), abdominal (AC), and femur length (FL), were assessed. Estimated fetal weight was calculated using a validated equation [50]. Gestational age was calculated using the date of last menstrual period. All ultrasounds were reviewed by a study physician, and any concerns identified were communicated to the participant's personal physician. Birth weight and gestational age at birth were abstracted from medical records or reported by participants if medical records were not available. Sex-specific birth weight-for-gestational age percentiles were determined using the INTERGROWTH-21st standards [51].

2.4. Analysis

Eating behavior variables were centered to disaggregate within- and between-person effects and to aid interpretation. Between-person effects were assessed using baseline levels of each eating behavior centered on the population mean, so that the centered baseline values represented each individual's deviation from the average baseline level across all women in the study. Within-person effects were assessed using change scores, such that each monthly eating behavior score represented a woman's change from her own baseline. Though survey data (i.e., eating behavior) and fetal ultrasounds were collected monthly, the timing of these measurements relative to one another was variable due to variation in gestational age at enrollment. For the purpose of this analysis, ultrasound data was paired with the most recent eating behavior survey completed prior to the scan. In seven instances across four participants, no eating behavior data preceded the scan, thus, survey data completed within seven days after the scan were used. The mean (SD) time between survey completion and ultrasound scan was 12.2 (8.9) days.

Statistical analysis was performed in SAS 9.4 (SAS Institute, Cary, NC, USA). Data were analyzed using multilevel modeling (PROC MIXED). All models utilized the restricted maximum likelihood (REML) estimation method, Satterthwaite method for computing df, and an unstructured covariance structure. Determination of model fit was based on several criteria including model convergence, a positive definite G matrix, and Akaike information criteria. Intraclass correlation coefficients (ICCs) were calculated as the ratio of between-subjects variance to total variance. Change over time in each of the three eating behavior subscales was modeled as a linear function of gestational week with a random intercept and slope. For cognitive restraint, the random gestational week slope term resulted in a nonpositive definite G matrix, so it was removed from the model. Interactions of gestational week with study group and prepregnancy BMI were tested to determine if the rate of change over time differed by these variables.

Estimated fetal weight was also modeled using multilevel modeling. Linear, quadratic, and cubic effects of gestational week were considered. Once the model describing fetal weight over time was finalized, study group and maternal prepregnancy BMI were added, and interactions with gestational week were tested. Next, a set of models were run for each of the three eating behavior constructs. These models tested (a) main effect of the baseline eating behavior, (b) interaction between baseline eating behavior and gestational week, (c) main effect of change in eating behavior, (d) interaction between change in eating behavior and gestational week, (e) interaction between baseline and change in eating behavior, and (f) three-way interaction between baseline and change in eating behavior and gestational week. For all interactions with gestational week, interactions with both the linear and quadratic gestational week terms were tested; the interaction with the quadratic term was dropped when not significant and was not reported here. Finally, two sets of models were tested to evaluate interactions between cognitive restraint and each of the two disinhibited eating behavior constructs (uncontrolled eating and emotional eating). These models tested (a) interaction between baseline cognitive restraint and baseline disinhibited eating; (b) three-way interactions between baseline cognitive restraint, baseline disinhibited eating, and gestational week; (c) interactions of baseline cognitive restraint and disinhibited eating with change in cognitive restraint and disinhibited eating; (d) three-way interactions between change in cognitive restraint, change in disinhibited eating, and gestational week; and (e) interactions described in (a)–(d) but with the quadratic gestational age term in place of the linear term. Two final models, one including cognitive restraint × uncontrolled eating and one including cognitive restraint × emotional eating, were evaluated using backward and forward selections.

3. Results

3.1. Descriptive Characteristics

Descriptive statistics are located in Table 1. Mean age of women at study entry was 30.6 (SD = 2.9) ranging from 24 to 37 years. Mean gestational age at the start of the study was 10.6 weeks (median 10.7 weeks). The majority of women were married and pregnant with their first baby (66.7% no prior live births; 33.3% had 1 previous live birth). Women were, on average, well-educated and fairly affluent. Moreover, the majority (81.5%) of women were employed full-time. Forty-one percent of women had obesity prior to pregnancy. Infants were born at a mean of 39.5 (1.4) weeks gestation. Mean birth weight was 3381 g (SD 617, range 2050–4570), reflecting an average birth weight for gestational age percentile of 56.6 (SD 31.9, range 2.6–99.4). Three infants (11.1%, 2 intervention, 1 control) were macrosomic (birth weight > 4000 g), and four infants (14.8%, 2 intervention, 2 control) were small for gestational age (SGA, <10th percentile) at birth. Out of 162 ultrasound scans, eight (4.9%) were flagged by the study physician, including five for growth restriction concerns. For two of these participants (both intervention), the concern resolved on its own at subsequent scans.

Mean (SD) baseline values were 2.79 (0.47) for cognitive restraint (range = 2.00–3.67), 2.15 (0.39) for uncontrolled eating (range = 1.33–3.00), and 2.14 (0.59) for emotional eating (range = 1.00–3.33). Baseline values did not differ by study group. Baseline cognitive restraint and uncontrolled eating did not differ by prepregnancy BMI status, but women with obesity had significantly higher baseline emotional eating than those with overweight (2.41 (0.52) vs. 1.95 (0.57), $p = 0.047$). Cognitive restraint significantly decreased across gestation (restraint = 2.90–0.014 × gestational age in weeks, $p = 0.0008$), as did uncontrolled eating (uncontrolled eating = 2.25–0.007 × gestational age in weeks, $p = 0.006$). Emotional eating did not significantly change across gestation (emotional eating = 2.10–0.006 × gestational age in weeks, $p = 0.14$), and there were no differences in rate of change by study group or prepregnancy BMI status for any of the eating behavior subscales. Baseline cognitive restraint was negatively associated with change in cognitive restraint ($p = 0.03$), such that women with higher baseline restraint experienced greater decreases in restraint across pregnancy. However, there was no association between baseline scores and change scores for uncontrolled eating and emotional eating. Intraclass correlations were 0.59 for cognitive restraint, 0.85 for uncontrolled eating, and 0.79 for emotional eating, suggesting that the majority of variance in all three constructs was between individuals rather than within individuals.

Table 1. Characteristics of women at baseline ($n = 27$).

	Overall ($n = 27$)	Intervention ($n = 13$)	Control ($n = 14$)
	M (SD) Range	M (SD) Range	M (SD) Range
Age (years)	30.6 (2.9) 24–37	30.3 (2.8) 24–35	30.9 (3.2) 27–37
Gestational Age at Study Start (weeks)	10.6 (1.6) 7.3–13.0	10.5 (1.7) 7.3–13.0	10.6 (1.6) 7.9–12.9
Weight at Study Start (lbs.)	190.4 (46.8) 134–294	182.2 (40.5) 145–294	197.9 (52.3) 134–292
Body Mass Index	31.6 (7.0) 24.1–48.9	30.7 (6.7) 24.1–48.9	32.4 (7.5) 24.5–42.8
	%	%	%
% BMI = 24.1–29.9	59.3	69.2	50
% Obese (BMI ≥ 30)	40.7	30.8	50
Parity			
Primiparous (no prior live birth)	66.7	53.9	78.6
Multiparous: 1 previous live birth	33.3	46.2	21.4
Education			
Graduate/professional	48.2	38.5	57.1
College	48.2	61.5	35.7
High school	3.7	0	7.1
Marital Status			
Married	92.6	92.3	92.9
Single	3.7	0	7.1
Divorced	3.7	7.7	0
Race			
Non-Hispanic White	96.3	92.3	100
Asian	3.7	7.7	0
Family Income			
>$100,000	33.3	30.8	35.7
$40–$100,000	44.4	61.5	28.6
$20–$40,000	18.5	0	35.7
$10–$20,000	3.7	7.7	0
Employment			
Full-time	81.5	84.6	78.6
Part-time	11.1	15.4	7.1
Self-employed	3.7	0	7.1
Other	3.7	0	7.1

Abbreviations: M = Mean; SD = Standard deviation; BMI = body mass index.

3.2. Unconditional Polynomial Models for Fetal Weight Change

Fetal growth from 14–34 weeks was modeled using a multilevel model. The first model included a random intercept and random and fixed linear effects of gestational week. The random intercept resulted in a nonpositive definite G matrix and was removed from the model. Next, a quadratic fixed effect of gestational week was added, which improved the fit of the model (likelihood ratio test $p < 0.0001$). Quadratic random effects and cubic fixed and random effects were tested but did not improve model fit.

3.3. Conditional Polynomial Models for Fetal Weight Change

3.3.1. Treatment Effect of Intervention and Fetal Growth

For Model 1, there was no significant main effect of study group on fetal weight. There was a trend for an interaction between study group and gestational week ($p = 0.095$, Model 1, Table 2), such that fetuses of women in the control group tended to grow faster and be larger in later gestation than those in the intervention group (Figure 1), before and after adjusting for prepregnancy BMI.

Table 2. Multilevel model parameter estimates predicting change in fetal weight (g) based on randomized intervention vs. control and interactions with eating ($n = 27$).

Term	Model 1: Study Group		
	Est	**SE**	**p**
Intercept	1232.8	151.2	<0.0001
Gestational week	−161.9	10.34	<0.0001
Gestational week^2	5.80	0.21	<0.0001
Prepregnancy BMI	0.30	2.74	0.91
Study group: Intervention *	89.46	61.56	0.14
Study group * × gestational week	−4.94	2.93	0.095

* Study group reference level is control. Abbreviations: Est = estimate; SE = standard error; BMI = body mass index.

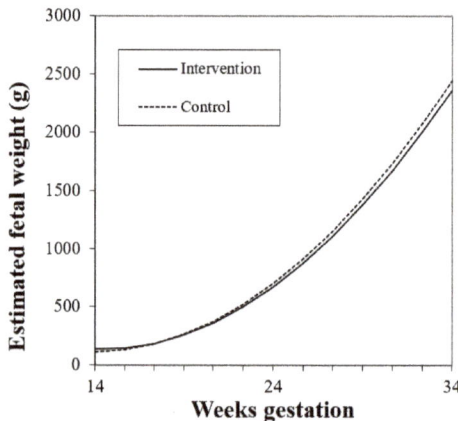

Figure 1. Estimated fetal weight trajectory in intervention and control group (study group × gestational week, $p = 0.095$). Estimates were generated using multilevel modeling (SAS PROC MIXED).

3.3.2. Individual Eating Behaviors and Fetal Growth

Eating characteristics (uncontrolled eating, emotional eating, and cognitive restraint) were each added, individually, to Model 1 to test higher order interactions and main effects. When higher-order interactions were not significant, models were simplified. For uncontrolled eating, there was a significant two-way interaction between baseline uncontrolled eating and gestational age^2 (Table 3, Model 2, $p = 0.03$). As shown in Figure 2, women with higher baseline levels of uncontrolled eating had faster rates of fetal growth in late gestation. Next, change in uncontrolled eating from baseline was added to the model. There was a significant interaction between change in uncontrolled eating and gestational age (Table 4, Model 3, $p = 0.04$). In contrast to the effect of baseline uncontrolled eating, women who experienced greater decreases in uncontrolled eating had faster rates of fetal growth. In both Models 2 and 3, the gestational week × study group effect was statistically significant. In addition, the interaction between baseline uncontrolled eating and gestational week^2 from Model 2 remained statistically significant with the addition of change in uncontrolled eating, suggesting independent effects of baseline and change in this eating behavior. Neither baseline levels nor change across gestation in cognitive restraint or emotional eating were associated with fetal growth.

Table 3. Multilevel model parameter estimates predicting change in fetal weight (g) based on randomized intervention vs. control and baseline uncontrolled eating ($n = 26$).

	Model 2: Study group + Baseline Uncontrolled Eating		
Term	Est	SE	*p*
Intercept	1195.7	154.5	<0.0001
Gestational week	−160.9	10.34	<0.0001
Gestational week^2	5.80	0.21	<0.0001
Prepregnancy BMI	0.89	2.86	0.76
Study group: Intervention *	122.0	63.02	0.06
Study group * × gestational week	−6.59	3.03	0.03
Baseline uncontrolled eating	475.3	330.5	0.15
Baseline uncontrolled eating × gestational week	−49.04	28.21	0.08
Baseline uncontrolled eating × gestational week^2	1.21	0.57	0.03

* Study group reference level is control. Abbreviations: Est = estimate; SE = standard error; BMI = body mass index.

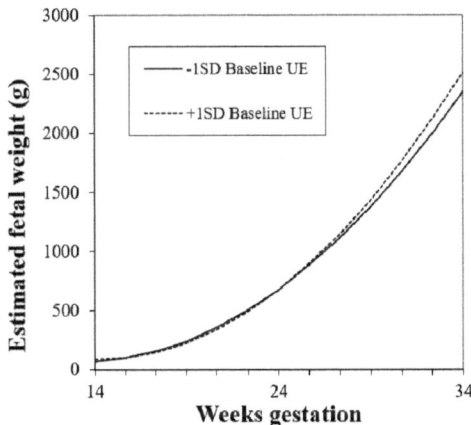

Figure 2. Estimated fetal weight trajectory at higher (mean + 1 SD) and lower (mean − 1 SD) levels of baseline uncontrolled eating (UE), (baseline uncontrolled eating × gestational age^2, $p = 0.03$). Estimates were generated using multilevel modeling (SAS PROC MIXED).

Table 4. Multilevel model parameter estimates predicting change in fetal weight (g) based on randomized intervention vs. control, baseline uncontrolled eating, and change in uncontrolled eating ($n = 26$).

Term	Model 3: Study group + Baseline Uncontrolled Eating + Change in Uncontrolled Eating		
	Est	SE	p
Intercept	1270.8	154.3	<0.0001
Gestational week	−159.6	10.63	<0.0001
Gestational week^2	5.74	0.21	<0.0001
Prepregnancy BMI	−1.45	2.71	0.60
Study group: Intervention *	110.3	64.65	0.09
Study group * × gestational week	−6.56	2.96	0.03
Baseline uncontrolled eating	532.6	330.8	0.11
Baseline uncontrolled eating × gestational week	−53.72	28.26	0.06
Baseline uncontrolled eating × gestational week^2	1.32	0.57	0.02
Change in uncontrolled eating	252.1	181.0	0.17
Change in uncontrolled eating × gestational week	−13.94	6.74	0.04

* Study group reference level is control. Abbreviations: Est = estimate; SE = standard error; BMI = body mass index.

3.3.3. Cognitive Restraint × Disinhibited Eating and Fetal Growth

Lastly, backward and forward selection were used to estimate a final model that included interactions between cognitive restraint and disinhibited eating (i.e., emotional eating and uncontrolled eating). All potential three-way interactions were considered. For cognitive restraint × uncontrolled eating (Table 5, Model 4), there were no significant three-way interactions. Baseline cognitive restraint × change in uncontrolled eating emerged as significant ($p = 0.02$). In women who experienced increases in uncontrolled eating across pregnancy, baseline cognitive restraint was negatively associated with fetal weight. This interaction was not specific to a particular time in gestation (i.e., there was no significant three-way interaction between baseline cognitive restraint, change in uncontrolled eating, and gestational week). The interaction between baseline uncontrolled eating and gestational age^2 remained significant, such that women with higher baseline levels of uncontrolled eating had faster rates of fetal growth in late gestation. However, the interaction between change in uncontrolled eating and gestational age observed in Model 3 did not emerge as significant in this final model. Lastly, the effect of study group on change in fetal growth remained significant (study group × gestational week $p = 0.02$).

For cognitive restraint × emotional eating, there were no statistically significant main effects or interactions between restraint and emotional eating.

3.3.4. Post-hoc Analysis: Study Group × Uncontrolled Eating Interaction

Although neither baseline nor change in uncontrolled eating differed by study group, inclusion of uncontrolled eating in Models 2–4 resulted in a statistically significant interaction between study group and gestational age. To further investigate why this relationship might emerge after accounting for baseline uncontrolled eating, we explored adding interactions between uncontrolled eating and study group to Model 4. A two-way interaction between uncontrolled eating and study group was not statistically significant, but there was evidence for a three-way interaction between baseline uncontrolled eating, study group, and gestational age ($p = 0.02$, Model 5, Table 6). Higher baseline level of uncontrolled eating was associated with faster fetal growth among control group participants but not among those receiving the intervention (Figure 3). Similar results were observed when these interactions were added to Models 2 and 3.

Table 5. Multilevel model parameter estimates predicting change in fetal weight (g) based on randomized intervention vs. control, uncontrolled eating, and cognitive restraint ($n = 26$).

Term	Model 4: Study Group + Uncontrolled Eating + Cognitive Restraint		
	Est	SE	*p*
Intercept	1283.4	156.4	<0.0001
Gestational week	−161.4	10.49	<0.0001
Gestational week^2	5.80	0.21	<0.0001
Prepregnancy BMI	−1.28	2.92	0.66
Study group: Intervention *	113.2	64.17	0.08
Study group * × gestational week	−7.45	3.05	0.02
Baseline uncontrolled eating	501.5	328.3	0.13
Baseline uncontrolled eating × gestational week	−49.78	28.03	0.08
Baseline uncontrolled eating × gestational week^2	1.25	0.56	0.03
Change in uncontrolled eating	−116.4	52.68	0.03
Baseline cognitive restraint	11.18	42.88	0.80
Baseline cognitive restraint × change in uncontrolled eating	−272.3	116.9	0.02

* Study group reference level is control. Abbreviations: Est = estimate; SE = standard error; BMI = body mass index.

Table 6. Multilevel model parameter estimates predicting change in fetal weight (g) including interactions between study group and uncontrolled eating ($n = 26$).

Term	Model 5: Study Group × Uncontrolled Eating + Cognitive Restraint		
	Est	SE	*p*
Intercept	1267.4	155.7	<0.0001
Gestational week	−160.6	10.40	<0.0001
Gestational week^2	5.80	0.21	<0.0001
Prepregnancy BMI	−1.21	2.93	0.68
Study group: Intervention *	117.0	63.58	0.07
Study group * × gestational week	−7.67	3.03	0.01
Baseline uncontrolled eating	256.8	352.8	0.47
Baseline uncontrolled eating × gestational week	−35.22	28.64	0.22
Baseline uncontrolled eating × gestational week^2	1.18	0.56	0.04
Change in uncontrolled eating	−111.3	52.60	0.04
Baseline cognitive restraint	10.36	42.70	0.81
Baseline cognitive restraint × change in uncontrolled eating	−259.0	116.1	0.03
Study group × baseline uncontrolled eating	310.6	175.3	0.08
Study group × baseline uncontrolled eating × gestational week	−17.00	8.25	0.04

* Study group reference level is control. Abbreviations: Est = estimate; SE = standard error; BMI = body mass index.

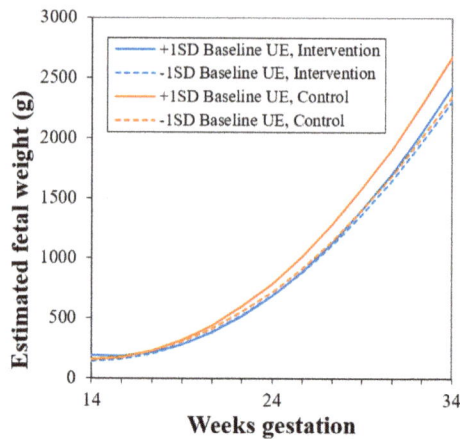

Figure 3. Estimated fetal weight trajectory at higher (mean + 1 SD) and lower (mean – 1 SD) levels of baseline uncontrolled eating (UE) in intervention and control group participants, (baseline uncontrolled eating × study group × gestational age, $p = 0.04$). Estimates were generated using multilevel modeling (SAS PROC MIXED).

4. Discussion

Healthy Mom Zone, an individually tailored, adaptive intervention designed to effectively manage gestational weight gain among women with overweight or obesity tended to result in a slower rate of fetal growth compared to fetuses of control women, with significant effects emerging in models that accounted for uncontrolled eating. Higher uncontrolled eating at study entry was associated with faster fetal growth. Findings also suggest that cognitive restraint on its own appears to have little effect on fetal growth, but may help moderate fetal growth among women who experience increases in uncontrolled eating during pregnancy. Several of these findings warrant discussion.

Though not statistically significant in Model 1, the Healthy Mom Zone intervention had a protective effect on fetal growth once uncontrolled eating was added (Models 2–4); intervention fetuses gained weight at a slower rate on average compared to control fetuses. It is important to note that the intervention did not increase risk for an infant being born small for gestational age. This finding has clinical relevance for several reasons. Women who are overweight or obese are 65% and 163% more likely, respectively, to deliver babies who are large for their gestational age, regardless of whether the mother develops gestational diabetes during pregnancy [52]. In addition, risk of fetal macrosomia, having a birth weight of more than 8 pounds or 4000 grams regardless of gestational age, is more likely among women with obesity. This is important because women with overweight or obesity are also more likely to exceed gestational weight gain guidelines, further elevating infant risk for macrosomia. The clinical relevance is that decreasing risk for faster weight gain may decrease health risks for the mother (e.g., labor and delivery complications) and the infant (e.g., childhood obesity, metabolic syndrome) both during pregnancy and after childbirth. Thus, our behavioral intervention has the potential to mediate this risk, in particular among pregnant women with overweight or obesity.

In the current study, both cognitive restraint and uncontrolled eating decreased on average across gestation, and there was no change in emotional eating. Similarly, previous studies have found that cognitive restraint is lower during pregnancy compared to prepregnancy levels or non-pregnant women [53,54]. Research on changes in disinhibition across pregnancy is scant. We did not observe any association between prepregnancy BMI and cognitive restraint or uncontrolled eating; however, this may be because our sample was limited to women with BMI > 24. Two previous studies have found that pregnant women with obesity tend to report higher levels of disinhibited eating [55,56] than their normal weight counterparts. Further research is needed to evaluate how these eating behaviors change throughout pregnancy.

As hypothesized, greater uncontrolled eating in late first trimester was associated with faster fetal growth in late pregnancy. However, there was no association between emotional eating and fetal growth. To our knowledge, this study is the first to describe relationships between maternal disinhibited eating during pregnancy and fetal growth in utero in a non-eating disorder population. One previous study has examined the relationship between maternal eating behavior and postnatal growth. Mothers' emotional eating, assessed at infant age 12 months, was positively associated with infant weight gain through 12 months, while external eating was negatively associated with infant weight gain. Maternal restraint was not associated with infant weight gain [57]. Further research is needed to clarify the relationship between maternal disinhibited eating and fetal/infant weight gain, and to determine the extent to which this relationship is reflective of modifiable factors, such as maternal diet, weight gain, and feeding practices, as opposed to nonmodifiable factors such as genetics.

Emotional eating and external eating have been associated with excess gestational weight gain [42]. Qualitative studies of pregnant women have found that emotional and external eating are often described as a barrier to healthy eating and management of gestational weight gain [58–60]. Our finding that uncontrolled eating is associated with fetal weight gain adds to this literature, suggesting that disinhibited eating behaviors may be an ideal target for gestational weight gain management interventions. Our Healthy Mom Zone intervention focused primarily on dietary intake, physical activity, and self-monitoring of weight. Expansion of intervention content focused on mindfulness, stress reduction, and emotion regulation, strategies that have been shown to reduce disinhibited eating in nonpregnant populations [61–63], may further benefit pregnant women with overweight or obesity.

Similar to the uncertain relationship between restraint and weight in nonpregnant adults, mixed findings have been reported for restraint and gestational weight gain [41,42,64,65]. Previous work in nonpregnant populations suggests that restraint may not be directly associated with weight but rather as a moderator of the effect of disinhibition [30,31,44,45]. A similar interaction effect may be involved in fetal growth. We observed a significant interaction between baseline cognitive restraint and change in uncontrolled eating, where among women who experienced increases in uncontrolled eating during pregnancy, those with higher baseline cognitive restraint had smaller fetuses than those with lower baseline cognitive restraint [31,44,45]. This suggests that a greater ability to consciously control food intake may help protect against adverse effects of increases in uncontrolled eating on fetal weight gain.

We did not have an a priori hypothesis about differential effects of the intervention on fetal growth by baseline eating behavior, but our post-hoc analysis of the interaction between study group and baseline uncontrolled eating suggests that the intervention may have been beneficial, particularly for women with higher baseline uncontrolled eating. It is possible that the intervention provided women who were already prone to uncontrolled eating with strategies to cope with inclinations to overeat, providing a buffering effect against excess fetal growth. Future analyses will examine whether baseline levels of eating behavior predict differential response to the Healthy Mom Zone intervention for other outcomes including maternal weight gain. Factors such as eating behavior that predict intervention response may be used to identify patients that are most likely to benefit from an intensive intervention.

There are also some limitations of this research. The small sample size, although adequate for an optimization trial within the MOST framework [46], precludes the ability to make assumptions at a population level. The intervention and control groups were unbalanced in terms of the proportion of women with obesity; however, all analyses controlled for prepregnancy BMI. In addition, the target population was a homogenous sample of women (e.g., married, middle to upper class, Caucasian) from largely rural and suburban areas in central Pennsylvania, thus limiting the extension of the study findings to more culturally diverse and urban populations of pregnant women with overweight or obesity. Future research will aim to broaden generalizability of our study findings. In addition, we do plan to understand the application of Healthy Mom Zone to a more diverse sample of pregnant women in the future, including those who are normal weight, more culturally diverse, and reside in varied communities across the US. Lastly, intensive longitudinal data collection may have affected behavior, and responses may have become biased due to repeated assessments throughout gestation. However, this longitudinal design is also a strength of this study, allowing us to better understand how individual differences in eating behavior are related to fetal growth.

5. Conclusions

Healthy Mom Zone, an individually tailored, intensive behavioral intervention designed to help women with overweight or obesity manage gestational weight gain shows promise for preventing rapid fetal growth, particularly for women prone to uncontrolled eating, though these findings must be replicated in a larger and more diverse sample. Our results also suggest that uncontrolled eating may increase risk for faster fetal growth, and that restraint on its own seems to have little influence. However, among women who experience increases in uncontrolled eating during pregnancy, higher restraint may help moderate fetal growth. Implications for these findings include that understanding participant eating behaviors at baseline may be important for future gestational weight gain interventions. Lastly, findings illustrate a need to develop intervention content and strategies specific to women with high vs. low levels of disinhibited eating.

Author Contributions: J.S.S. lead the fetal growth ancillary project, contributed to study conception and design, contributed to data analysis, interpreted results, and drafted the manuscript. E.E.H. contributed to study design, analyzed the data, interpreted results, and aided in drafting the manuscript. K.M.M., A.M.P., K.S.L., T.T., and J.M.P. contributed to study design and data collection. A.D.G. and D.E.R. contributed to study design and interpretation of results. D.S.D. lead the intervention study, contributed to study conception and design, and interpretation of results. All authors participated in critical revision of the manuscript and approved the final version.

Acknowledgments: We would like to acknowledge the assistance of the Healthy Mom Zone team who assisted with participant recruitment and data collection for this study. Support for this work has been provided by the National Heart, Lung, and Blood Institute (NHLBI) of the National Institutes of Health (NIH) through grants R56-HL126799 and R01-HL119245. REDCap support was received from the Penn State Clinical & Translational Sciences Institute through the National Center for Advancing Translational Sciences (NCATS) NIH grant UL1 TR002014. The opinions expressed in this article are the authors' own and do not necessarily reflect the views of NIH.

Conflicts of Interest: The authors declare no conflicts of interest.

References

1. Institute of Medicine (US) and National Research Council (US) Committee to Reexamine IOM Pregnancy Weight Guidelines. *Weight Gain during Pregnancy: Reexamining the Guidelines*; National Academies Press (US): Washington, DC, USA, 2009.
2. Ogden, C.L.; Carroll, M.D.; Kit, B.K.; Flegal, K.M. Prevalence of childhood and adult obesity in the United States, 2011–2012. *JAMA* **2014**, *311*, 806–814. [CrossRef]
3. Ogden, C.L.; Carroll, M.D.; Lawman, H.G.; Fryar, C.D.; Kruszon-Moran, D.; Kit, B.K.; Flegal, K.M. Trends in Obesity Prevalence Among Children and Adolescents in the United States, 1988–1994 through 2013–2014. *JAMA* **2016**, *315*, 2292–2299. [PubMed]
4. Hales, C.M.; Fryar, C.D.; Carroll, M.D.; Freedman, D.S.; Ogden, C.L. Trends in Obesity and Severe Obesity Prevalence in US Youth and Adults by Sex and Age, 2007–2008 to 2015–2016. *JAMA* **2018**, *319*, 1723–1725. [CrossRef] [PubMed]
5. Patti, M.E. Intergenerational programming of metabolic disease: Evidence from human populations and experimental animal models. *Cell. Mol. Life Sci.* **2013**, *70*, 1597–1608. [CrossRef]
6. Adamo, K.B.; Ferraro, Z.M.; Brett, K.E. Can we modify the intrauterine environment to halt the intergenerational cycle of obesity? *Int. J. Environ. Res. Public Health* **2012**, *9*, 1263–1307. [CrossRef] [PubMed]
7. Dong, Y.; Rivera, D.E.; Thomas, D.M.; Navarro-Barrientos, J.E.; Downs, D.S.; Savage, J.S.; Collins, L.M. A Dynamical Systems Model for Improving Gestational Weight Gain Behavioral Interventions. In Proceedings of the American Control Conference, Montreal, QC, Canada, 27–29 June 2012; pp. 4059–4064.

8. Fallucca, S.; Vasta, M.; Sciullo, E.; Balducci, S.; Fallucca, F. Birth weight: Genetic and intrauterine environment in normal pregnancy. *Diabetes Care* **2009**, *32*, e149. [CrossRef] [PubMed]
9. Reilly, J.J.; Armstrong, J.; Dorosty, A.R.; Emmett, P.M.; Ness, A.; Rogers, I.; Steer, C.; Sherriff, A. Early life risk factors for obesity in childhood: Cohort study. *BMJ* **2005**, *330*, 1357. [CrossRef]
10. Martin, J.A.; Hamilton, B.E.; Ventura, S.J.; Osterman, M.J.; Kirmeyer, S.; Mathews, T.J.; Wilson, E.C. Births: Final data for 2009. *Natl. Vital Stat. Rep.* **2011**, *60*, 1–70.
11. Nohr, E.A.; Vaeth, M.; Baker, J.L.; Sorensen, T.I.; Olsen, J.; Rasmussen, K.M. Pregnancy outcomes related to gestational weight gain in women defined by their body mass index, parity, height, and smoking status. *Am. J. Clin. Nutr.* **2009**, *90*, 1288–1294. [CrossRef] [PubMed]
12. Wei, J.-N.; Sung, F.-C.; Li, C.-Y.; Chang, C.-H.; Lin, R.-S.; Lin, C.-C.; Chiang, C.-C.; Chuang, L.-M. Low birth weight and high birth weight infants are both at an increased risk to have type 2 diabetes among schoolchildren in Taiwan. *Diabetes Care* **2003**, *26*, 343–348. [CrossRef]
13. Muktabhant, B.; Lawrie, T.A.; Lumbiganon, P.; Laopaiboon, M. Diet or exercise, or both, for preventing excessive weight gain in pregnancy. *Cochrane Database Syst. Rev.* **2015**. [CrossRef] [PubMed]
14. Pauley, A.M.; Hohman, E.; Savage, J.S.; Rivera, D.E.; Guo, P.; Leonard, K.S.; Downs, D.S. Gestational Weight Gain Intervention Impacts Determinants of Healthy Eating and Exercise in Overweight/Obese Pregnant Women. *J. Obes.* **2018**, *2018*, 6469170. [CrossRef]
15. Downs, D.S. Obesity in Special Populations: Pregnancy. *Prim. Care* **2016**, *43*, 109–120. [CrossRef] [PubMed]
16. Downs, D.S.; Savage, J.S.; Rauff, E.L. Falling Short of Guidelines? Nutrition and Weight Gain Knowledge in Pregnancy. *J. Women's Heal. Care* **2014**, *3*, 1–6.
17. Dong, Y.; Rivera, D.E.; Downs, D.S.; Savage, J.S.; Thomas, D.M.; Collins, L.M. Hybrid Model Predictive Control for Optimizing Gestational Weight Gain Behavioral Interventions. In Proceedings of the American Control Conference, Washington, DC, USA, 17–19 June 2013; pp. 1970–1975.
18. Symons Downs, D.; Savage, J.S.; Rivera, D.E.; Smyth, J.M.; Rolls, B.J.; Hohman, E.E.; McNitt, K.M.; Kunselman, A.R.; Stetter, C.; Pauley, A.M.; et al. Individually Tailored, Adaptive Intervention to Manage Gestational Weight Gain: Protocol for a Randomized Controlled Trial in Women with Overweight and Obesity. *JMIR Res. Protoc.* **2018**, *7*, e150. [CrossRef] [PubMed]
19. Collins, L.M.; Murphy, S.A.; Strecher, V. The multiphase optimization strategy (MOST) and the sequential multiple assignment randomized trial (SMART): New methods for more potent eHealth interventions. *Am. J. Prevent. Med.* **2007**, *32* (Suppl. 5), S112–S118. [CrossRef] [PubMed]
20. Christian, P.; Mullany, L.C.; Hurley, K.M.; Katz, J.; Black, R.E. Nutrition and maternal, neonatal, and child health. *Semin. Perinatol.* **2015**, *39*, 361–372. [CrossRef] [PubMed]
21. Livingstone, M.B.; Black, A.E. Markers of the validity of reported energy intake. *J. Nutr.* **2003**, *133* (Suppl. 3), 895S–920S. [CrossRef]
22. Sui, Z.; Cramp, C.; Moran, L.J.; McNaughton, S.A.; Deussen, A.R.; Grivell, R.M.; Dodd, J.M. The characterisation of overweight and obese women who are under reporting energy intake during pregnancy. *BMC Pregnancy Childbirth* **2018**, *18*, 204.
23. Guo, P.; Rivera, D.E.; Savage, J.S.; Hohman, E.E.; Pauley, A.M.; Leonard, K.S.; Symons Downs, D. System identification approaches for energy intake estimation: Enhancing interventions for managing gestational weight gain. *IEEE Trans. Control Syst. Technol.* **2018**. [CrossRef]
24. Cappelleri, J.C.; Bushmakin, A.G.; Gerber, R.; Leidy, N.K.; Sexton, C.C.; Lowe, M.R.; Karlsson, J. Psychometric analysis of the Three-Factor Eating Questionnaire-R21: Results from a large diverse sample of obese and non-obese participants. *Int. J. Obes.* **2009**, *33*, 611–620. [CrossRef] [PubMed]
25. Stunkard, A.J.; Messick, S. The three-factor eating questionnaire to measure dietary restraint, disinhibition and hunger. *J. Psychosom. Res.* **1985**, *29*, 71–83. [CrossRef]
26. Stunkard, A.J.; Messick, S. *Eating Inventory Manual*; The Psychological Corporation: San Antonio, TX, USA, 1988.
27. Lowe, M.R.; Kral, T.V. Stress-induced eating in restrained eaters may not be caused by stress or restraint. *Appetite* **2006**, *46*, 16–21. [CrossRef] [PubMed]
28. Lindroos, A.-K.; Mathiassen, M.E.; Bengtsson, C.; Lindroos, A.; Lissner, L.; Karlsson, J.; Sullivan, M.; Sjöström, L. Dietary intake in relation to restrained eating, disinhibition, and hunger in obese and nonobese Swedish women. *Obes. Res.* **1997**, *5*, 175–185. [CrossRef] [PubMed]

29. Hainer, V.; Kunešová, M.; Bellisle, F.; Parizkova, J.; Braunerova, R.; Wagenknecht, M.; Lajka, J.; Hill, M.; Stunkard, A. The eating inventory, body adiposity and prevalence of disease in a quota sample of Czech adults. *Int. J. Obesity (Lond.)* **2006**, *30*, 830–836. [CrossRef]

30. Williamson, D.A.; Lawson, O.J.; Brooks, E.R.; Wozniak, P.J.; Ryan, D.H.; Bray, G.A.; Duchmann, E.G. Association of body mass with dietary restraint and disinhibition. *Appetite* **1995**, *25*, 31–41. [CrossRef]

31. Hays, N.P.; Bathalon, G.P.; McCrory, M.A.; Roubenoff, R.; Lipman, R.; Roberts, S.B. Eating behavior correlates of adult weight gain and obesity in healthy women aged 55–65 y. *Am. J. Clin. Nutr.* **2002**, *75*, 476–483. [CrossRef]

32. Smith, C.F.; Geiselman, P.J.; Williamson, D.A.; Champagne, C.M.; Bray, G.A.; Ryan, D.H. Association of dietary restraint and disinhibition with eating behavior, body mass, and hunger. *Eat. Weight Disord. Ewd* **1998**, *3*, 7–15. [CrossRef]

33. Foster, G.D.; Wadden, T.A.; Swain, R.M.; Stunkard, A.J.; Platte, P.; Vogt, R.A. The eating inventory in obese women: Clinical correlates and relationship to weight loss. *Int. J. Obes.* **1998**, *22*, 778–785. [CrossRef]

34. Lauzon-Guillain, B.; Basdevant, A.; Romon, M.; Karlsson, J.; Borys, J.M.; Charles, M.A. Is restraint eating a risk factor for weight gain in a general population? *Am. J. Clin. Nutr.* **2006**, *83*, 132–138. [CrossRef]

35. Lowe, M.R.; Annunziato, R.A.; Markowitz, J.T.; Didie, E.; Bellace, D.L.; Riddell, L.; Maille, C.; McKinney, S.; Stice, E. Multiple types of dieting prospectively predict weight gain during the freshman year of college. *Appetite* **2006**, *47*, 83–90. [CrossRef]

36. Olea Lopez, A.L.; Johnson, L. Associations between Restrained Eating and the Size and Frequency of Overall Intake, Meal, Snack and Drink Occasions in the UK Adult National Diet and Nutrition Survey. *PLoS ONE* **2016**, *11*, e0156320. [CrossRef] [PubMed]

37. Bongers, P.; Jansen, A. Emotional Eating Is Not What You Think It Is and Emotional Eating Scales Do Not Measure What You Think They Measure. *Front. Psychol.* **2016**, *7*, 1932. [CrossRef] [PubMed]

38. Hill, J.O.; Wyatt, H.R.; Reed, G.W.; Peters, J.C. Obesity and the environment: Where do we go from here? *Science* **2003**, *299*, 853–855. [CrossRef] [PubMed]

39. Keränen, A.-M.; Savolainen, M.J.; Reponen, A.H.; Kujari, M.-L.; Lindeman, S.M.; Bloigu, R.S.; Laitinen, J.H.; Teeriniemi, A.-M. The effect of eating behavior on weight loss and maintenance during a lifestyle intervention. *Prev. Med.* **2009**, *49*, 32–38. [CrossRef] [PubMed]

40. Mumford, S.L.; Siega-Riz, A.M.; Herring, A.; Evenson, K.R. Dietary restraint and gestational weight gain. *J. Am. Diet. Assoc.* **2008**, *108*, 1646–1653. [CrossRef]

41. Heery, E.; Wall, P.G.; Kelleher, C.C.; McAuliffe, F.M. Effects of dietary restraint and weight gain attitudes on gestational weight gain. *Appetite* **2016**, *107*, 501–510. [CrossRef]

42. Blau, L.E.; Orloff, N.C.; Flammer, A.; Slatch, C.; Hormes, J.M. Food craving frequency mediates the relationship between emotional eating and excess weight gain in pregnancy. *Eat. Behav.* **2018**, *31*, 120–124. [CrossRef]

43. Hutchinson, A.D.; Charters, M.; Prichard, I.; Fletcher, C.; Wilson, C. Understanding maternal dietary choices during pregnancy: The role of social norms and mindful eating. *Appetite* **2017**, *112*, 227–234. [CrossRef]

44. Lawson, O.J.; Champagne, C.M.; Brooks, E.R.; Howat, P.M.; Wozniak, P.J.; Williamson, D.A.; Delany, J.P.; Bray, G.A.; Ryan, D.H. The association of body weight, dietary intake, and energy expenditure with dietary restraint and disinhibition. *Obes. Res.* **1995**, *3*, 153–161. [CrossRef]

45. Savage, J.S.; Hoffman, L.; Birch, L.L. Dieting, restraint, and disinhibition predict women's weight change over 6 y. *Am. J. Clin. Nutr.* **2009**, *90*, 33–40. [CrossRef]

46. Rivera, D.E.; Hekler, E.; Savage, J.S.; Symons Downs, D. Intensively adaptive interventions using control systems engineering: Two illustrative examples. In *Optimization of Behavioral, Biobehavioral, and Biomedical Interventions*; Collins, L.M., Ed.; Springer: Cham, Switzerland, 2018; pp. 21–173.

47. Ajzen, I. The theory of planned behavior. *Organ. Behav. Hum. Decis. Process.* **1991**, *50*, 179–211. [CrossRef]

48. Carver, C.; Scheier, M. *On the Self-Regulation of Behavior*; Cambridge University Press: Cambridge, UK, 1998.

49. Harris, P.A.; Taylor, R.; Thielke, R.; Payne, J.; Gonzalez, N.; Conde, J.G. Research electronic data capture (REDCap)—A metadata-driven methodology and workflow process for providing translational research informatics support. *J. Biomed. Inf.* **2009**, *42*, 377–381. [CrossRef]

50. Hadlock, F.P.; Harrist, R.B.; Carpenter, R.J.; Deter, R.L.; Park, S.K. Sonographic estimation of fetal weight. The value of femur length in addition to head and abdomen measurements. *Radiology* **1984**, *150*, 535–540. [CrossRef]

51. Villar, J.; Ismail, L.C.; Victora, C.G.; Ohuma, E.; Bertino, E.; Altman, D.G.; Lambert, A.; Papageorghiou, A.T.; Carvalho, M.; Jaffer, Y.; et al. International standards for newborn weight, length, and head circumference by gestational age and sex: The Newborn Cross-Sectional Study of the INTERGROWTH-21st Project. *Lancet* **2014**, *384*, 857–868. [CrossRef]

52. Black, M.H.; Sacks, D.A.; Xiang, A.H.; Lawrence, J.M. The relative contribution of prepregnancy overweight and obesity, gestational weight gain, and IADPSG-defined gestational diabetes mellitus to fetal overgrowth. *Diabetes Care* **2013**, *36*, 56–62. [CrossRef]

53. Clark, M.; Ogden, J. The impact of pregnancy on eating behaviour and aspects of weight concern. *Int. J. Obes. Relat. Metab. Disord.* **1999**, *23*, 18–24. [CrossRef]

54. Fairburn, C.G.; Stein, A.; Jones, R. Eating habits and eating disorders during pregnancy. *Psychosom. Med.* **1992**, *54*, 665–672. [CrossRef] [PubMed]

55. Shloim, N.; Hetherington, M.M.; Rudolf, M.; Feltbower, R.G. Relationship between body mass index and women's body image, self-esteem and eating behaviours in pregnancy: A cross-cultural study. *J. Health Psychol.* **2015**, *20*, 413–426. [CrossRef] [PubMed]

56. Slane, J.D.; Levine, M.D. Association of Restraint and Disinhibition to Gestational Weight Gain among Pregnant Former Smokers. *Women's Health Issues* **2015**, *25*, 390–395. [CrossRef]

57. Wright, C.M.; Parkinson, K.N.; Drewett, R.F. The influence of maternal socioeconomic and emotional factors on infant weight gain and weight faltering (failure to thrive): Data from a prospective birth cohort. *Arch. Dis. Child.* **2006**, *91*, 312–317. [CrossRef] [PubMed]

58. Wang, M.L.; Arroyo, J.; Druker, S.; Sankey, H.Z.; Rosal, M.C. Knowledge, Attitudes and Provider Advice by Pre-Pregnancy Weight Status: A Qualitative Study of Pregnant Latinas with Excessive Gestational Weight Gain. *Women Health* **2015**, *55*, 805–828. [CrossRef] [PubMed]

59. Chang, M.W.; Nitzke, S.; Buist, D.; Cain, D.; Horning, S.; Eghtedary, K. I am pregnant and want to do better but I can't: Focus groups with low-income overweight and obese pregnant women. *Matern. Child Health J.* **2015**, *19*, 1060–1070. [CrossRef] [PubMed]

60. Chang, M.W.; Nitzke, S.; Guilford, E.; Adair, C.H.; Hazard, D.L. Motivators and barriers to healthful eating and physical activity among low-income overweight and obese mothers. *J. Am. Diet Assoc.* **2008**, *108*, 1023–1028. [CrossRef] [PubMed]

61. Levoy, E.; Lazaridou, A.; Brewer, J.; Fulwiler, C. An exploratory study of Mindfulness Based Stress Reduction for emotional eating. *Appetite* **2017**, *109*, 124–130. [CrossRef]

62. Armitage, C.J. Randomized test of a brief psychological intervention to reduce and prevent emotional eating in a community sample. *J. Public Health (Oxf.)* **2015**, *37*, 438–444. [CrossRef]

63. Katterman, S.N.; Kleinman, B.M.; Hood, M.M.; Nackers, L.M.; Corsica, J.A. Mindfulness meditation as an intervention for binge eating, emotional eating, and weight loss: A systematic review. *Eat. Behav.* **2014**, *15*, 197–204. [CrossRef]

64. Conway, R.; Reddy, S.; Davies, J. Dietary restraint and weight gain during pregnancy. *Eur. J. Clin. Nutr.* **1999**, *53*, 849–853. [CrossRef]

65. Laraia, B.; Epel, E.; Siega-Riz, A.M. Food insecurity with past experience of restrained eating is a recipe for increased gestational weight gain. *Appetite* **2013**, *65*, 178–184. [CrossRef] [PubMed]

MDPI

Article

Prenatal Intervention with Partial Meal Replacement Improves Micronutrient Intake of Pregnant Women with Obesity

Suzanne Phelan [1,*], Barbara Abrams [2] and Rena R. Wing [3]

[1] Department of Kinesiology & Public Health, California Polytechnic State University, 1 Grand Ave, San Luis Obispo, CA 93407, USA
[2] Division of Epidemiology, University of California at Berkeley School of Public Health, 2121 Berkeley Way #5302, Berkeley, CA 94720-7360, USA; babrams@berkeley.edu
[3] Warren Alpert Medical School at Brown University Department of Psychiatry and Human Behavior, 197 Richmond Street, Providence, RI 02906, USA; rwing@lifespan.org
* Correspondence: sphelan@calpoly.edu; Tel.: +805-756-2087; Fax: +805-756-7273

Received: 30 April 2019; Accepted: 10 May 2019; Published: 14 May 2019

Abstract: A behavioral lifestyle intervention with partial meal replacement reduced excess gestational weight gain in ethnically diverse women with overweight/obesity, but the effects on micronutrient intake remained unknown. A secondary analysis of a randomized, controlled trial tested whether the intervention improved micronutrient intake relative to usual care. Pregnant women (n = 211; 30.5 years of age, body mass index, BMI, of 32.0 kg/m^2) were enrolled and randomized within site and ethnicity (40% were Hispanic) into intervention (n = 102) or usual care (n = 109) groups. Two 24 h dietary recalls were conducted on random days at study entry and late pregnancy (35–36 weeks gestation). Nutrient adequacy was defined using the Estimated Average Requirement cut-point method. At study entry and including prenatal vitamins, ≥90% of participants reported inadequate intake of vitamins D and E and iron; 40–50% reported inadequate intake of calcium, protein, vitamins A, C, B$_6$, folate, magnesium, and zinc. From study entry to late pregnancy, the behavioral intervention with partial meal replacement increased the overall intake of vitamins A, E, and D and copper and reduced the odds of inadequate intake of calcium (odds ratio (OR) = 0.37 (0.18, 0.76)), vitamins A (OR = 0.39 (0.21, 0.72)) and E (OR = 0.17 (0.06, 0.48)), and magnesium (OR = 0.36 (0.20, 0.65)). A behavioral intervention with partial meal replacement during pregnancy improved the intake of several micronutrients in Hispanic and non-Hispanic women with overweight/obesity.

Keywords: prenatal intervention; meal replacements; randomized clinical trial; lifestyle intervention; obesity; RDA; micronutrients

1. Introduction

Maternal pre-pregnancy obesity and excess gestational weight gain are well established risk factors for several adverse short- and long-term maternal and child health outcomes, including pregnancy complications, diabetes, obesity, and cardio-metabolic comorbidities [1]. Compounding risks, pregnant women with obesity are more likely than those with normal weight to have micronutrient insufficiency [2–4], conferring additional potential risks of pregnancy complications and chronic conditions in later life [5,6].

The Academy of Nutrition and Dietetics recommends that the optimal prenatal diet should limit overconsumption yet prevent micronutrient insufficiency [5]. A varied and balanced diet, rich in fruits, vegetables, and whole grains is recommended. Several studies have shown that comprehensive lifestyle interventions that target reduced calorie intake and balanced nutrition can effectively reduce excess gestational weight gain in women with obesity [7], but less is known about intervention effects on micronutrient intake. Micronutrient needs increase during pregnancy, particularly for folic acid, iron, zinc, calcium, vitamin C, and vitamin D [8]. While supplementation can reduce micronutrient deficiencies and some associated maternal and fetal complications [9–12], adherence to prenatal multivitamins may be limited by gastrointestinal distress or nausea [13]. The provision of micronutrient-rich foods [14,15] and nutrition educational interventions [16] can also improve dietary quality and pregnancy outcomes, but few randomized clinical trials have been done, particularly in pregnant women with obesity who are at risk of micronutrient insufficiency [15].

Healthy Beginnings/Cominezos Saludables was a randomized clinical trial designed to test the efficacy of a partial meal replacement program versus usual care to reduce excessive gestational weight gain in 257 pregnant women with overweight or obesity. The intervention significantly reduced excess gestational weight gain, which was related to increased use of meal replacements during pregnancy [17]. No significant treatment effect was seen on energy or macronutrient composition [17]. Since the provided meal replacements were fortified with vitamins and minerals, it is possible that they could address underlying micronutrient inadequacies in pregnant women with overweight or obesity. The purpose of this study was to determine if the behavioral intervention with partial meal replacement compared with usual care improved micronutrient intake of pregnant women with overweight and obesity.

2. Materials and Methods

2.1. Design

Healthy Beginnings/Comienzo Saludables was a randomized controlled trial conducted at California Polytechnic State University, San Luis Obispo, California, and at the Miriam Hospital with Women & Infants Hospital in Providence, Rhode Island, and was part of the Lifestyle Interventions for Expectant Moms (LIFE-Moms) consortium [18]. Clinical Trial Registry Number: ClinicalTrials.gov, www.clinicaltrials.gov, NCT01545934.

2.2. Participants

The study was conducted in accordance with the ethical principles of research; the protocol was approved by the Institutional Review Boards of California Polytechnic State University (2018-264) and the Miriam Hospital (2144-11), and all participants provided written informed consent. As previously described [17,19], recruitment occurred November 2012 and October 2015 in California and Rhode Island. Eligibility criteria included gestational age between 9 and 16 weeks, body mass index (BMI) ≥25, being English or Spanish-speaking, age ≥18 years, and singleton pregnancy. Participants were excluded if they had glycosylated hemoglobin (Hb A1c) ≥ 6.5 or self-reported major health diseases (e.g., heart disease, cancer, renal disease, and diabetes), and other (Figure 1). Of the 5381 screened women, 24% were excluded due to BMI < 25, and 24% were excluded due to gestational age >16 weeks. Other prevalent reasons for exclusions are shown in Figure 1.

Figure 1.

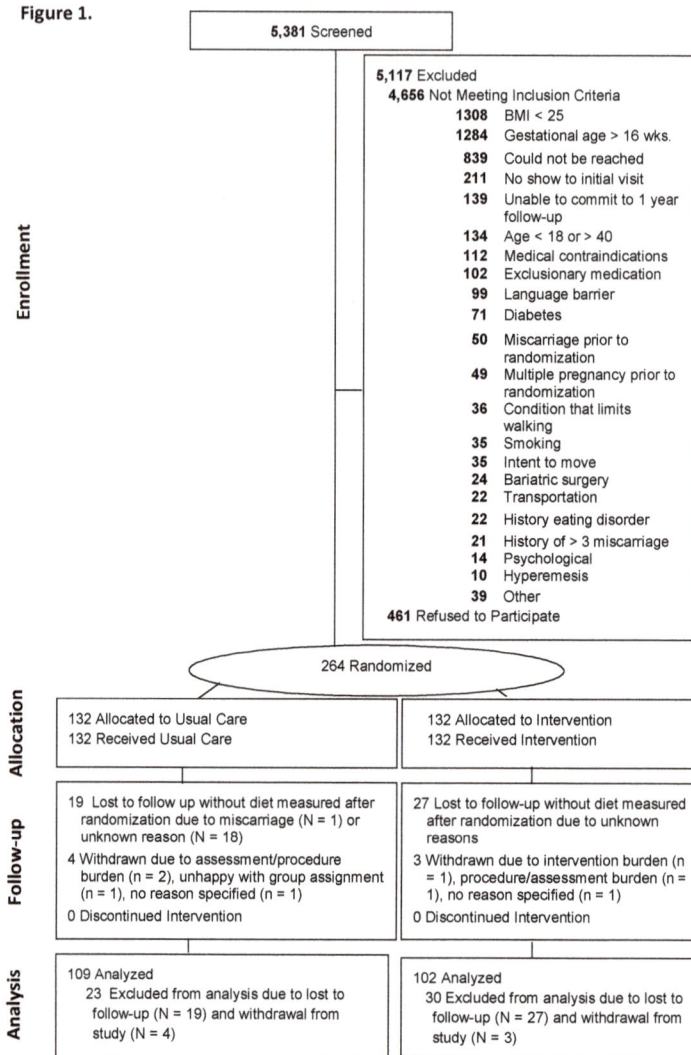

Figure 1. Participant flow and Retention in Healthy Beginnings/Comienzos Saludables. BMI = body mass index.

2.3. Interventions

In this two-site trial, randomization was computer-generated by the study statistician, and women were randomly assigned within site and ethnicity (Hispanic versus non-Hispanic) to one of the two treatment conditions: (1) usual care or (2) behavioral lifestyle intervention with partial meal replacement.

2.4. Usual Care

Participants in the usual care group received all aspects of usual care offered by their prenatal care providers [20]. Usual prenatal care visits typically occur monthly until 28 weeks of gestation, bi-weekly between 28 and 36 weeks of gestation, and weekly until delivery. Also, in this group at the time of study randomization, participants attended a ~20 min welcome visit with a study interventionist, providing general information about healthy eating, physical activity, and the Institte of Medicine (IOM)recommendations for total gestational weight gain [21]. Study interventionists were bilingual registered dietitians or counselors with degrees in nutrition, community health, psychology, kinesiology, or a related field. Participants received study newsletters with general information about pregnancy-related health, including consuming prenatal vitamins, quitting smoking, planning to breastfeed, and fetal growth.

2.5. Behavioral Lifestyle Intervention with Partial Meal Replacement during Pregnancy

Participants in the intervention group received all aspects of usual care plus a behavioral lifestyle intervention designed to prevent excessive weight gain during pregnancy. As described previously [17], the intervention targeted healthy eating, activity, and behavioral strategies. Each woman received ~20 min, individual, face-to-face counseling sessions with a study interventionist every two weeks until 20 weeks of gestation and then monthly visits until delivery. Women were encouraged to gain approximately one-half pound (0.23 kg) per week, on the basis of the 2009 IOM guidelines [21]. To promote adherence to weight gain guidelines, women were provided with a structured meal plan [22] that was individually tailored to meet each participant's self-reported dietary needs, including food aversions, cravings, lactose intolerance, and specialized diets, such as vegetarianism. The plan provided a caloric prescription of ~18 kcal/kg of body weight at study entry [23] and consisted of 30% of calories from fat, 15–20% from protein, and 50–55% from carbohydrates [24]. Women were instructed to replace two meals with a provided meal replacement shake or bar and to consume at least one meal of regular foods and two to four healthy snacks each day. The meal replacement products were provided free of charge at every intervention visit and in the quantities needed until the next scheduled intervention visit. The study's meal replacement options were selected at study onset by the investigators after an analysis of various meal plan scenarios that considered the micronutrient and macronutrient composition of specific meal replacement products, the participant's use of prenatal vitamins, the intervention's calorie and nutritional goals, and the current micronutrient and macronutrient recommendations for pregnant women [24]. Options included organic and lactose-free drinks and bars in a variety of flavors and brands (Supplemental Table S1).

2.6. Outcome Assessments

Assessments were conducted early in pregnancy (between 9 and 16 weeks) and at 35–36 weeks of gestation. The participants received $25 for completing each assessment. The assessment staff was masked to randomization to minimize potential bias. Dietary intake was assessed at study entry and 35–36 weeks of gestation using interview-administered 24 h recalls on two random days over a week and completed using the National Cancer Institute Automated Self-Administered 24 h recall (ASA-24; http://riskfactor.cancer.gov/tools/instruments/asa24.html) [25]. The ASA-24 provided values (combined from food, beverages, and supplements) for thiamin, riboflavin, niacin, vitamin B6, folate, vitamin B12, vitamin C, vitamin A, vitamin E, iron, zinc, calcium, magnesium, phosphorous, copper, selenium, water, energy, carbohydrate, total fat, and protein and included intake of supplements [26]. Dietary intake was categorized as meeting or not meeting the Recommended Daily Allowance (RDA) based on the Estimated Average Requirement (EAR) of the Institute of Medicine (IOM) for pregnant women [27–32]. The RDA represents an estimate of the average daily intake level sufficient to meet the nutrient requirements for 97–98% of healthy individuals. To assess the prevalence of inadequacy, the cut-point method was used [29,33] to classify individuals with intakes below versus at or above the median EAR considered needed for half of the individuals in the population (Table 1) [27–32,34,35]. For micronutrients in which an RDA and EAR had not been established (i.e., vitamin K, choline, potassium, sodium), cutoffs based on Adequate Intakes (AI) were used, albeit interpreted with less confidence [27–32]. The ASA-24 was also used to measure meal replacement intake, quantified as the total number of meal replacement products including shakes and bars that were consumed each day, on average, during the assessment period.

Table 1. Estimated average requirements (EAR) [1] for pregnancy.

Nutrient	Recommended Amount
Total Water (L/d) **	3.0 **
Calcium (mg/d)	800
CHO (g/d)	135
Fiber (g/d) **	28 **
Added sugars	≤25% of TE
Protein (g/kg/d)	0.88
Vit A µg, RAE/d	550
Vit C (mg/d)	70
Vit D (µg/d)	10
Vit E (mg/d)	12
Vit K µg/d **	90 **
Thiamin (mg/d)	1.2
Riboflavin (mg/d)	1.2
Niacin (mg/d)	14
Vit B6 (mg/d)	1.6
Folate (µg/d)	520
Vit B12 (µg/d)	2.2
Choline mg/d **	450 **
Copper (µg/d)	800
Iron (mg/d)	22
Magnesium (mg/d)	290
Phosphorus (mg/d)	580
Selenium (µg/d)	49
Zinc (mg/d)	9.5
Potassium mg/d **	4.7 **
Sodium g/d **	1.5 **

Abbreviations: Vit: vitamin; TE: total energy; d: day; RAE: retinol activity equivalents; CHO: carbohydrates: [1] EAR is the average daily nutrient intake level estimated to meet the requirements of half of the healthy individuals who are pregnant and aged 19–30 years [27–32]. Vitamin A based on RAEs; vitamin E based on α-tocopherol; niacin expressed as niacin equivalents; folate expressed as folate equivalents. Food and Nutrition Board, Institute of Medicine, National Academies reports may be accessed via www.nap.edu [27–32], ** Adequate Intakes (AI) because EARs have not been established; this cutoff is made with less confidence [27–32].

Weight and height were assessed in duplicate to the nearest 0.1 kg or 0.1 cm using a calibrated standard digital scale and stadiometer with the participant in lightweight clothing without shoes. Heritage and ethnicity were assessed by self-report using questionnaires with fixed categories. Marital status, income, education, employment status, and childbearing history were also assessed by self-report questionnaires. Gestational age in weeks at study entry was measured via clinical ultrasound.

2.7. Statistical Methods

Analysis Plan

To compare participants in the two groups and completers versus non-completers, Independent *t*-test for continuous variables and Pearson χ2 test or exact tests for categorical variables were used. To test if the intervention versus usual care affected micronutrient values, a repeated measures analysis of variance was used. The models included the terms treatment group and group × time interactions (fixed effect) and a priori defined covariates that included weeks of gestation at randomization, age, ethnicity (Hispanic versus non-Hispanic), parity (multiparity versus primiparity), study entry BMI category (overweight versus obese), household family income (≥50,000/year versus <50,000/year), and baseline value of variable of interest [36]; site (California versus Rhode Island) was also included as a fixed effect. (A sensitivity analysis that divided the income into four categories and also included education did not alter the results). To test whether the intervention versus usual care improved micronutrient adequacy (meeting versus not meeting RDA's EAR cutpoint), logistic regression was used that included the treatment group and the same covariates. Within the intervention group alone, number of meal replacement products/day and changes in micronutrients status were also analyzed via Pearson's partial correlations, adjusting for the same covariates. Statistical significance was set to $p < 0.05$. The SPSS (23.0.0; IBM Corporation; Armonk, NY, USA) statistical package was used for all analyses.

3. Results

Figure 1 summarizes the participant flow and retention into Healthy Beginnings/Comienzos Saludables. Participant characteristics were well balanced by randomized group (Table 2). At the 35–36 weeks of gestation visit, 82.1% (211/257) of participants completed the dietary assessment, including 85.2% (109/128) of usual care and 79.1% (102/129) of intervention participants, with no statistically significant ($p = 0.55$) differences in retention by group. The demographic characteristics (site, group, BMI, age, education, parity, weeks of gestation at randomization) did not significantly differ between participants who completed and those who did not complete the 35–36 weeks gestation visit. Completers were more likely than non-completers to have adequate intake of phosphorus (38/43 versus 209/211; chi square test $p = 0.002$), niacin (32/46 versus 187/211; $p = 0.002$), riboflavin (33/46 versus 185/211; $p = 0.011$), and thiamin (23/46 versus 144/211; $p = 0.03$); no other significant differences were observed.

Table 2. Baseline characteristics of the participants by condition.

Characteristic	Total $n = 211$	Usual Care $n = 109$	Intervention $n = 102$
Age, years, Mean (SD)	30.5 (5.3)	30.0 (5.6)	31.0 (4.9)
Hispanic/Latino, No. (%)			
Yes	85 (40.3)	43 (39.4)	42 (41.2)
No	126 (59.7)	66 (60.6)	60 (58.8)
Heritage, No. (%) (participants could select multiple)			
American Indian or Alaskan Native	7 (3.3)	3 (2.8)	4 (3.9)
Asian	2 (0.9)	0 (0)	2 (2.0)
Black or African American	15 (7.1)	7 (6.4)	8 (7.8)
Native Hawaiian or Pacific Islander	4 (1.9)	3 (2.8)	1 (1.0)
White	131 (62.1)	67 (61.5)	64 (62.7)
Other	60 (28.4)	30 (27.5)	30 (29.4)
Marital Status, No. (%)			
Married or living with significant other	148 (70.1)	97 (89.0)	51 (50.0)
Never married/divorced/widowed	63 (29.9)	12 (11.0)	51 (50.0)
Annual household Income $, No. (%)			
<$24,999	49 (23.2)	28 (25.7)	21 (20.6)
$25,000–49,999	62 (29.4)	30 (27.5)	32 (31.4)
$50,000–99,999	60 (28.4)	30 (27.5)	30 (29.4)
≥$100,000	40 (19.0)	21 (19.3)	19 (18.6)
Education, No. (%)			
High school or less	48 (22.7)	29 (26.6)	19 (18.6)
Some college/College	130 (61.6)	62 (56.9)	68 (66.7)
Post-graduate work	33 (15.6)	18 (16.5)	15 (14.7)
Employment, No. (%)			
Employed Full Time (at least 35 hours/week)	121 (57.3)	62 (56.9)	59 (57.8)
Employed Part-Time (less than 35 hours/week)	38 (18.0)	22 (20.2)	16 (15.7)
Unemployed	52 (24.6)	25 (22.9)	27 (26.5)
Childbearing history, No. (%)			
Primiparous	53 (25.1)	25 (22.9)	28 (27.5)
Multiparous	154 (73.0)	82 (75.2)	72 (70.6)
Weeks of gestation at study entry, Mean (SD)	13.6 (1.7)	13.5 (1.9)	13.8 (1.5)
Weight, kg, at study entry, Mean (SD)	84.9 (16.5)	86.1 (17.9)	83.6 (14.8)
BMI, kg/m^2, at study entry, Mean (SD)	32. (5.3)	32.5 (5.4)	32.1 (5.3)

Table 2. *Cont.*

Characteristic	Total *n* = 211	Usual Care *n* = 109	Intervention *n* = 102
Weight status			
Overweight, No. (%)	86 (40.8)	42 (38.5)	44 (43.1)
Obese, No. (%)	125 (59.2)	67 (61.5)	58 (56.9)
Preconception weight, Mean (SD)	83.0 (16.5)	81.9 (14.8)	84.1 (17.9)
Preconception weight status			
Overweight, No. (%)	94 (44.5)	44 (40.4)	50 (49.0)
Obese, No. (%)	114 (54.0)	63 (57.8)	51 (50.0)
Weight gain from preconception to study entry, kg, Mean (SD)	1.9 (4.4)	1.8 (3.3)	19 (5.0)
Daily prenatal vitamin intake, No. (%)	105 (96.7%)	105 (96.3)	99 (97.0)

Abbreviations: SD: standard deviation; BMI is calculated as weight in kilograms divided by the square of height in meters. Y: years. No.: number.

3.1. Intervention Effects on Micronutrient Intake

When examining the average changes in micronutrients from study entry to 35–36 weeks gestation (Table 3), significant group × time interactions indicated that the intervention relative to usual care increased the average intake of vitamins A (178.3 versus 34.6, µg/day, respectively; p = 0.0001), E (1.9 versus −0.3 mg/day; p = 0.0001), and K (23.4 versus −11.7 µg/day; p = 0.04). In addition, the intervention significantly increased (relative to usual care) the intakes of vitamin D (1.5 versus 0.5 µg/day; p = 0.045) and copper (259.2 versus −34.5, µg/day; p = 0.001) and significantly decreased the intake of selenium (−11.0 versus −7.2 µg/day; p = 0.002).

Table 3. Micronutrient intake from early pregnancy (baseline) to 35 weeks of gestation by treatment group.

	EAR	Usual Care; *n* = 109 Mean (SD)		Intervention; *n* = 102 Mean (SD)		Statistical Results [1]	
		Baseline	35 Weeks	Baseline	35 Weeks	T	G × T
Total Water, L/d	3.0 **	2.6 (0.9)	2.8 (1.1)	2.5 (0.8)	2.8 (0.9)	0.20	0.72
Calcium, mg/d	800	973.5 (355.8)	1026.0 (387.8)	928.9 (352.1)	1097.2 (378.0)	0.06	0.14
CHO, g/d	135	219.6 (68.3)	225.1 (71.9)	219.7 (75.6)	217.7 (63.3)	0.001	0.783
Fiber, g/d	28 **	16.5 (6.3)	15.6 (6.6)	16.7 (6.9)	14.2 (6.4)	0.02	0.07
Added sugars, % TE	<25% of TE	9.6 (4.6)	10.3 (7.0)	10.5 (6.1)	12.3 (10.3)	0.28	0.35
Protein, g/kg/d,	0.88	0.9 (0.3)	0.8 (0.3)	0.9 (.31665)	0.8 (0.3)	0.92	0.76
Vit A µg, RAE/d·	550	631.2 (336.6)	667.6 (372.3)	720.0 (378.9)	**898.0 (402.9)**	0.10	**0.0001**
Vit C, mg/d	70	90.1 (62.4)	78.0 (57.3)	103.3 (70.5)	94.6 (64.8)	0.03	0.09
Vit D, µg/d	10	4.1 (2.6)	4.6 (2.9)	4.0 (2.5)	**5.5 (3.3)**	0.55	**0.045**
Vit E, mg/d	12	7.2 (4.4)	6.9 (3.1)	7.3 (3.7)	**9.2 (5.0)**	0.03	**0.0001**
Vit K, µg/d	90 *	111.8 (108.6)	100.0 (105.0)	116.1 (114.3)	**140.0 (175.3)**	0.07	**0.04**
Thiamin, mg/d	1.2	1.5 (0.5)	1.6 (0.6)	1.5 (0.6)	1.5 (0.62)	0.04	0.91
Riboflavin, mg/d	1.2	1.9 (0.6)	2.1 (0.7)	1.9 (0.8)	2.0 (0.8)	0.63	0.77
Niacin, mg/d	14	22.2 (7.6)	22.3 (8.6)	22.2 (7.8)	22.4 (7.4)	0.01	0.95
Vit B6, mg/d	1.6	1.9 (0.9)	2.0 (0.9)	1.9 (0.8)	2.0 (2.0)	0.02	0.72
Folate, µg/d	520	519.4 (218.2)	554.1 (248.7)	557.9 (267.5)	550.3 (287.3)	0.01	0.71
Vit B12, µg/d	2.2	4.8 (3.1)	5.3 (2.8)	5.4 (4.2)	5.2 (2.9)	0.9	0.70
**Choline, mg/d ** **	450 **	285.1 (160.4)	294.0 (150.3)	291.5 (208.6)	**230.4 (138.5)**	0.11	**0.005**
Copper, µg/d	800	1270 (471)	1236 (390)	1208 (363)	**1467 (473)**	0.14	**0.0001**
Iron, mg/d	22	14.0 (5.3)	15.1 (5.5)	15.1 (6.0)	15.7 (6.2)	0.003	0.70
Magnesium, mg/d	290	280.7 (90.5)	282.4 (93.7)	269.6 (79.4)	**326.5 (99.3)**	0.09	**0.001**
Phosphorus, mg/d	580	1246.6 (391.9)	1262.9 (407.4)	1172.6 (360.1)	1255.3 (353.4)	0.035	0.91
Selenium, µg/d	49	108.9 (37.9)	101.6 (28.3)	100.4 (38.4)	**88.1 (28.9)**	0.012	**0.002**
Zinc, mg/d	9.5	11.2 (4.6)	11.0 (4.0)	10.6 (4.0)	12.2 (4.8)	0.635	0.06
Potassium mg/d	4.7**	2.4 (0.7)	2.4 (0.8)	2.4 (0.8)	2.4 (0.7)	0.003	0.63
Sodium g/d	1.5 **	3.2 (1.0)	3.1 (0.9)	3.1 (1.0)	**2.8 (1.1)**	0.118	**0.04**

Bold font used to highlight nutrients that statistically differed by randomized group. Abbreviations: TE: total energy; T: time; G × T: group by time, Vitamin A based on RAEs; [1] Repeated measures ANOVA adjusted for weeks gestation at randomization, age, ethnicity (Hispanic versus non-Hispanic), parity (multiparity versus primiparity), study entry BMI category (overweight verssu obese), household family income (>50,000/year versus <50,000/year), and baseline value of the variable of interest. However, the mean (SD) values in table are shown unadjusted. ** AI because EARs have not been established; this cutoff is made with less confidence [27–32].

3.2. Intervention Effects on Micronutrient Adequacy Based on the RDAs

Study Entry. Despite the prevalent intake of prenatal vitamins (97%, Table 2), significant proportions of participants reported inadequate intakes of nearly every micronutrient on the basis of the EARs for pregnant women (Table 4). The vast majority (≥90%) reported inadequate intakes of fiber, vitamin D, Vitamin E, iron. Nearly half (between 40 and 50%) of the participants also reported inadequate intakes of calcium, protein, vitamin A, vitamin C, vitamin B$_6$, folate, magnesium, and zinc. On the basis of AI cutoffs, inadequate intakes of choline and vitamin K were also quite prevalent (Table 4). Few participants reported intakes that were at or above the recommended tolerable limit for the micronutrients. Exceptions were that a majority of participants had higher than recommended as tolerable levels of sodium (170/211; 81.0%), and a minority had excess magnesium (39/211; 18.5%), folate (12/211; 5.7%), and niacin (19/211; 9.0%) at study entry (Table 5).

Table 4. Proportions with micronutrient inadequacy at 35 weeks of gestation by treatment group.

	EAR	Usual Care; *n* = 109 No. (%)		Intervention; *n* = 102 No. (%)		Sig [1]	OR [1]	95% CI [1]	
		Baseline	35 Weeks	Baseline	35 Weeks			Lower	Upper
Total Water, L/d; No. (%)	3.0 **	74 (67.9)	72 (66.1)	85 (83.3)	64 (62.8)	0.27	0.70	0.37	1.33
Calcium, mg/d; No. (%)	800	43 (39.5)	34 (31.2)	41 (40.2)	18 (17.7)	0.007	0.37	0.18	0.76
CHO, g/d; No. (%)	135	9 (8.3)	9 (8.3)	11 (10.8)	5 (4.9)	0.34	0.55	0.16	1.88
Fiber, g/d; No. (%)	28 **	102 (93.6)	104 (95.4)	98 (96.1)	100 (98.0)	0.32	2.54	0.41	15.91
Added sugars, % TE; No. (%)	<25% of TE	0 (0.00)	5 (4.6)	2 (2.0)	6 (5.9)	0.53	0.52	0.07	4.10
Protein, g/kg/d; No. (%)	0.88	53 (48.6)	71 (65.2)	52 (51.0)	71 (69.6)	0.42	1.30	0.69	2.45
Vit A, μg_RAE/d; No. (%)	550	53 (48.6)	50 (45.9)	44 (43.1)	26 (25.5)	0.003	0.39	0.21	0.72
Vit C, mg/d; No. (%)	70	45 (41.3)	56 (51.4)	39 (38.2)	41 (40.2)	0.17	0.66	0.37	1.18
Vit D, μg/d; No. (%)	10	106 (97.3)	103 (94.5)	99 (97.1)	95 (93.1)	0.53	0.68	0.20	2.27
Vit E, mg/d; No. (%)	12	98 (89.9)	104 (95.4)	91 (89.2)	79 (77.5)	0.001	0.17	0.06	0.48
Vit K, μg/d; No. (%)	90 **	65 (59.6)	70 (64.2)	57 (55.9)	49 (48.0)	0.023	0.49	0.26	0.91
Thiamin, mg/d; No. (%)	1.2	29 (26.6)	24 (22.0)	38 (37.3)	31 (30.4)	0.08	1.61	0.95	2.73
Riboflavin (mg/d; No. (%)	1.2	10 (9.2)	9 (8.3)	16 (15.7)	10 (9.8)	0.25	1.43	0.77	2.66
Niacin (mg/d; No. (%)	14	12 (11.0)	15 (13.8)	12 (11.8)	11 (10.8)	0.58	1.18	0.66	2.11
Vit B6 (mg/d; No. (%)	1.6	46 (42.2)	44 (40.4)	36 (35.3)	27 (26.5)	0.06	0.56	0.31	1.03
Folate (μg/d; No. (%)	520	63 (57.8)	59 (54.1)	52 (51.0)	55 (53.9)	0.85	1.06	0.59	1.89
Vit B12 (μg/d; No. (%)	2.2	13 (11.9)	8 (7.3)	11 (10.8)	10 (9.8)	0.37	1.62	0.57	4.60
Choline mg/d; No. (%)	450 **	98 (89.9)	99 (90.8)	91 (89.2)	100 (98.0)	0.048	5.00	1.02	24.60
Copper (μg/d; No. (%)	800	13 (11.9)	10 (9.2)	8 (7.8)	7 (6.7)	0.85	1.11	0.37	3.32
Iron (mg/d; No. (%)	22	102 (93.6)	102 (93.6)	90 (88.2)	90 (88.2)	0.55	0.72	0.25	2.08
Magnesium (mg/d; No. (%))	290	69 (63.3)	64 (58.7)	70 (68.6)	38 (37.3)	0.001	0.36	0.20	0.65
Phosphorus (mg/d; No. (%)	580	1 (0.9)	3 (2.8)	1 (1.0)	3 (2.9)	0.99	1.01	0.18	5.74
Selenium (μg/d; No. (%)	49	2 (1.8)	3 (2.8)	8 (7.8)	9 (8.8)	0.08	3.67	0.88	15.30
Zinc (mg/d; No. (%)	9.5	42 (38.5)	47 (43.1)	44 (43.1)	36 (35.3)	0.25	0.71	0.39	1.27
Potassium mg/d; No. (%)	4.7 **	0 (0)	0 (0)	0 (0)	0 (0)	–	–	–	–
Sodium, g/d; No. (%)	1.5 **	2 (1.8)	2 (1.8)	5 (4.9)	7 (6.9)	0.68	0.91	0.57	1.45

Bold font used to highlight nutrients that statistically differed by randomized group. Abbreviations: [1] Logistic regression analysis adjusted for weeks of gestation at randomization, age, ethnicity (Hispanic versus non-Hispanic), parity (multiparity versus primiparity), study entry BMI category (overweight versus obese), household family income (>50,000/year versus <50,000/year), and baseline value of the variable of interest. ** AI because EARs have not been established; this cutoff is made with less confidence [27–32].

Study Entry to 35–36 Weeks of Gestation. Significant group x time interactions were observed that indicated that the intervention significantly reduced the odds at 35–36 weeks of gestation of inadequate intake based on the EARs for calcium (odds ratio (OR) = 0.34 (0.18, 0.76) *p* = 0.007), vitamin A (OR = 0.39 (0.21, 0.72) *p* = 0.003;), vitamin E (OR = 0.17 (0.06, 0.48) *p* = 0.001), and magnesium (OR = 0.36 (0.20, 0.65) *p* = 0.001), as shown in Table 4. Based on AIs, the intervention also decreased the odds of inadequate intake of vitamin K (OR = 0.49 (0.26, 0.91)) and increased the odds of inadequate intake of choline (OR = 5.0 (1.0, 24.6) *p* = 0.04).

From study entry to 35–36 weeks of gestation, the intervention reduced the odds of intake above tolerable limits for sodium (OR = 0.47 (0.24, 0.91) *p* = 0.026) and increased the odds of intake above tolerable limits for magnesium (OR = 2.0 (1.0, 3.7); *p* = 0.038) (Table 5).

Table 5. Proportions with micronutrient levels above recommended tolerable limits.

	Upper Limit [1]	Usual Care n = 109		Intervention n = 102		Sig [2]	OR [2]	95% CI	
		Baseline	35 Weeks	Baseline	35 Weeks			Lower	Upper
Calcium no., %	2500 (mg/d)	0	0	0	0	–	–	–	–
Vit A no., %	3000 (µg/d)	0	0	0	0	–	–	–	–
Vit C no., %	2000 (mg/d)	0	0	0	0	–	–	–	–
Vit E no., %	1000 (mg/d)	0	0	0	0	–	–	–	–
Niacin no., %	35 (mg/d)	9 (8.2)	5 (4.6)	10 (9.8)	5 (4.9)	0.89	0.91	0.23	3.52
Vit B6 no., %	1000 (mg/d)	0	0	0	0	–	–	–	–
Folate no., %	100 (µg/d)	4 (3.7)	7 (6.4)	8	4 (3.9)	0.10	0.27	0.05	1.30
Choline no., %	3500 mg/d	0	0	0	0	–	–	–	–
Copper no., %	10,000 (µg/d)	0	0	0	0	–	–	–	–
Magnesium no., %	350 (mg/d)	23 (21.1)	24 (22.0)	16 (15.7)	36 (35.3)	0.038	1.97	1.04	3.74
Phosphorus no., %	3500(mg/d)	0	0	0	0	–	–	–	–
Selenium no., %	400 (µg/d)	0	0	0	0	–	–	–	–
Zinc no., %	40 (mg/d)	0	0	0	0	–	–	–	–
Sodium no., %	2.3 g/d	90 (82.6)	85 (78.0)	80 (78.4)	64 (62.7)	0.026	0.47	0.24	0.91

Abbreviations: [1] Upper limit (UL) is the highest level of daily nutrient intake that is likely to pose no risk of adverse health effects to almost all pregnant women aged 19–30 years. As intake increases above the UL, the risk for adverse effects increases. ULs are not intended to be a recommended level of intake, rather a level of intake that most individuals can likely tolerate. [2] Logistic regression analysis adjusted for weeks of gestation at randomization, age, ethnicity (Hispanic versus non-Hispanic), parity (multiparity versus primiparity), study entry BMI category (overweight versus obese), household family income (>50,000/year versus <50,000/year), baseline value of the variable of interest.

3.3. Intervention Adherence

From baseline to weeks 35–36, the intervention increased the average number of meal replacement products reported each day by an additional 0.63 (SD 0.83) products/day. Increased reported intake of meal replacement products/day was significantly related to increased intake of micronutrients (Figure 2), including vitamin A ($r = 0.33$; $p = 0.002$), vitamin E ($r = 0.30$; $p = 0.004$), niacin ($r = 0.21$; $p = 0.043$), thiamin ($r = 0.31$; $p = 0.003$), copper ($r = 0.50$; $p = 0.0001$), iron ($r = 0.28$; $p = 0.008$), magnesium ($r = 0.47$; $p = 0.0001$), phosphorus ($r = 0.22$; $p = 0.04$), and zinc ($r = 0.241$ $p = 0.02$). Trend positive partial correlations were observed for calcium ($r = 0.21$; $p = 0.05$), riboflavin ($r = 0.20$; $p = 0.058$), vitamin B6 ($r = 0.20$; $p = 0.06$), and choline ($r = 0.20$; $p = 0.06$), and a trend inverse correlation was observed for vitamin K ($r = -0.18$; $p = 0.08$) (Figure 2). No significant correlations were observed for changes in meal replacements and vitamin C, K, folic acid, B$_{12}$, selenium, potassium, or sodium.

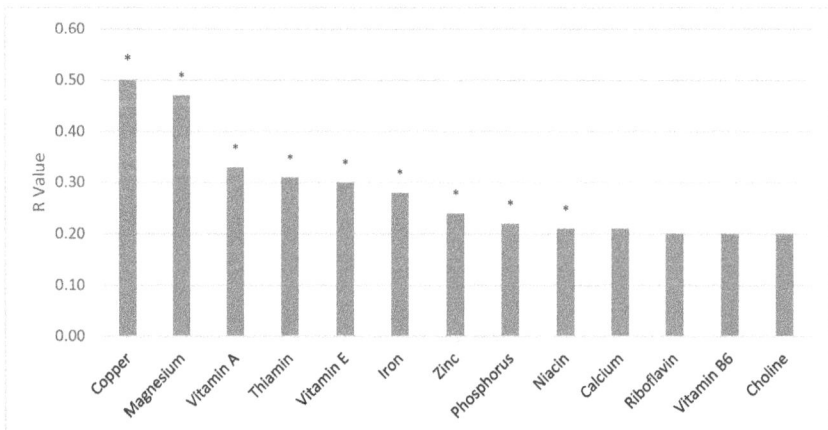

Figure 2. R values for changes in daily meal replacement servings/day and micronutrients from study entry to 35 weeks gestation among the intervention participants. * $p < 0.05$.

4. Discussion

A prenatal meal replacement intervention that reduced excess gestational weight gain also improved the micronutrient intake of pregnant women with overweight and obesity. At study entry, despite the intake of prenatal vitamins, more than 90% of participants reported inadequate intakes of vitamin E, and about 50% reported inadequate intakes of calcium, vitamins A and K, and magnesium. By 35 weeks of gestation, the prenatal lifestyle intervention with partial meal replacement had cut by more than half the odds of these micronutrient insufficiencies. Given the association between inadequate levels of nutrients during pregnancy and later adverse maternal and child health conditions [5], these findings highlight an additional potential benefit of the lifestyle intervention that reduced excess gestational weight gain [17].

At study entry, the vast majority of participants (81%) had higher than recommended intakes for sodium. By 35 weeks gestation, the intervention reduced by nearly 90% the odds of excess sodium intake based on AI cutoffs. The effects of excess sodium intake during pregnancy remain unclear [37]. Animal and some human studies have suggested that excess sodium may negatively affect the immune system [38] and placental functioning [37,39,40] and contribute to the development of high blood pressure and pre-eclampsia [41]. However, mixed findings and lack of evidence that restricting sodium during pregnancy has any long-term health benefits to mothers or children have led the current guidelines not to recommend sodium restriction during pregnancy, unless a woman has high blood pressure [42,43]. In the current trial, the intervention did not have a significant effect on lowering blood pressure [17], despite the reductions in sodium intake reported here.

The intervention-related increases in micronutrient intake generally occurred without promoting an excess in micronutrients—with the exception of magnesium. The intervention increased the odds of exceeding the recommended tolerable limit for intake of magnesium (i.e., 350 mg/day). The upper limits are not intended to be a recommended "cap" but rather a level of intake that most individuals could likely tolerate [29,30,34,35]. Nevertheless, high amounts of magnesium could result in diarrhea, nausea, and abdominal cramping [29,30,34,35], which were not measured in the current study.

Also of note, the intervention increased the odds of inadequate intake of choline. Choline does not yet have an established EAR, so cutoffs were based on AI, which should be interpreted with less confidence [27–32]. Nevertheless, the reasons for an intervention-related decline in choline remain unclear. The meal replacements likely did not contain adequate amounts of choline and might have displaced typical consumption of choline-containing foods, such as eggs and milk. However, group differences in milk intake were not observed (data not shown). Adequate intake of dietary choline may be important for optimal fetal outcome (birth defects, brain development) and for maternal liver and placental function [44]. Future research of the intervention approach should consider the promotion of both egg and milk consumption or other choline-rich foods or a diet supplement that might assist women in the intervention in meeting the daily choline intake recommendations.

This study is the first to examine whether a prenatal lifestyle intervention with partial meal replacement that reduced excess gestational weight gain had positive effects on micronutrient intake of pregnant women with overweight and obesity. The study's strengths include its randomized design, diverse population, blinded assessors, and two repeated, random, and interview-administered 24 h recalls to measure the dietary intake. Limitations of this study include the fact that the assessment of micronutrients was solely based on self-report, which could have overestimated the proportions with inadequate intake [45]. Also, data were lacking for some micronutrients, including molybdenum and iodine. Also, for micronutrients in which an RDA and EAR had not been established (i.e., vitamin K, choline, potassium, sodium), cutoffs were based on AIs, which should be interpreted with less confidence, since the amount and quality of data currently available may not be sufficient to make reliable estimates [46]. The study design tested a treatment "package" and did not allow for isolation of the independent contribution of meal replacements from other intervention components. Future randomized clinical trials are needed to tease apart the intervention "package" and identify the independent contribution of meal replacements and other intervention components on improving micronutrient adequacy. Only 82% of the original sample completed the dietary assessment, although the characteristics of completers versus non-completers were similar. Also, the study's sample size was not powered to examine the effect of changes in micronutrients on maternal and fetal complications.

5. Conclusions

A comprehensive prenatal lifestyle intervention that reduced excess gestational weight gain [17] also improved micronutrient intake and reduced the odds of inadequate intake of several micronutrients. A low nutritional quality of the diet and inadequate intakes of micronutrients can have significant consequences for both the mother and the developing fetus. Future research is now needed to examine the generalizability and effectiveness of this prenatal lifestyle modification program in other populations and settings.

Supplementary Materials: The following are available online at http://www.mdpi.com/2072-6643/11/5/1071/s1, Table S1: Nutritional Composition of Healthy Beginnings/Comienzos Saludables Meal Replacements.

Author Contributions: S.P., B.A., R.R.W. designed the research; S.P., R.R.W. conducted the research; S.P. analyzed the data; S.P., R.R.W., B.A. wrote the paper; S.P. had primary responsibility for the final content. All authors read and approved the final manuscript.

Funding: This research was funded by the National Institutes of Health National Heart, Lung, and Blood Institute (HL114377). LIFE-Moms is supported by the National Institutes of Health through the National Institute of Diabetes and Digestive and Kidney Diseases (NIDDK, U01 DK094418, U01 DK094463, U01 DK094416, 5U01 DK094466 [RCU]), the National Heart, Lung, and Blood Institute (NHLBI, U01 HL114344, U01 HL114377), the *Eunice Kennedy Shriver* National Institute of Child Health and Human Development (NICHD, U01 HD072834), the National Center for Complementary and Integrative Health (NCCIH), the NIH Office of Research in Women's Health (ORWH), the Office of Behavioral and Social Science Research (OBSSR), the Indian Health Service, and the Intramural Research Program of the NIDDK.

Acknowledgments: We thank the participants in the study. We thank our unpaid recruitment sites, including in California, Pacific Central Coast Health Centers, including Santa Maria Women's Health Center, Bishop's Peak Women's Health Center, and Templeton Women's Health Center and the Community Health Centers (CHC) in Templeton, San Luis Obispo, Santa Maria, and Nipomo, and Creating Harmony Women's Healthcare, French Hospital Medical Center, and Marian Regional Medical Center, and San Luis Obispo and Santa Barbara WIC clinics, and, in Rhode Island, the Women's Primary Care Center at Women and Infants Hospital, Center for OB-GYN, Women's Care, Inc. in Providence, Pawtucket and East Greenwich, Beitle, Bayside OB-GYN, Broadway OB/GYN, RI WIC Clinics. We thank the paid research team members in CA, including Adilene Quintana-Diaz, Noemi Alarcon, Martha La Spina, Maria Legato, Samantha Lalush, Adrian Mercado, Nick Katsantones, Megan Ershov, Natali Valdez, Ana Stewart, Vanessa Rodriguez, Jill Jacoby, Hannah Feldman, and in RI, including Juliana Duszlak (Luciani), Erica Ferguson-Robichaud, Denise Fernandes-Pierre, Kristen DeLayo, Kathryn Story, Stephanie Guerra, Patricia Sandoval, Zeely Sylvia-Denmat, Isabella Cassell, Julie Krol, Sarah Morris, Briana Borgolini, Molly La Rue, Kaitlyn Dahlborg, Genevieve Ramos, Whitney Howie, and Leah Sabatino. We thank the LIFE-Moms consortium members for their paid contributions to the development and oversight of the common measures and procedures shared across the trials. We thank Orgain and Pure Fit bar for providing product purchase discounts in support of this study.

Conflicts of Interest: The authors report grants from the National Institutes of Health during the conduct of the study. SP reports a grant from Weight Watchers International, outside the submitted work. The authors have no other conflicts of interest to report.

References

1. Koletzko, B.; Godfrey, K.M.; Poston, L.; Szajewska, H.; van Goudoever, J.B.; de Waard, M.; Brands, B.; Grivell, R.M.; Deussen, A.R.; Dodd, J.M.; et al. Nutrition during pregnancy, lactation and early childhood and its implications for maternal and long-term child health: The early nutrition project recommendations. *Ann. Nutr. Metab.* **2019**, *74*, 93–106. [CrossRef] [PubMed]

2. Nesby-O'Dell, S.; Scanlon, K.S.; Cogswell, M.E.; Gillespie, C.; Hollis, B.W.; Looker, A.C.; Allen, C.; Doughertly, C.; Gunter, E.W.; Bowman, B.A. Hypovitaminosis D prevalence and determinants among African American and white women of reproductive age: Third National Health and Nutrition Examination Survey, 1988–1994. *Am. J. Clin. Nutr.* **2002**, *76*, 187–192. [CrossRef]

3. Agarwal, S.; Reider, C.; Brooks, J.R.; Fulgoni, V.L. Comparison of prevalence of inadequate nutrient intake based on body weight status of adults in the United States: An analysis of NHANES 2001–2008. *J. Am. Coll. Nutr.* **2015**, *34*, 126–134. [CrossRef]

4. Astrup, A.; Bugel, S. Overfed but undernourished: recognizing nutritional inadequacies/deficiencies in patients with overweight or obesity. *Int. J. Obes.* **2018**. [CrossRef] [PubMed]

5. Procter, S.B.; Campbell, C.G. Position of the Academy of Nutrition and Dietetics: Nutrition and lifestyle for a healthy pregnancy outcome. *J. Acad. Nutr. Diet.* **2014**, *114*, 1099–1103. [CrossRef] [PubMed]

6. Shapira, N. Prenatal nutrition: A critical window of opportunity for mother and child. *Womens Health* **2008**, *4*, 639–656. [CrossRef]

7. Peaceman, A.M.; Clifton, R.G.; Phelan, S.; Gallagher, D.; Evans, M.; Redman, L.M.; Knowler, W.C.; Joshipura, K.; Haire-Joshu, D.; Yanovski, S.Z.; et al. Lifestyle Interventions Limit Gestational Weight Gain in Women with Overweight or Obesity: LIFE-Moms Prospective Meta-Analysis. *Obesity (Silver Spring)* **2018**, *26*, 1396–1404. [CrossRef]

8. Marangoni, F.; Cetin, I.; Verduci, E.; Canzone, G.; Giovannini, M.; Scollo, P.; Corsello, G.; Poli, A. Maternal diet and nutrient requirements in pregnancy and breastfeeding. An italian consensus document. *Nutrients* **2016**, *8*, 629. [CrossRef]

9. Milman, N.; Bergholt, T.; Byg, K.E.; Eriksen, L.; Graudal, N. Iron status and iron balance during pregnancy. A critical reappraisal of iron supplementation. *Acta. Obstet. Gynecol. Scand.* **1999**, *78*, 749–757. [CrossRef]

10. Czeizel, A.E. Periconceptional folic acid containing multivitamin supplementation. *Eur. J. Obstet. Gynecol. Reprod. Biol.* **1998**, *78*, 151–161. [CrossRef]

11. Murphy, M.M.; Scott, J.M.; McPartlin, J.M.; Fernandez-Ballart, J.D. The pregnancy-related decrease in fasting plasma homocysteine is not explained by folic acid supplementation, hemodilution, or a decrease in albumin in a longitudinal study. *Am. J. Clin. Nutr.* **2002**, *76*, 614–619. [CrossRef] [PubMed]

12. Zeghoud, F.; Vervel, C.; Guillozo, H.; Walrant-Debray, O.; Boutignon, H.; Garabedian, M. Subclinical vitamin D deficiency in neonates: Definition and response to vitamin D supplements. *Am. J. Clin. Nutr.* **1997**, *65*, 771–778. [CrossRef] [PubMed]

13. Koren, G.; Pairaideau, N. Compliance with prenatal vitamins. Patients with morning sickness sometimes find it difficult. *Can. Fam. Phys.* **2006**, *52*, 1392–1393. [PubMed]

14. Dobbing, J.; Rybar Laboratories. *Prevention of Spina Bifida and Other Neural Tube Defects*; Academic Press: London, UK; New York, NY, USA, 1983; p. xiv. 251p.

15. Fall, C.H.; Yajnik, C.S.; Rao, S.; Davies, A.A.; Brown, N.; Farrant, H.J. Micronutrients and fetal growth. *J. Nutr.* **2003**, *133*, S1747–S1756. [CrossRef]

16. Long, V.A.; Martin, T.; Janson-Sand, C. The great beginnings program: Impact of a nutrition curriculum on nutrition knowledge, diet quality, and birth outcomes in pregnant and parenting teens. *J. Am. Diet. Assoc.* **2002**, *102*, S86–S89. [CrossRef]

17. Phelan, S.; Wing, R.R.; Brannen, A.; McHugh, A.; Hagobian, T.A.; Schaffner, A.; Jelalian, E.; Hart, C.N.; Scholl, T.O.; Munoz-Christian, K.; et al. Randomized controlled clinical trial of behavioral lifestyle intervention with partial meal replacement to reduce excessive gestational weight gain. *Am. J. Clin. Nutr.* **2018**, *107*, 183–194. [CrossRef]

18. Clifton, R.G.; Evans, M.; Cahill, A.G.; Franks, P.W.; Gallagher, D.; Phelan, S.; Pomeroy, J.; Redman, L.M.; Van Horn, L.; Group, L.I.-M.R. Design of lifestyle intervention trials to prevent excessive gestational weight gain in women with overweight or obesity. *Obesity (Silver Spring)* **2016**, *24*, 305–313. [CrossRef]

19. Phelan, S.; Wing, R.R.; Brannen, A.; McHugh, A.; Hagobian, T.; Schaffner, A.; Jelalian, E.; Hart, C.N.; Scholl, T.O.; Munoz-Christian, K.; et al. Does Partial Meal Replacement During Pregnancy Reduce 12-Month Postpartum Weight Retention? *Obesity (Silver Spring)* **2019**, *27*, 226–236. [CrossRef]

20. Conner, P.; Bartlett, S.; Mendelson, M.; Condon, K.; Sutcliffe, C. (Eds.) *WIC Participant and Program Characteristics 2008, WIC-08-PC*; U.S. Department of Agriculture, Food and Nutrition Service, Office of Research and Analysis: Alexandria, VA, USA, 2010.

21. Phelan, S.; Phipps, M.G.; Abrams, B.; Darroch, F.; Schaffner, A.; Wing, R.R. Randomized trial of a behavioral intervention to prevent excessive gestational weight gain: The Fit for Delivery Study. *Am. J. Clin. Nutr.* **2011**, *93*, 772–779. [CrossRef]

22. Wing, R.R.; Jeffery, R.W.; Burton, L.R.; Thorson, C.; Nissinoff, K.S.; Baxter, J.E. Food provision vs structured meal plans in the behavioral treatment of obesity. *Int. J. Obes. Relat. Metab. Disord.* **1996**, *20*, 56–62.

23. Artal, R.; Catanzaro, R.B.; Gavard, J.A.; Mostello, D.J.; Friganza, J.C. A lifestyle intervention of weight-gain restriction: Diet and exercise in obese women with gestational diabetes mellitus. *Appl. Physiol. Nutr. Metab.* **2007**, *32*, 596–601. [CrossRef]

24. Otten, J.J.; Hellwig, J.P.; Meyers, L.D. *DRI, Dietary Reference Intakes: The Essential Guide to Nutrient Requirements*; National Academies Press: Washington, DC, USA, 2006; p. xiii. 543p.

25. Kirkpatrick, S.I.; Subar, A.F.; Douglass, D.; Zimmerman, T.P.; Thompson, F.E.; Kahle, L.L.; George, S.M.; Dodd, K.W.; Potischman, N. Performance of the Automated Self-Administered 24-h Recall relative to a measure of true intakes and to an interviewer-administered 24-h recall. *Am. J. Clin. Nutr.* **2014**, *100*, 233–240. [CrossRef]

26. Subar, A.F.; Kirkpatrick, S.I.; Mittl, B.; Zimmerman, T.P.; Thompson, F.E.; Bingley, C.; Willis, G.; Islam, N.G.; Baranowski, T.; McNutt, S.; et al. The Automated Self-Administered 24-h dietary recall (ASA24): A resource for researchers, clinicians, and educators from the National Cancer Institute. *J. Acad. Nutr. Diet.* **2012**, *112*, 1134–1137. [CrossRef] [PubMed]

27. Institute of Medicine (U.S.). Standing Committee on the Scientific Evaluation of Dietary Reference Intakes. *Dietary Reference Intakes: For Calcium, Phosphorus, Magnesium, Vitamin D, and Fluoride*; National Academy Press: Washington, DC, USA, 1997; p. xv. 432p.

28. Institute of Medicine (U.S.). Standing Committee on the Scientific Evaluation of Dietary Reference Intakes; Institute of Medicine (U.S.). Panel on Folate Other B Vitamins and Choline; Institute of Medicine (U.S.). Subcommittee on Upper Reference Levels of Nutrients. *Dietary Reference Intakes for Thiamin, Riboflavin, Niacin, Vitamin B$_6$, Folate, Vitamin B$_{12}$, Pantothenic Acid, Biotin, and Choline*; National Academy Press: Washington, DC, USA, 1998; p. xxii. 564p.

29. Institute of Medicine (U.S.). Panel on Dietary Antioxidants and Related Compounds. *Dietary Reference Intakes for Vitamin C, Vitamin E, Selenium, and Carotenoids: A Report of the Panel on Dietary Antioxidants and Related Compounds, Subcommittees on Upper Reference Levels of Nutrients and of Interpretation and Use of Dietary Reference Intakes, and the Standing Committee on the Scientific Evaluation of Dietary Reference Intakes, Food and Nutrition Board, Institute of Medicine*; National Academy Press: Washington, DC, USA, 2000; p. xx, 506p.

30. Institute of Medicine (U.S.). Panel on Micronutrients. *DRI: Dietary Reference Intakes for Vitamin A, Vitamin K, Arsenic, Boron, Chromium, Copper, Iodine, Iron, Manganese, Molybdenum, Nickel, Silicon, Vanadium, and Zinc: A Report of the Panel on Micronutrients and the Standing Committee on the Scientific Evaluation of Dietary Reference Intakes, Food and Nutrition Board, Institute of Medicine*; National Academy Press: Washington, DC, USA, 2001; p. xxii, 773p.

31. Institute of Medicine (U.S.). Panel on Macronutrients; Institute of Medicine (U.S.). Standing Committee on the Scientific Evaluation of Dietary Reference Intakes. *Dietary Reference Intakes for Energy, Carbohydrate, Fiber, Fat, Fatty Acids, Cholesterol, Protein, and Amino Acids*; National Academies Press: Washington, DC, USA, 2005; p. xxv, 1331p.

32. Ross, A.C. Dietary Reference Intakes for Adequacy: Calcium and Vitamin D. In *Dietary Reference Intakes for Calcium and Vitamin D*; National Academies Press: Washington, DC, USA, 2011; pp. 345–402.

33. Beaton, G.H. Approaches to analysis of dietary data: relationship between planned analyses and choice of methodology. *Am. J. Clin. Nutr.* **1994**, *59*, 253S–261S. [CrossRef]

34. Institute of Medicine (U.S.). Food and Nutrition Board; Suitor, C.W.; Meyers, L.D. *Dietary Reference Intakes Research Synthesis Workshop Summary*; National Academies Press: Washington, DC, USA, 2007; p. xii, 297p.

35. Institute of Medicine (U.S.). Panel on Dietary Reference Intakes for Electrolytes and Water. *DRI, Dietary Reference Intakes for Water, Potassium, Sodium, Chloride, and Sulfate*; National Academies Press: Washington, DC, USA, 2005; p. xviii, 617p.

36. Rasmussen, K.M.; Yaktine, A.L.; Institute of Medicine (U.S.). Committee to Reexamine IOM Pregnancy Weight Guidelines. *Weight Gain during Pregnancy: Reexamining the Guidelines*; National Academies Press: Washington, DC, USA, 2009.

37. Asayama, K.; Imai, Y. The impact of salt intake during and after pregnancy. *Hypertens. Res.* **2018**, *41*, 1–5. [CrossRef]

38. Luo, T.; Ji, W.J.; Yuan, F.; Guo, Z.Z.; Li, Y.X.; Dong, Y.; Ma, Y.Q.; Zhou, X.; Li, Y.M. Th17/Treg Imbalance Induced by Dietary Salt Variation Indicates Inflammation of Target Organs in Humans. *Sci. Rep.* **2016**, *6*, 26767. [CrossRef]

39. Lenda, D.M.; Boegehold, M.A. Effect of a high salt diet on microvascular antioxidant enzymes. *J. Vasc. Res.* **2002**, *39*, 41–50. [CrossRef]

40. Lenda, D.M.; Boegehold, M.A. Effect of a high-salt diet on oxidant enzyme activity in skeletal muscle microcirculation. *Am. J. Physiol. Heart. Circ. Physiol.* **2002**, *282*, H395–H402. [CrossRef] [PubMed]

41. Dishy, V.; Sofowora, G.G.; Imamura, H.; Nishimi, Y.; Xie, H.G.; Wood, A.J.; Stein, C.M. Nitric oxide production decreases after salt loading but is not related to blood pressure changes or nitric oxide-mediated vascular responses. *J. Hypertens.* **2003**, *21*, 153–157. [CrossRef] [PubMed]

42. Visintin, C.; Mugglestone, M.A.; Almerie, M.Q.; Nherera, L.M.; James, D.; Walkinshaw, S.; Guideline Development Group. Hypertension in Pregnancy: The Management of Hypertensive Disorders During Pregnancy. *BMJ* **2010**. [CrossRef] [PubMed]

43. Hypertension in pregnancy. Report of the American College of Obstetricians and Gynecologists' Task Force on Hypertension in Pregnancy. Hypertension in pregnancy. Report of the American College of Obstetricians and Gynecologists' Task Force on Hypertension in Pregnancy. *Obstet. Gynecol.* **2013**, *122*, 1122–1131.

44. Zeisel, S.H. Nutrition in pregnancy: The argument for including a source of choline. *Int. J. Womens Health* **2013**, *5*, 193–199. [CrossRef] [PubMed]

45. Garriguet, D. Impact of identifying plausible respondents on the under-reporting of energy intake in the Canadian Community Health Survey. *Health Rep.* **2008**, *19*, 47–55. [PubMed]

46. Institute of Medicine (U.S.). *DRI Dietary Reference Intakes: Applications in Dietary Assessment*; National Academies Press: Washington, DC, USA, 2000. [CrossRef]

nutrients

MDPI

Article

Behavioral Determinants of Objectively Assessed Diet Quality in Obese Pregnancy

Jasper Most [1], Candida J. Rebello [2], Abby D. Altazan [1], Corby K. Martin [3], Marshall St Amant [4] and Leanne M. Redman [1,*]

[1] Reproductive Endocrinology and Women's Health, Pennington Biomedical Research Center, Baton Rouge, LA 70817, USA
[2] Clinical Trials Unit, Pennington Biomedical Research Center, Baton Rouge, LA 70817, USA
[3] Ingestive Behavior, Weight Management, and Health Promotion Laboratory, Pennington Biomedical Research Center, Baton Rouge, LA 70817, USA
[4] Woman's Hospital, 100 Woman's Way, Baton Rouge, LA 70817, USA
*   Correspondence: leanne.redman@pbrc.edu; Tel.: +1-225-763-0947

Received: 30 April 2019; Accepted: 24 June 2019; Published: 26 June 2019

Abstract: Interventions to promote healthy pregnancy in women with obesity by improving diet quality have been widely unsuccessful. We hypothesized that diet quality is determined by eating behaviors, but evidence in women with obesity is lacking. We evaluated diet quality and eating behavior in 56 women with obesity (mean ± SEM, 36.7 ± 0.7 kg/m^2, 46% White, 50% nulliparous) early in pregnancy (14.9 ± 0.1 weeks). Diet quality was objectively assessed with food photography over six days and defined by Healthy Eating Index. Eating behaviors were assessed by validated questionnaires. Women reported consuming diets high in fat (38 ± 1% of energy) and the HEI was considered "poor" on average (46.7 ± 1.3), and for 71% of women. Diet quality was independently associated with education level ($p = 0.01$), food cravings ($p < 0.01$), and awareness towards eating ($p = 0.01$). Cravings for sweets and fast foods were positively correlated with respective intakes of these foods ($p < 0.01$ and $p = 0.04$, respectively), whereas cravings for fruits and vegetables did not relate to diet intake. We provide evidence of the determinants of poor diet quality in pregnant women with obesity. Based on this observational study, strategies to improve diet quality and pregnancy outcomes are to satisfy cravings for healthy snacks and foods, and to promote awareness towards eating behaviors.

Keywords: Pregnancy; obesity; diet quality; Healthy Eating Index; food cravings; mindful eating; education; race; food photography

1. Introduction

Poor quality of maternal diet is considered one of the most significant predictors of adverse pregnancy outcomes [1–5] and poor infant health [5–9]. Both poor diet quality and adverse pregnancy outcomes are more frequent among women with obesity [10,11].

Among women with normal weight or who are overweight, improvements in maternal diet quality reduced the prevalence of adverse outcomes including excess gestational weight gain [12,13], gestational diabetes, and hypertensive disorders in mothers [2] and macrosomia in infants [14]. However, in women with obesity, successful dietary interventions are seldom reported, likely due to small effects on diet quality [14–17].

To develop more successful behavioral interventions for women with obesity requires an understanding of determinants of poor diet quality, i.e., consumption of specific food groups, mindful eating, or food cravings. Others have shown in 24 h recalls, that poor diet quality of women with obesity was determined by low intake of fruits [10], but not due to intake of other food groups. Eating behaviors such as mindful eating or food cravings associate with poor diet quality in nonpregnant subjects [18]. Women with obesity appear more likely to report indulgence in food cravings and less mindfulness towards eating, but such data has only been reported in women who were not pregnant [19]. Furthermore, the available evidence from these studies is limited by their use of self-reported diet data, varying metrics for diet quality, and lack of information on eating behaviors to understand how behaviors could be targeted to improve diet quality. To our knowledge, no study has defined the behavioral determinants of diet quality in pregnant women with obesity.

The aim of this study was to characterize diet quality in pregnant women with obesity and identify maternal eating patterns and behaviors that contribute to the quality of the diet in early pregnancy. In addition, we assessed if maternal eating attitudes and behaviors are influenced by race, education level, or nausea and vomiting. To this end, dietary intake, including diet quality and eating patterns, and eating attitudes and behaviors were simultaneously assessed between 13 and 16 weeks gestation. We hypothesized that maternal eating behavior such as food cravings and mindfulness would contribute to maternal diet quality objectively assessed by food photography.

2. Materials and Methods

2.1. Study Design

Participants were enrolled in a prospective observational study at the Pennington Biomedical Research Center to assess the determinants of gestational weight gain [20,21]. Between 13 and 16 weeks of gestation (14.9 ± 0.1 weeks), dietary intake and eating patterns were assessed over six consecutive days by a validated food photography method [22]. Eating behaviors were assessed with self-report questionnaires. The study was approved by the Pennington Biomedical Research Center Institutional Review Board and written informed consent was obtained from all participants prior to the initiation of procedures.

2.2. Participants

Enrolled women were pregnant, 18 to 40 years of age, with a measured BMI $\geq$30 kg/m^2 at the screening visit (gestational age, <15 weeks). Obesity was defined as class I: 30$\leq$BMI<35 kg/m^2, class II: 35$\leq$BMI<40 kg/m^2, and class III: BMI$\geq$40 kg/m^2. Women who reported smoking, alcohol, or drug use or had pre-existing hypertension, diabetes (HbA1c $\geq$6.5%), psychological or eating disorders or reported medications/supplements that might affect body weight, planned termination of pregnancy, and bariatric surgery were excluded. Fasting plasma glucose was measured in a venous blood sample, collected after an overnight fast at the research center. Sociodemographic parameters including age, race, parity, family income, education, and employment were obtained from self-report questionnaires. Poverty to income ratio was determined as the ratio of individual income to poverty threshold based on family size.

2.3. Dietary Assessment

Dietary intake was assessed over six days with the SmartIntake® smartphone application, which is a validated food photography method for adult energy intake and diet quality, as assessed by macronutrient intake (kcal from protein, fat, or carbohydrate) and intake of vitamin C, calcium, iron, sodium, or fiber [22]. Prior to analysis, routine data handling procedures were followed and days in which reported caloric intake was <60% of total daily energy expenditure measured by doubly labeled water, were excluded to improve accuracy and eliminate reporting bias by BMI [20]. Total daily energy expenditure was measured for seven days simultaneous to dietary intake [20]. When using SmartIntake®, participants captured images of meals or food items consumed and plate (food) waste. Participants defined the type of meal (breakfast, lunch, dinner, snacks) and every image was automatically time-stamped to allow calculation of eating patterns (first and last meal of a day, eating duration). Images were transmitted automatically in real-time via the phone application and were reviewed as a set by the participant and study staff at the end of the assessment period to document potential missing data. Dietitians determined portion sizes using validated visual comparison procedures [22], and the nutritional characteristics of each food were determined from the United States Department of Agriculture (USDA) Food and Nutrient Database for Dietary Studies 2011–2012 [23] and manufacturers' nutrient information.

Diet quality was defined by macronutrient composition expressed relative to total energy intake (%EI), by the 2015 Healthy Eating Index (HEI) [24] and by macronutrient and micronutrient intake as compared to the intake recommendations by the Institute of Medicine [25,26]. To calculate HEI, foods and beverages were converted into the 37 USDA food pattern components [27] using the Food Patterns Equivalents Database (FPED). The FPED serves as a unique research tool to evaluate food and beverage intakes of Americans with respect to the 2015–2020 Dietary Guidelines for Americans recommendations [28]. The 37 USDA scores were converted into 12 HEI subscores measured as cup equivalents of fruit, vegetables, and dairy; ounce equivalents of grains and protein foods; teaspoon equivalents of added sugars; gram equivalents of solid fats and oils; and the number of alcoholic drinks. The HEI is a composite score of these subscores. The Healthy Eating Index 2015 ranges from 0 to 100, and HEI <51 indicates "Poor", a score between 51 and 80 indicates "Needs Improvement" and a HEI score >80 indicates "Good".

2.4. Nausea and Vomiting of Pregnancy

The Modified-Pregnancy-Unique Quantification of Emesis and Nausea Index (PUQE) questionnaire [29] is a validated instrument that assesses the severity of nausea and vomiting of pregnancy [29,30]. A PUQE Summary Score of ≤6 is considered as having no or mild symptoms of nausea and vomiting of pregnancy, a score between 7 and 12 is considered moderate and a score of ≥13 is defined as severe.

2.5. Eating Inventory

The Eating Inventory (EI), previously known as the Three Factor Eating Questionnaire [31,32], evaluates eating behavior in three factor domains: dietary restraint, disinhibition, and perceived hunger. A low score for dietary restraint, and high scores for disinhibition and perceived hunger, indicate less control over eating behavior and more episodic overeating.

2.6. Food Cravings

Food cravings were evaluated with the Food Craving Inventory (FCI), which measures the frequency of food cravings over the previous month. The FCI measures general cravings as well as cravings for high-fat foods, sweets, carbohydrates/starches, and fast food fats [33]. The version used in this study also included a scale to measure cravings for fruits and vegetables, and cravings for specific foods measured with the FCI have been found to correlate with intake of those foods [34].

2.7. Mindfulness

Mindfulness towards eating was assessed with the Mindful Eating Questionnaire (MEQ) which has been validated in pregnant women [35]. The MEQ explores mindfulness across five subscales, including disinhibition, awareness, external cues, emotional response, and distraction [36]. Mindful eating refers to an unbiased awareness of sensations around eating. Disinhibition measures the inability to stop eating even when full. The awareness subscale measures an individual's awareness of the sensory aspects of eating. Distraction refers to the tendency to think about other things and rush while eating. The external cues subscale refers to eating in response to environmental cues, and emotional response refers to eating in response to negative emotions. The emotional response and distraction subscales are reverse scored, and five questions on the disinhibition are reverse scored. High scores are indicative of mindful eating.

2.8. Statistical Analysis

The sample size was attained from available data of the prospective observational study to assess determinants of gestational weight gain (clinicaltrials.gov: NCT01954342), with energy intake measured by energy intake-balance method [37] as the primary outcome. At the given sample size (n = 56), the minimal detectable simple correlation for a power of 80%, at a two-sided significance level of 0.05, was r = 0.37. For the regression analysis, we assessed the power (1-β) of independent predictors post-hoc. Relationships between diet quality, eating patterns, and constructs of eating behavior were first evaluated by simple Pearson correlation coefficients. Significant correlations were linear; using power other than 1 did not increase the explained variability (change in R^2 < 0.01). To test independent effects of eating behaviors on diet quality adjusted for demographic factors, we performed linear regression with HEI as the dependent variable and significant individual correlates of HEI as independent variables. Frequencies were compared using Chi-Square tests. Comparisons between African-American and White women were performed by independent Student's *t*-test. Statistical analyses were completed using SPSS Inc. software, Version 24 for Windows (IBM Corp, Armonk, NY). Data are presented as mean ± SEM. All tests were performed with significance level α = 0.05, and findings were considered significant when $p < α$.

3. Results

Fifty-six women with obesity (36.7 ± 0.7 kg/m^2, 46% White, 50% nulliparous) who collected dietary intake data that satisfied the criteria for inclusion were included in this analysis (Table 1).

Table 1. Subject Characteristics.

	All, n = 56	Association with Diet Quality, Healthy Eating Index
	Mean ± SEM	r, p
Race, n (African-American, White, Others)	25, 26, 5	0.11, 0.43
Gestational Age, weeks	11.3 ± 0.3	-
Maternal Age, years	28.3 ± 0.6	0.19, 0.16
Body weight, kg	97.8 ± 2.2	-
Body mass index, kg/m^2	36.7 ± 0.7	0.01, 0.96
Obesity Class, n (1, 2, 3)	23, 21, 12	-
Fasting Plasma Glucose, mmol/L (range)	4.9 ± 0.1 (4.2, 6.1)	0.17, 0.20
Education, n (1, 2, 3)	6, 39, 11	0.31, 0.02
Employment, n (1, 2, 3, 4)	1, 14, 9, 32	0.18, 0.18
Household Income *	3.6 ± 0.3	0.14, 0.30
Parity, n (0, 1, ≥2)	28, 16, 12	−0.10, 0.46
Healthy Eating Index	46.7 ± 1.3	-

Education is categorized into High School (1), college (2), postgraduate work (3). Employment into medically disabled (1), unemployed (2), part-time employment (3), and full-time employment (4). * Household income was computed as a percent of federal poverty line ("poverty" < 1.0) according to family size. Parity is defined as number of previous pregnancies of viable infant >20 weeks gestation.

3.1. Dietary Assessment

Participants reported consuming 3.8 ± 0.1 meals per day, of which 2.7 ± 0.05 meals were reported as main meals (breakfast, lunch, or dinner) and 1.1 ± 0.1 meals as snacks. The average reported eating duration (time between first and last photograph of any meal eaten) was 8:28 ± 0:31 h.

The mean reported energy intake was 2165 ± 43 kcal/day, which is equivalent to 84 ± 2% of energy requirements assessed by doubly labeled water during the same period. On average, participants consumed 46 ± 1% of energy as carbohydrates, 38 ± 1% as fat (41% saturated fatty acids, 35% mono-unsaturated fatty acids, 24% poly-unsaturated fatty acids), and 16 ± 1% as protein. The HEI of the diet was 46.7 ± 1.3, which is considered to reflect a "Poor" (<51) diet quality (Table 1). The HEI was positively associated with education level, but was not associated with BMI (as continuous or categorical variable; Class 1: 46.2 ± 1.9, Class 2: 45.9 ± 2.4, Class 3: 49.1 ± 2.6, $p = 0.47$), parity, race, age, income, or employment (Table 1). Breakfast consumption ($p = 0.04$) and an earlier last meal of the day ($p < 0.01$) were associated with a higher HEI.

On an individual level, 71% of women in the cohort ($n = 40$) had a HEI <51 that is "poor", 29% had a HEI between 50 and 80 indicating that diet quality "needs improvement", and none of the participants had a HEI >80 which is "good" diet quality (Supplemental Table S1). Scores of the individual components within the HEI are summarized in Figure 1A and Table S1. Among these, only mean intake of protein foods exceeded 80% of the HEI maximum protein score. Vegetables, dairy, refined grains, and added sugars were between 50% and 80% of the maximum scores within the respective categories, and intake of fruits (total fruits including juices, and whole fruits), greens and beans, whole-grains, seafood, and plant protein, fatty acids and saturated fats, and sodium were below 50%, and thereby "poor".

Macronutrient and fat composition associated significantly with diet quality. HEI associated with intakes of protein (as % of total energy intake, $r = 0.29$, $p = 0.03$), but not carbohydrates. Moreover, HEI was not associated with overall fat intake, but with intakes of saturated fatty acids ($r = -0.37$, $p = 0.005$), mono-unsaturated fatty acids ($r = 0.32$, $p = 0.02$), and the ratio of unsaturated to saturated fatty acids ($r = 0.37$, $p = 0.005$). Macronutrient and micronutrient intakes as compared to the recommendations by the Institute of Medicine are presented in Figure 1B. For majority of nutrients, intakes were close to or exceeding the recommendations, but intakes of fiber (62.5 ± 3.5%), vitamin B12 (41.5 ± 8.0%), vitamin E (69.2 ± 3.8%), and iron (66.0 ± 2.9%) were particularly low.

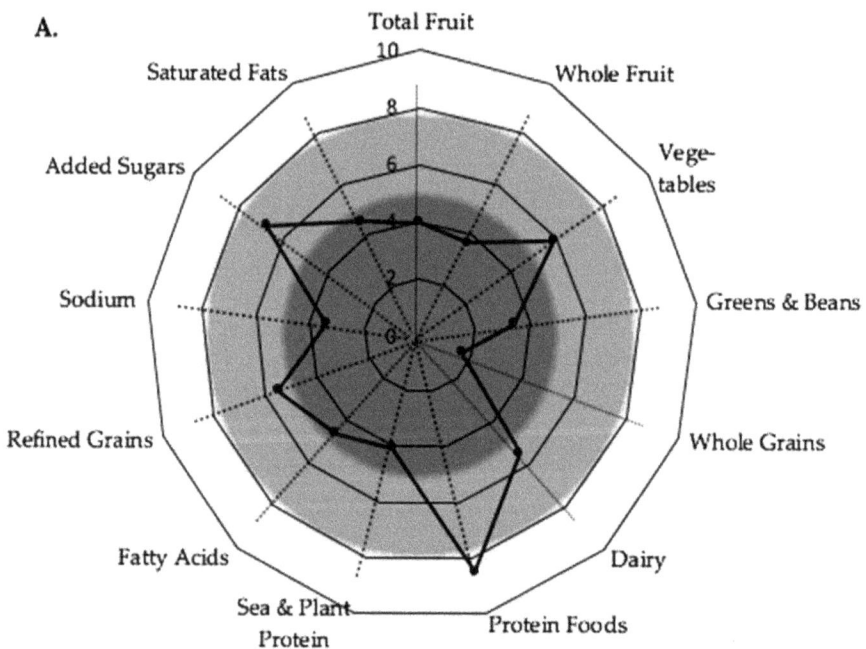

Figure 1. *Cont.*

B

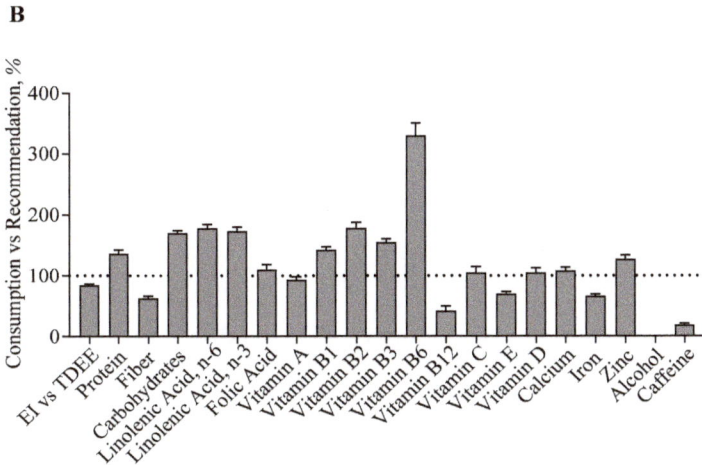

Figure 1. Diet Quality per Healthy Eating Index Components, and Recommendations for Intakes of Macronutrients and Micronutrients. Data are given as mean and mean ± SEM. (**A**) Values in the dark grey area indicate the respective Healthy Eating Index (HEI) food group to be consumed "poorly" (0–5), values in the light grey area as "needs improvement" (5–8) and in the white area as "sufficient" (8–10), according to the HEI guidelines [25]. HEI components scaled from 0–10 are reported as assessed and components scaled 0–5 were multiplied by two to facilitate comparability between factors. (**B**) Values are expressed as intake as compared to the minimal macronutrient and micronutrient intake recommendation by the WHO (maximum intake recommendation for alcohol and caffeine). Macronutrient and Micronutrient intake values are adjusted for the degree of underreporting per individual (−16 ± 2% vs. TDEE, adjusted intake=reported intake/reporting accuracy ["EI vs. TDEE"]).

3.2. Food Cravings

The total food craving score was 2.19 ± 0.07, ranging from 1.09 to 3.33. The most frequent cravings were reported for "fast food fats" and "fruits and vegetables", whereas "high-fat foods" and "sweets" were less frequently craved. In Figure 2, the associations between food cravings and diet quality were reported. Cravings for "high fat foods", "sweets", "carbohydrates and starches", and "fast food fats" associated negatively with HEI, but cravings for "fruit and vegetables" did not associate with HEI.

Cravings for "sweets" correlated significantly with added sugar intake ($r = 0.39$, $p < 0.01$), but were not associated with the reported frequency of snack consumption ($p = 0.65$). Cravings for "fast food fats" were significantly associated with increased consumption of carbohydrate ($r = 0.27$, $p = 0.04$) and less consumption of fats ($r = −0.30$, $p = 0.03$). Increased cravings for both "sweets" and "fast food fats" were associated with poor consumption of whole grains, seafood, and plant proteins ("sweets": $p = 0.02$ and $p = 0.04$, and "fast food fats": $p = 0.04$ and $p = 0.01$, respectively). Cravings for "Fruit and Vegetables" did not correlate with intake of any HEI component.

3.3. Eating Inventory

The average scores from the Eating Inventory were 8.1 ± 0.6 for cognitive restraint, 5.4 ± 0.4 for disinhibition and 3.9 ± 0.3 for hunger. Fourteen (25%) women in the cohort were considered restrained eaters and one (2%) a disinhibited eater (respective subscores >12). The HEI score for diet quality was not associated cognitive restraint ($p = 0.09$), disinhibition ($p = 0.73$), or hunger ($p = 0.43$).

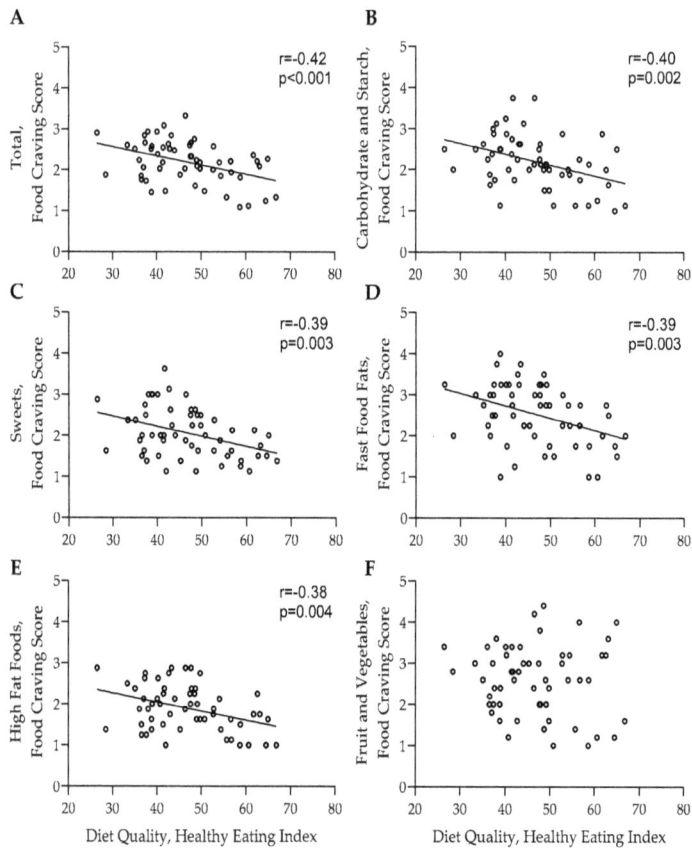

Figure 2. Food Cravings and Diet Quality. Food cravings scores are reported per individual in relation to their respective Healthy Eating Index. Cravings are reported as (**A**) total cravings and as cravings for (**B**) Carbohydrates and Starch, (**C**) Sweets, (**D**) Fast Food Fats, (**E**) High Fat Foods and (**F**) Fruit and Vegetables. Significant associations are indicated by linear regression.

3.4. Mindful Eating

The mindful eating summary score was 2.93 ± 0.04. Overall mindfulness as assessed by the summary score was not associated with HEI ($p = 0.22$), meal/snack frequency, or timing. Subscores higher than the overall summary score indicating mindfulness towards these eating behaviors were reported for eating in response to negative emotions ("emotional eating"), inability to stop eating even when full ("disinhibition") and the tendency to think about other things and rush while eating ("distraction"). Of the mindful eating subscores, only awareness correlated with HEI ($r = 0.34$, $p = 0.01$). Awareness associated significantly with consumption of "Greens and Beans" ($r = 0.30$, $p = 0.02$), but not with other HEI components.

3.5. Nausea and Vomiting and the Influence on Diet Quality

Forty-six percent of women reported no or only mild symptoms of nausea and vomiting whereas 54% experienced moderate symptoms, and none reported severe symptoms. Nausea and vomiting severity did not correlate with demographic characteristics or HEI but was associated with a lower energy intake ($p = 0.02$), skipping breakfast ($p = 0.02$), eating later in the day ($p < 0.01$), and consuming all meals during a shorter period of time ($p < 0.01$).

3.6. The Influence of Maternal Education on Diet Quality and Eating Behaviors

Education level was positively associated with HEI ($r = 0.31$, $p = 0.02$). Women with higher education consumed more whole grains ($p = 0.03$), greens and beans ($p = 0.02$), and less added sugars ($p = 0.04$). Women with higher education also reported eating more main meals ($r = 0.30$, $p = 0.02$), and increased awareness towards the sensory aspects of eating ($r = 0.33$, $p = 0.01$). Education level was positively associated with disinhibition scores, which indicates that higher educated women had less control over eating behavior ($r = 0.31$, $p = 0.02$). The intent to restrict food intake also was associated with education level (cognitive restraint, $r = 0.29$, $p = 0.03$) and could be in response to the tendency to overeat, as indicated by the association between disinhibition and education level.

3.7. Independent Effects of Awareness, Cravings, and Education

In the linear stepwise regression model with HEI as the dependent variable and significant individual correlates with HEI ($n = 8$) as independent variables, we observed that an early last meal of the day (standardized $\beta = -0.40$, $p < 0.001$, $1-\beta = 0.94$), cravings for sweets (standardized $\beta = -0.35$, $p = 0.004$, $1-\beta = 0.84$), and education (standardized $\beta = 0.27$, $p = 0.02$, $1-\beta = 0.65$) were independent predictors of HEI ($R^2 = 0.38$, $p < 0.001$), but not other cravings, awareness towards eating or breakfast consumption.

3.8. The Influence of Maternal Race on Diet Quality and Eating Behaviors

African-American women reported lower education, less income and higher parity than White women. The HEI was not significantly different between African-American and White women (AA: 45.2 ± 1.8 vs. White: 47.2 ± 1.8, $p = 0.44$). Reported energy intake as compared to energy requirements was also comparable between races. However, eating patterns and eating behaviors differed significantly by maternal race. Compared to White women, African-American women consumed their first meal later in the day (+1:28 ± 0:36 h, $p = 0.03$), and reported eating fewer snacks (African-American vs. White, −0.6 ± 0.2 snacks/day, $p = 0.01$). African-American women reported more cravings for high-fat foods and carbohydrates and starches ($p = 0.02$ and $p = 0.03$, respectively), yet less disinhibition ($p < 0.01$), indicating better control over their eating behavior compared to White women (Table 2).

Table 2. Eating Behavior Assessment.

	African-American, $n = 26$	White, $n = 25$	p for Race	Range	High Values Indicative of
Food Cravings					
High Fat	2.12 ± 0.11	1.76 ± 0.10	0.02	1–5	More Craving
Sweets	2.12 ± 0.12	2.04 ± 0.12	0.64	1–5	More Craving
Carbohydrates and Starch	2.42 ± 0.11	2.03 ± 0.13	0.03	1–5	More Craving
Fast Food Fats	2.65 ± 0.12	2.47 ± 0.18	0.41	1–5	More Craving
Fruit and Vegetables	2.79 ± 0.16	2.45 ± 0.18	0.16	1–5	More Craving
Total Score	2.36 ± 0.09	2.08 ± 0.11	0.06	1–5	More Craving
Mindful Eating					
Awareness	2.58 ± 0.12	2.53 ± 0.10	0.77	1–4	Mindfulness
Distraction	3.24 ± 0.10	2.95 ± 0.13	0.09	1–4	Mindfulness
Disinhibition	3.34 ± 0.08	3.00 ± 0.10	0.01	1–4	Mindfulness
Emotional Cues	3.54 ± 0.09	3.32 ± 0.10	0.11	1–4	Mindfulness
External	2.29 ± 0.11	2.69 ± 0.10	0.01	1–4	Mindfulness
Summary Score	2.99 ± 0.06	2.89 ± 0.06	0.24	1–4	Mindfulness
Eating Inventory					
Cognitive Restraint	8.08 ± 1.04	7.69 ± 0.91	0.78	0–21	Greater Control
Disinhibition	4.20 ± 0.47	6.69 ± 0.66	0.004	0–16	Less Control
Hunger	3.52 ± 0.42	4.50 ± 0.40	0.10	0–14	Less Control
PUQE					
Summary Score	6.48 ± 0.43	5.54 ± 0.4	0.12	3–15	Severe Symptoms

PUQE, Modified-Pregnancy-Unique Quantification of Emesis and Nausea Index.

4. Discussion

Improvements in diet quality during pregnancy can prevent poor outcomes in nonobese women [1–9], but dietary interventions have been largely unsuccessful in women with obesity [12,14,16,17,38,39]. We speculate that this may be explained in part by intervention approaches not being specific to the eating patterns and behaviors in women with obesity and thus achieving only small effect sizes on diet quality. Identifying maternal eating patterns that relate to diet quality in women with obesity will help inform the development of future dietary interventions with greater specificity for this group. To this end, we performed an observational cohort study and measured diet quality and eating behavior in 56 healthy women with obesity. In support of our hypothesis, diet quality in early pregnancy was poor, which was related to increased food cravings and less awareness towards eating. Nausea and vomiting severity were not associated with diet quality, but were inversely related to breakfast consumption and energy intake. Lastly, we observed that diet quality and eating behaviors differed by education and race, respectively.

Diet quality in our cohort of pregnant women with obesity was poor (HEI = 47, fat content = 38%) and lower than prior reports of nonpregnancy cohorts in the US (49–64 [40–43]). Interventions in pregnant cohorts with poor diet quality may be more successful in improving maternal and infant outcomes as compared to previous lifestyle intervention studies, because the opportunity for improvement is larger. Diet quality was higher in pregnant women in Australia (HEI = 72 [38]) and the UK (fat content = 31% [15]), and diet quality improved by only 2% [12,38] to 5% [32], as assessed by healthy eating index [38], adherence score to the recommended diet [12], Glycemic Index and Glycemic Load [15].

We observed that poor diet quality is due to poor consumption of all food groups, except for protein intake, and thus appropriate interventions require approaches to target all components of the diet, e.g., Mediterranean diet, DASH diet. Nevertheless, our data also identify specific food groups and behaviors that may affect overall diet quality more than others. For example, strategies to reduce high fat intake, e.g., by reducing fast food intake, will likely reduce saturated fatty acid intake, but may also result in less refined grain consumption. Moreover, increasing fruit and vegetable consumption would likely increase dietary fiber and unsaturated fatty acid intake, thus providing a strategy to improve multiple diet quality components. Increasing fiber intake may not only improve satiety and reduce energy intake [44,45], but also improve gastrointestinal health and glucose homeostasis [46].

Food cravings increase the desire to eat and the satisfaction associated with eating [47]. In the present study, cravings for healthy foods, e.g., fruits and vegetables, were reported more frequently compared to cravings for unhealthy foods, e.g., sweets, fast food, high fat foods. Despite the frequent cravings, fruit was consumed in poor quantities according to the 2015–2020 Dietary Guidelines for Americans [29]. The inability to increase consumption of healthy foods in order to satisfy food cravings for such foods may be due to a lack of availability, perceived cost, or misperception about the health of certain foods [48]. Facilitating indulgence in cravings for healthy foods may also prevent cravings and consumption of unhealthy foods [49].

Conversely to cravings for healthy foods, cravings for foods with poor nutritional value were often indulged in and thereby contributed to poor diet quality, which is consistent with other studies [50]. Specifically, cravings for sweets were linked to intake of added sugar. This data supports and connects evidence for a causal relationship between cravings for sugar [51], consumption of sugar [52] and excess gestational weight gain. We also observed that fast food fats were frequently craved and correlated with poor diet quality, whereas high fat foods were less frequently craved. The fast food fats subscale is comprised solely of pizza, hamburgers, french fries, and potato chips, while the high fat subscale includes eight foods, including bacon, gravy, and sausage. Thus, an important factor distinguishing the two fat subscales is ease of access being that fast food fats are easier to acquire compared to foods on the high fat subscale. Hence, these data support the notion that food accessibility may influence the perception of cravings and the likelihood of eating for indulgence.

Increasing consciousness about diet choices may benefit diet quality [53,54]. We observed that awareness towards eating was associated with better diet quality. This confirms previous reports of mindfulness in pregnant women with overweight and obesity [36,55]. In addition, we show that mindfulness associates specifically with consumption of healthy foods, i.e., 'Greens and Beans', but not with other HEI components. Thus, whereas increasing awareness would likely increase consumption of healthy food items, no direct effect of reducing intake of unhealthy foods would be expected. To our knowledge, only one study has successfully implemented a mindfulness-based intervention in overweight, pregnant women [56,57]; however, the effect of the intervention on diet quality or pregnancy outcomes was not reported.

We hypothesized that diet quality and eating behaviors would be affected by maternal demographic characteristics such as education and race. Diet quality was poor in women with a college degree or postgraduate education (HEI = 49.2) and in women with lower education levels (HEI = 44.4, $p = 0.06$), which confirms similar findings of previous studies [58]. Women with less education were less aware of their eating behavior and had a lower intent to restrict food intake, which may reflect lacking or misguided knowledge on the risks of excess weight gain [59]. Qualitative studies in women enrolled in the Special Supplemental Nutrition Program for Women, Infants, and Children (WIC) program who generally have poor socio-economic status, report that they indeed feel uninformed about weight gain in pregnancy and that receiving more information may increase the intent to restrict food intake to optimize pregnancy outcomes [60]. Women with higher education levels had lower disinhibition scores, indicating poorer self-control. To understand the main independent effects of individual correlates with diet quality, we performed linear regression analysis with HEI as the dependent variable. We identified education, cravings for sweets and an early last meal of the day as independent predictors of diet quality. Awareness towards eating was not a significant predictor in the model, which suggests that the beneficial effect of increased awareness towards eating on diet quality is mediated by eating the last meal of the day earlier, or that education, cravings for sweets, and awareness were low/high in the same women and were thus not independent of each other. Indeed, awareness correlated significantly with both education and cravings for sweets.

We observed significant differences in eating behaviors, but not in diet quality between African-American and White women in the present study, indicating the need for different approaches to improve poor diet quality in all women. African-American women, who reported lower education levels, less income, and higher parity, consumed the first meal later in the day but not the last meal, so eating duration tended to be shorter. The severity of morning sickness was not reported to be different between African-American and White women. The observation that African-American women ate breakfast less frequently, and thus ate later in the day may have contributed to a reduced dietary intake, which we reported previously [21]. In line with these findings, African-American women also reported better control over their eating behavior and reported less snacking despite more cravings.

The strength of this study is the use of food photography. Food photography methods are more accurate in measuring portion sizes and the composition of meals since these are analyzed in an objective manner [22]. Furthermore, the food photography method is performed in real time and is not prone to recall bias as with instruments reliant on recall, yet possibly to "attention bias" for the reporting of snacks [20]. With respect to energy intake, implementation of data quality criteria in the analytical stages minimizes the degree of under-reporting to less than 20% compared to doubly labeled water and eliminates increased reporting bias typically observed with higher energy intake [20]. The major weakness of this study is the lack of prepregnancy data. Thereby, we were unable to determine whether poor diet quality and associated eating behaviors can be targeted prior to pregnancy. Further, the observation period (six days) was relatively short as compared to other instruments such as a food frequency questionnaire which recalls intake over a period of 1 to 12 months. Eating behaviors were assessed by questionnaires and are prone to self-reporting bias. Importantly, all questionnaires were previously validated in pregnancy. In addition, micronutrient, but not macronutrient, intakes may be higher than reported due to prenatal vitamin supplement intake. Lastly, this study did not enroll women without obesity, and therefore we were unable to compare diet quality and eating behaviors between women with normal weight, overweight, or obesity. Early pregnancy diabetes was excluded by HbA1c, but no glucose tolerance test was performed to determine gestational diabetes.

5. Conclusions

The main behavioral determinants of poor diet quality in pregnant women with obesity include the low consumption of fruits and vegetables, the indulgence into cravings for sweets, and eating late in the day. Future intervention strategies developed with a specificity for improving diet quality are therefore to increase the availability of fruits and vegetables rather than sweets or fast foods, which may facilitate indulgence in the observed cravings for healthy foods, and to reduce late-night snacking. In addition, women with lower education levels may benefit from educational interventions on the risks of overeating, whereas women with higher education may benefit from strategies to increase self-control. Future studies are required to show whether strategies to change those behaviors can be successfully implemented and improve pregnancy outcomes in women with obesity.

Supplementary Materials: The following are available online at http://www.mdpi.com/2072-6643/11/7/1446/s1, Table S1: HEI components.

Author Contributions: Conceptualization, J.M., L.M.R.; methodology, J.M., A.D.A., C.J.R.; software, J.M.; validation, A.D.A., C.J.R. and L.M.R.; formal analysis, J.M., L.M.R.; investigation, J.M., C.J.R., A.D.A., and L.M.R.; resources, L.M.R.; data curation, J.M.; writing—original draft preparation, J.M.; writing—review and editing, J.M., C.J.R., A.D.A., M.S.A. C.K.M. L.M.R.; visualization, J.M.; supervision, L.M.R.; project administration, L.M.R.; funding acquisition, C.K.M. and L.M.R.

Funding: This study was funded by the National Institutes of Health (R01DK099175, U54GM104940 and P30DK072476).

Acknowledgments: We are indebted to the participants and clinical/technical support from Porsha Vallo, Elizabeth Sutton, Karissa Elsass, Taylor Ayers and Ray Allen.

Conflicts of Interest: The intellectual property surrounding the Remote Food Photography Method© and SmartIntake® application are owned by Louisiana State University/Pennington Biomedical Research Center and CKM is an inventor. J.M., C.J.R., A.D.A., M.S.A., L.M.R. have no conflict of interest. The funders had no role in the design of the study; in the collection, analyses, or interpretation of data; in the writing of the manuscript, or in the decision to publish the results.

References

1. Mijatovic-Vukas, J.; Capling, L.; Cheng, S.; Stamatakis, E.; Louie, J.; Cheung, N.W.; Markovic, T.; Ross, G.; Senior, A.; Brand-Miller, J.C.; et al. Associations of Diet and Physical Activity with Risk for Gestational Diabetes Mellitus: A Systematic Review and Meta-Analysis. *Nutrients* **2018**, *10*, 698. [CrossRef] [PubMed]

2. Schoenaker, D.A.; Soedamah-Muthu, S.S.; Mishra, G.D. Quantifying the mediating effect of body mass index on the relation between a Mediterranean diet and development of maternal pregnancy complications: The Australian Longitudinal Study on Women's Health. *Am. J. Clin. Nutr.* **2016**, *104*, 638–645. [CrossRef] [PubMed]

3. Hillesund, E.R.; Overby, N.C.; Engel, S.M.; Klungsoyr, K.; Harmon, Q.E.; Haugen, M.; Bere, E. Associations of adherence to the New Nordic Diet with risk of preeclampsia and preterm delivery in the Norwegian Mother and Child Cohort Study (MoBa). *Eur. J. Epidemiol.* **2014**, *29*, 753–765. [CrossRef] [PubMed]

4. Zhang, C.; Tobias, D.K.; Chavarro, J.E.; Bao, W.; Wang, D.; Ley, S.H.; Hu, F.B. Adherence to healthy lifestyle and risk of gestational diabetes mellitus: Prospective cohort study. *BMJ* **2014**, *349*, g5450. [CrossRef] [PubMed]

5. Gresham, E.; Collins, C.E.; Mishra, G.D.; Byles, J.E.; Hure, A.J. Diet quality before or during pregnancy and the relationship with pregnancy and birth outcomes: The Australian Longitudinal Study on Women's Health. *Public Health Nutr.* **2016**, *19*, 2975–2983. [CrossRef] [PubMed]

6. Koletzko, B.; Brands, B.; Grote, V.; Kirchberg, F.F.; Prell, C.; Rzehak, P.; Uhl, O.; Weber, M. Early Nutrition Programming Project. Long-Term Health Impact of Early Nutrition: The Power of Programming. *Ann. Nutr. Metab.* **2017**, *70*, 161–169. [CrossRef]

7.   Shapiro, A.L.; Kaar, J.L.; Crume, T.L.; Starling, A.P.; Siega-Riz, A.M.; Ringham, B.M.; Glueck, D.H.; Norris, J.M.; Barbour, L.A.; Friedman, J.E.; et al. Maternal diet quality in pregnancy and neonatal adiposity: The Healthy Start Study. *Int. J. Obes.* **2016**, *40*, 1056–1062. [CrossRef]

8.   Grandy, M.; Snowden, J.M.; Boone-Heinonen, J.; Purnell, J.Q.; Thornburg, K.L.; Marshall, N.E. Poorer maternal diet quality and increased birth weight. *J. Matern. Fetal Neonatal Med.* **2018**, *31*, 1613–1619. [CrossRef]

9.   Brei, C.; Stecher, L.; Meyer, D.M.; Young, V.; Much, D.; Brunner, S.; Hauner, H. Impact of Dietary Macronutrient Intake during Early and Late Gestation on Offspring Body Composition at Birth, 1, 3, and 5 Years of Age. *Nutrients* **2018**, *10*, 579. [CrossRef]

10. Shin, D.; Lee, K.W.; Song, W.O. Pre-Pregnancy Weight Status Is Associated with Diet Quality and Nutritional Biomarkers during Pregnancy. *Nutrients* **2016**, *8*, 162. [CrossRef]

11. Catalano, P.M.; Shankar, K. Obesity and pregnancy: Mechanisms of short term and long term adverse consequences for mother and child. *BMJ* **2017**, *356*, j1. [CrossRef] [PubMed]

12. Koivusalo, S.B.; Rono, K.; Klemetti, M.M.; Roine, R.P.; Lindstrom, J.; Erkkola, M.; Kaaja, R.J.; Poyhonen-Alho, M.; Tiitinen, A.; Huvinen, E.; et al. Gestational Diabetes Mellitus Can Be Prevented by Lifestyle Intervention: The Finnish Gestational Diabetes Prevention Study (RADIEL): A Randomized Controlled Trial. *Diabetes Care* **2016**, *39*, 24–30. [CrossRef] [PubMed]

13. Shieh, C.; Cullen, D.L.; Pike, C.; Pressler, S.J. Intervention strategies for preventing excessive gestational weight gain: Systematic review and meta-analysis. *Obes. Rev.* **2018**. [CrossRef] [PubMed]

14. Dodd, J.M.; McPhee, A.J.; Turnbull, D.; Yelland, L.N.; Deussen, A.R.; Grivell, R.M.; Crowther, C.A.; Wittert, G.; Owens, J.A.; Robinson, J.S.; et al. The effects of antenatal dietary and lifestyle advice for women who are overweight or obese on neonatal health outcomes: The LIMIT randomised trial. *BMC Med.* **2014**, *12*, 163. [CrossRef] [PubMed]

15. Poston, L.; Bell, R.; Croker, H.; Flynn, A.C.; Godfrey, K.M.; Goff, L.; Hayes, L.; Khazaezadeh, N.; Nelson, S.M.; Oteng-Ntim, E.; et al. Effect of a behavioural intervention in obese pregnant women (the UPBEAT study): A multicentre, randomised controlled trial. *Lancet Diabetes Endocrinol.* **2015**, *3*, 767–777. [CrossRef]

16. Simmons, D.; Devlieger, R.; van Assche, A.; Jans, G.; Galjaard, S.; Corcoy, R.; Adelantado, J.M.; Dunne, F.; Desoye, G.; Harreiter, J.; et al. Effect of Physical Activity and/or Healthy Eating on GDM Risk: The DALI Lifestyle Study. *J. Clin. Endocrinol. Metab.* **2017**, *102*, 903–913. [CrossRef]

17. Rono, K.; Grotenfelt, N.E.; Klemetti, M.M.; Stach-Lempinen, B.; Huvinen, E.; Meinila, J.; Valkama, A.; Tiitinen, A.; Roine, R.P.; Poyhonen-Alho, M.; et al. Effect of a lifestyle intervention during pregnancy-findings from the Finnish gestational diabetes prevention trial (RADIEL). *J. Perinatol.* **2018**. [CrossRef]

18. Boswell, R.G.; Kober, H. Food cue reactivity and craving predict eating and weight gain: A meta-analytic review. *Obes. Rev.* **2016**, *17*, 159–177. [CrossRef]

19. Delahanty, L.M.; Meigs, J.B.; Hayden, D.; Williamson, D.A.; Nathan, D.M.; Diabetes Prevenion Program Research Group. Psychological and behavioral correlates of baseline BMI in the diabetes prevention program (DPP). *Diabetes Care* **2002**, *25*, 1992–1998. [CrossRef]

20. Most, J.; Vallo, P.M.; Altazan, A.D.; Gilmore, L.A.; Sutton, E.F.; Cain, L.E.; Burton, J.H.; Martin, C.K.; Redman, L.M. Food Photography Is Not an Accurate Measure of Energy Intake in Obese, Pregnant Women. *J. Nutr.* **2018**, *148*, 658–663. [CrossRef]

21. Most, J.; Gilmore, L.A.; Altazan, A.D.; St Amant, M.; Beyl, R.A.; Ravussin, E.; Redman, L.M. Propensity for adverse pregnancy outcomes in African-American women may be explained by low energy expenditure in early pregnancy. *Am. J. Clin. Nutr.* **2018**, *107*, 957–964. [CrossRef] [PubMed]

22. Martin, C.K.; Correa, J.B.; Han, H.; Allen, H.R.; Rood, J.C.; Champagne, C.M.; Gunturk, B.K.; Bray, G.A. Validity of the Remote Food Photography Method (RFPM) for estimating energy and nutrient intake in near real-time. *Obesity* **2012**, *20*, 891–899. [CrossRef] [PubMed]

23. Martin, C.L.; Montville, J.B.; Steinfeldt, L.C.; Omolewa-Tomobi, G.; Heendeniya, K.Y.; Adler, M.E.; Moshfegh, A.J. *The USDA Food and Nutrient Database for Dietary Studies 2011–2012*; U.S. Department of Agriculture: Beltsville, MD, USA, 2014.

24. The Epidemiology and Genomics Research Program. Overview & Background of the Healthy Eating Index. Available online: https://epi.grants.cancer.gov/hei/ (accessed on 28 May 2019).

25. Mousa, A.; Naqash, A.; Lim, S. Macronutrient and Micronutrient Intake during Pregnancy: An Overview of Recent Evidence. *Nutrients* **2019**, *11*, 443. [CrossRef] [PubMed]

26. Institute of Medicine (US); National Research Council (US) Committee to Reexamine IOM Pregnancy Weight Guidelines. Determining Optimal Weight Gain. In *Weight Gain During Pregnancy: Reexamining the Guidelines*; Rasmussen, K.M., Yaktine, A.L., Eds.; National Academies Press: Washington, DC, USA, 2009. [CrossRef]

27. Bowman, S.A.; Clemens, J.C.; Friday, J.E.; Lynch, K.L.; Moshfegh, A.J. Food Patterns Equivalents Database 2013–2014: Methodology and User Guide. Available online: http://www.ars.usda.gov/nea/bhnrc/fsrg (accessed on 28 May 2019).

28. Millen, B.; Lichtenstein, A.H.; Abrams, S.; Adams-Campbell, L.; Anderson, C.; Brenna, J.T.; Campbell, W.; Clinton, S.; Foster, G.; Hu, F.; et al. 2015–2020 Dietary Guidelines for Americans. Available online: http://health.gov/dietaryguidelines/2015/guidelines/ (accessed on 28 May 2019).

29. Lacasse, A.; Rey, E.; Ferreira, E.; Morin, C.; Berard, A. Validity of a modified Pregnancy-Unique Quantification of Emesis and Nausea (PUQE) scoring index to assess severity of nausea and vomiting of pregnancy. *Am. J. Obstet. Gynecol.* **2008**, *198*, 71.e1–71.e7. [CrossRef] [PubMed]

30. Birkeland, E.; Stokke, G.; Tangvik, R.J.; Torkildsen, E.A.; Boateng, J.; Wollen, A.L.; Albrechtsen, S.; Flaatten, H.; Trovik, J. Norwegian PUQE (Pregnancy-Unique Quantification of Emesis and nausea) identifies patients with hyperemesis gravidarum and poor nutritional intake: A prospective cohort validation study. *PLoS ONE* **2015**, *10*, e0119962. [CrossRef] [PubMed]

31. Stunkard, A.J.; Messick, S. *Eating Inventory Manual*; Harcourt Brace & Company: San Antonio, TX, USA, 1988.

32. Stunkard, A.J.; Waterland, R.A. The Three-Factor Eating Questionnaire-Eating Inventory. In *Obesity Assessment: Tools, Methods, Interpretation*; St Jeor, S.T., Ed.; Chapman and Hall: New York, NY, USA, 1997; pp. 343–370.

33. White, M.A.; Whisenhunt, B.L.; Williamson, D.A.; Greenway, F.L.; Netemeyer, R.G. Development and validation of the food-craving inventory. *Obes. Res.* **2002**, *10*, 107–114. [CrossRef] [PubMed]

34. Martin, C.K.; O'Neil, P.M.; Tollefson, G.; Greenway, F.L.; White, M.A. The association between food cravings and consumption of specific foods in a laboratory taste test. *Appetite* **2008**, *51*, 324–326. [CrossRef]

35. Apolzan, J.W.; Myers, C.A.; Cowley, A.D.; Brady, H.; Hsia, D.S.; Stewart, T.M.; Redman, L.M.; Martin, C.K. Examination of the reliability and validity of the Mindful Eating Questionnaire in pregnant women. *Appetite* **2016**, *100*, 142–151. [CrossRef]

36. Framson, C.; Kristal, A.R.; Schenk, J.M.; Littman, A.J.; Zeliadt, S.; Benitez, D. Development and validation of the mindful eating questionnaire. *J. Am. Diet. Assoc.* **2009**, *109*, 1439–1444. [CrossRef]

37. Thomas, D.M.; Navarro-Barrientos, J.E.; Rivera, D.E.; Heymsfield, S.B.; Bredlau, C.; Redman, L.M.; Martin, C.K.; Lederman, S.A.; L, M.C.; Butte, N.F. Dynamic energy-balance model predicting gestational weight gain. *Am. J. Clin. Nutr.* **2012**, *95*, 115–122. [CrossRef]

38. Dodd, J.M.; Cramp, C.; Sui, Z.; Yelland, L.N.; Deussen, A.R.; Grivell, R.M.; Moran, L.J.; Crowther, C.A.; Turnbull, D.; McPhee, A.J.; et al. The effects of antenatal dietary and lifestyle advice for women who are overweight or obese on maternal diet and physical activity: The LIMIT randomised trial. *BMC Med.* **2014**, *12*, 161. [CrossRef] [PubMed]

39. Dodd, J.M.; Turnbull, D.; McPhee, A.J.; Deussen, A.R.; Grivell, R.M.; Yelland, L.N.; Crowther, C.A.; Wittert, G.; Owens, J.A.; Robinson, J.S.; et al. Antenatal lifestyle advice for women who are overweight or obese: LIMIT randomised trial. *BMJ* **2014**, *348*, g1285. [CrossRef] [PubMed]

40. McCullough, M.L.; Feskanich, D.; Stampfer, M.J.; Rosner, B.A.; Hu, F.B.; Hunter, D.J.; Variyam, J.N.; Colditz, G.A.; Willett, W.C. Adherence to the Dietary Guidelines for Americans and risk of major chronic disease in women. *Am. J. Clin. Nutr.* **2000**, *72*, 1214–1222. [CrossRef] [PubMed]

41. Wilson, M.M.; Reedy, J.; Krebs-Smith, S.M. American Diet Quality: Where It Is, Where It Is Heading, and What It Could Be. *J. Acad. Nutr. Diet.* **2016**, *116*, 302–310. [CrossRef] [PubMed]

42. Gao, S.K.; Beresford, S.A.; Frank, L.L.; Schreiner, P.J.; Burke, G.L.; Fitzpatrick, A.L. Modifications to the Healthy Eating Index and its ability to predict obesity: The Multi-Ethnic Study of Atherosclerosis. *Am. J. Clin. Nutr.* **2008**, *88*, 64–69. [CrossRef] [PubMed]

43. Camhi, S.M.; Whitney Evans, E.; Hayman, L.L.; Lichtenstein, A.H.; Must, A. Healthy eating index and metabolically healthy obesity in U.S. adolescents and adults. *Prev. Med.* **2015**, *77*, 23–27. [CrossRef] [PubMed]

44. Wanders, A.J.; van den Borne, J.J.; de Graaf, C.; Hulshof, T.; Jonathan, M.C.; Kristensen, M.; Mars, M.; Schols, H.A.; Feskens, E.J. Effects of dietary fibre on subjective appetite, energy intake and body weight: A systematic review of randomized controlled trials. *Obes. Rev.* **2011**, *12*, 724–739. [CrossRef] [PubMed]

45. Hervik, A.K.; Svihus, B. The Role of Fiber in Energy Balance. *J. Nutr. Metab.* **2019**, *2019*, 4983657. [CrossRef]

46. Muller, M.; Canfora, E.E.; Blaak, E.E. Gastrointestinal Transit Time, Glucose Homeostasis and Metabolic Health: Modulation by Dietary Fibers. *Nutrients* **2018**, *10*, 275. [CrossRef]

47. Berthoud, H.R. Metabolic and hedonic drives in the neural control of appetite: Who is the boss? *Curr. Opin. Neurobiol.* **2011**, *21*, 888–896. [CrossRef]

48. Reyes, N.R.; Klotz, A.A.; Herring, S.J. A qualitative study of motivators and barriers to healthy eating in pregnancy for low-income, overweight, African-American mothers. *J. Acad. Nutr. Diet.* **2013**, *113*, 1175–1181. [CrossRef]

49. van Kleef, E.; Otten, K.; van Trijp, H.C. Healthy snacks at the checkout counter: A lab and field study on the impact of shelf arrangement and assortment structure on consumer choices. *BMC Public Health* **2012**, *12*, 1072. [CrossRef]

50. Farland, L.V.; Rifas-Shiman, S.L.; Gillman, M.W. Early Pregnancy Cravings, Dietary Intake, and Development of Abnormal Glucose Tolerance. *J. Acad. Nutr. Diet.* **2015**, *115*, 1958–1964. [CrossRef]

51. Orloff, N.C.; Flammer, A.; Hartnett, J.; Liquorman, S.; Samelson, R.; Hormes, J.M. Food cravings in pregnancy: Preliminary evidence for a role in excess gestational weight gain. *Appetite* **2016**, *105*, 259–265. [CrossRef]

52. Renault, K.M.; Carlsen, E.M.; Norgaard, K.; Nilas, L.; Pryds, O.; Secher, N.J.; Olsen, S.F.; Halldorsson, T.I. Intake of Sweets, Snacks and Soft Drinks Predicts Weight Gain in Obese Pregnant Women: Detailed Analysis of the Results of a Randomised Controlled Trial. *PLoS ONE* **2015**, *10*, e0133041. [CrossRef]

53. Warren, J.M.; Smith, N.; Ashwell, M. A structured literature review on the role of mindfulness, mindful eating and intuitive eating in changing eating behaviours: Effectiveness and associated potential mechanisms. *Nutr. Res. Rev.* **2017**, *30*, 272–283. [CrossRef]

54. O'Reilly, G.A.; Cook, L.; Spruijt-Metz, D.; Black, D.S. Mindfulness-based interventions for obesity-related eating behaviours: A literature review. *Obes. Rev.* **2014**, *15*, 453–461. [CrossRef]

55. Hutchinson, A.D.; Charters, M.; Prichard, I.; Fletcher, C.; Wilson, C. Understanding maternal dietary choices during pregnancy: The role of social norms and mindful eating. *Appetite* **2017**, *112*, 227–234. [CrossRef]

56. Vieten, C.; Laraia, B.A.; Kristeller, J.; Adler, N.; Coleman-Phox, K.; Bush, N.R.; Wahbeh, H.; Duncan, L.G.; Epel, E. The mindful moms training: Development of a mindfulness-based intervention to reduce stress and overeating during pregnancy. *BMC Pregnancy Childbirth* **2018**, *18*, 201. [CrossRef]

57. Laraia, B.A.; Adler, N.E.; Coleman-Phox, K.; Vieten, C.; Mellin, L.; Kristeller, J.L.; Thomas, M.; Stotland, N.E.; Lustig, R.H.; Dallman, M.F.; et al. Novel Interventions to Reduce Stress and Overeating in Overweight Pregnant Women: A Feasibility Study. *Matern. Child Health J.* **2018**, *22*, 670–678. [CrossRef]

58. Bodnar, L.M.; Simhan, H.N.; Parker, C.B.; Meier, H.; Mercer, B.M.; Grobman, W.A.; Haas, D.M.; Wing, D.A.; Hoffman, M.K.; Parry, S.; et al. Racial or Ethnic and Socioeconomic Inequalities in Adherence to National Dietary Guidance in a Large Cohort of US Pregnant Women. *J. Acad. Nutr. Diet.* **2017**, *117*, 867–877. [CrossRef]

59. Willcox, J.C.; Ball, K.; Campbell, K.J.; Crawford, D.A.; Wilkinson, S.A. Correlates of pregnant women's gestational weight gain knowledge. *Midwifery* **2017**, *49*, 32–39. [CrossRef]

60. Lindsay, A.C.; Wallington, S.F.; Greaney, M.L.; Tavares Machado, M.M.; De Andrade, G.P. Patient-Provider Communication and Counseling about Gestational Weight Gain and Physical Activity: A Qualitative Study of the Perceptions and Experiences of Latinas Pregnant with their First Child. *Int. J. Environ. Res. Public Health* **2017**, *14*, 1412. [CrossRef]

nutrients

MDPI

Article

Effectiveness of an Intervention of Dietary Counseling for Overweight and Obese Pregnant Women in the Consumption of Sugars and Energy

Elisa Anleu [1], Marcela Reyes [1], Marcela Araya B [2], Marcela Flores [3], Ricardo Uauy [1,4] and María Luisa Garmendia [1,*]

1   Institute of Nutrition and Food Technology (INTA), University of Chile, 7830490 Santiago, Chile; anleuelisa@gmail.com (E.A.); mreyes@inta.uchile.cl (M.R.); druauy@gmail.com (R.U.)
2   Department of Women and Newborn Health Promotion, Faculty of Medicine, University of Chile, 8380453 Santiago, Chile; marbannout@uchile.cl
3   Corporación de Salud Municipal de Puente Alto, 8210269 Santiago, Chile; marcela.flores@cmpuentealto.cl
4   Department of Pediatrics, School of Medicine, Pontificia Universidad Católica de Chile, 8330023 Santiago, Chile
*   Correspondence: mgarmendia@inta.uchile.cl; Tel.: +56-2-297-81-402

Received: 29 December 2018; Accepted: 4 February 2019; Published: 13 February 2019

Abstract: Objective: Evaluate if an intervention based on nutritional counseling decreases total sugars and energy consumption in overweight and obese pregnant women, compared to their previous consumption and compared to women who only received routine counseling. Methods: Randomized study of two groups: dietary counseling (Intervention Group: IG) and routine counseling (Control Group: CG). The intervention consisted of three educational sessions focused on decreasing intake of foods that most contribute to sugars consumption. Changes in sugars and energy consumption were evaluated by a food frequency questionnaire before and after the intervention. Results: We evaluated 433 pregnant women, 272 in IG and 161 in CG, who before intervention had a mean consumption of 140 g total sugars and 2134 kcal energy per day. At the end of the intervention, the IG showed 15 g/day lower consumption of total sugars (95% CI: −25 and −5 g/day), 2% less total energy from sugars (95% CI: −3% and −1% g/day), and 125 kcal/day less energy than the CG (95% CI: −239 and −10 kcal/day). Table sugar, sweets, and soft drinks had the greatest reduction in consumption. Conclusions: The intervention focused on counseling on the decrease in consumption of the foods that most contribute to sugars consumption in overweight and obese pregnant women was effective in decreasing total sugars and energy consumption, mainly in the food groups high in sugars. Future studies should examine if this intervention has an effect on maternal and fetal outcomes.

Keywords: nutritional intervention; pregnant women; overweight; obesity; total sugars; energy

1. Introduction

In recent years, Chile has had one of the highest prevalence rates of overweight and obesity in Latin America [1]. Overweight has 39% prevalence and obesity has 29%; 51% of women in reproductive age have a state of malnutrition due to excess [2]. This situation is of concern for the country, given the consequences it represents for maternal and infant health. Excessive weight during pregnancy and maternal obesity are associated with complications such as fetal trauma, congenital malformations, recurrent abortion, gestational hypertension, preeclampsia, macrosomia (birth weight >4 kg at birth), Caesarean births, and gestational diabetes, among others [3,4]. Long-term complications include the appearance of obesity and noncommunicable diseases in both mother and child.

The appearance of pregnancy complications depends on the pre-gestation nutritional status, but also on gestational weight gain (GWG). One of the most important contributors to GWG are eating habits. Elevated sugars consumption of foods with high energy value and high added sugars predisposes these women to excessive GWG [5,6]. The study by Renault et al. [6] revealed that large GWG is more related to intake of added sugars than to saturated fats. It is of interest to consider the nutritional state of pregnant women, emphasizing the habitual consumption of these foods.

Interventions focused on diet and physical exercise during pregnancy have shown to be efficacious in preventing excessive GWG and in other maternal–fetal outcomes, such as lower incidence of gestational diabetes and Caesarean births [7–9]. The dietary component of the interventions has usually been based on acquiring healthier eating habits (reduction of the glycemic index of foods, restricting the consumption of saturated fats, and increased consumption of fruits and vegetables [10–12]. Others have used recommendations based on food guides, considering preferences and alimentary beliefs [8]. However, there are no studies that show the effect of a nutritional intervention focused on the reduction in consumption of those foods that most contribute to total sugars consumption.

We developed a home intervention strategy in Chile based on diet counseling and omega-3 supplementation in order to decrease the incidence of gestational diabetes mellitus in women who began their pregnancy overweight or obese. The study protocol has been published elsewhere [13] (Trial registration: NCT02574767). Briefly, the intervention attempts to achieve adequate metabolic control for pregnant women and their children resulted in lower incidence of gestational diabetes mellitus and lower incidence of macrosomia in the newborn. The present study evaluates whether the dietary counseling decreased the consumption of total sugars and energy, compared to their consumption before the intervention and compared to women who only received routine counseling.

2. Materials and Methods

2.1. Study Design

The prospective experimental study evaluated the frequency of consumption before and after nutritional counseling of overweight and obese pregnant women, comparing the intervention group (IG) to a control group (CG). The IG received dietary counseling, while the CG received the routine control recommendations provided in the primary health centers of the Chilean Health-Care System.

2.2. Study Population

The study population consisted of pregnant women recruited from March 2016 to May 2018 in health centers of the Puente Alto county of Santiago, Chile: 272 participants in IG and 161 in CG. The inclusion criteria were: gestation of 15 weeks or less in the first prenatal control, 18 years of age or older, single pregnancy, and overweight or obese in the first control. The exclusion criteria were: previous diagnosis of diabetes or treatment with metformin or insulin, eating disorders (bulimia or anorexia), or a risk pregnancy defined in the health center using the definition in the Chilean Health Ministry guide [14].

The original group included 1002 women, 500 randomly assigned to IG and 502 to CG. In the present study we included the subsample of participants who answered the food frequency questionnaire (FFQ) at the beginning and end of the intervention between September 2016 and October 2018. Before the intervention, 408 and 264 participants answered the FFQ, respectively. At 35–37 weeks of pregnancy, 272 participants in the IG and 161 in the CG answered a second FFQ. The main reasons the second FFQs were not obtained are given in the flowchart in Figure 1. No significant differences were found in age, education, or nutritional status between the women who answered two questionnaires with those who answered only one, nor with the women who did not answer the FFQ ($p > 0.05$).

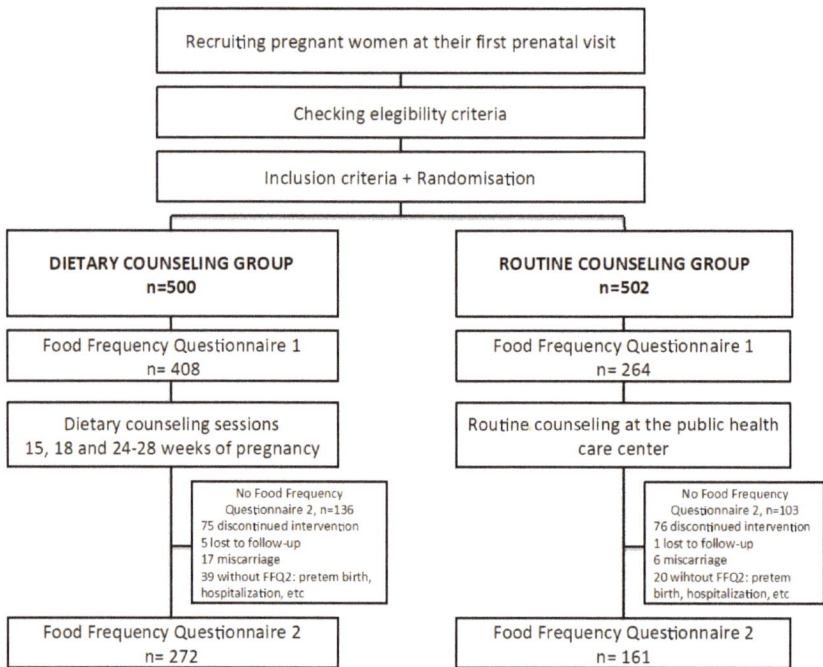

Figure 1. Flow diagram of the recruitment, randomization, and dietary measurements of trial participants.

2.3. Dietary Counseling

The general objective of the dietary intervention is to reduce the consumption of the main sources of total sugars through a culturally tailored intervention culturally with recommendations simple and easy to follow by pregnant women. Nutritional intervention consisted of three teaching sessions focused on decreasing consumption of foods that are a relevant dietary source of sugars as well as the use of behavioral techniques to allow participants to achieve eating behavioral changes. There are no interventions similar to ours in the literature, but several clinical trials have also focused on decreasing the consumption of foods with high glycemic index [15–18]. The consumption of high carbohydrate foods influences maternal glucose levels, which results excessive gestational weight gain and fetoplacental growth [17]. Some interventions focused on reducing foods with high glycemic index have shown beneficial effects on maternal outcomes (gestational weight gain, fasting, and postprandial glucose levels) and offspring birth weight [19].

We previously used a multiple pass 24-h dietary recall (R24h) applied to 114 pregnant women who attended in the same health centers to identify the seven foods that most contributed to their sugars consumption (43.4% of total sugars consumption). The "Top 7" were sweetened soft drinks, juice with added sugar, powdered juices with sugar, cookies, table sugar, sweetened milk products, and bread. The first session took place when the participants had less than 15 weeks gestation, the second session at week 18, and the third between 24–28 weeks.

Session 1: "Introduction to gestational diabetes: sugars consumption during pregnancy and consequences for the baby".

This consisted of showing an animated video with relevant information on gestational diabetes mellitus, illustrating the possible consequences of high sugars consumption on maternal–infant health and recommending general care in this phase of life, including healthy eating and physical activity. Participants were then given 6 photos with the Top 7 foods and a set of sugar cubes equivalent to

5 g each (one teaspoon). Participants were asked to place the sugar cubes that they thought were contained by each food on top of the corresponding photo. Then they were told the real amounts of sugar and given recommendations according to what each participant observed and commented.

Session 2: "Learning to substitute intelligently".

This consisted of presenting information on options of healthy food substitutes for those with high sugars content. Each participant received a magnetic board with images of the Top 7, as well as a number of photos of healthy and unhealthy foods. The participants were asked to choose two alternatives that best substituted for the high-sugars foods in the same meal in which they eat those foods, and to place the two photos of each pair together. Afterwards, the selected choices for each food were analyzed, and recommendations to reinforce the consumption of healthy foods were given.

Session 3: "Identifying my eating habits".

There were two activities in this session. The first activity used a board in the form of a traffic light; the participant was shown five photographs of eating habits of a pregnant woman, and a brief description of exactly what each photograph represented was read to her. The objective was for the participant to place the photograph on the color to which she considered it belonged, where green = healthy, yellow = caution, and red = risky.

The second activity was a roulette game with four colors: red, green, blue, and yellow. Each color represented one of the topics treated in the previous two sessions. The participant spun the wheel five times. According to the color that resulted for each time, she took a card from the stack with that color that had a question on the corresponding topic. There were direct questions and multiple-choice questions with three options, but only one correct.

According to the responses of the participant in the two activities her knowledge was reinforced, emphasizing her weak points, and a general feedback of all the educational sessions was given.

2.4. Routine Counseling

The CG only received the routine counseling given in primary health centers in prenatal controls.

2.5. Data Collection

When participants were enrolled, sociodemographic (age, marital status, occupation, income, composition of the home), obstetric (previous pregnancies and abortions, etc.), and morbidity (personal and family history of depression, type 2 diabetes, hypertension) information was collected, and blood pressure, height, and weight were measured. We also recorded information about the group of omega-3 fatty acids in which women were randomly allocated to receive (200 mg/day or 800 mg/day). By far, this information is not known to the participants or the researchers; thus, we classified this variable for the analyses as group 1 and group 2. GWG was calculated as the difference between the weight at the last visit during pregnancy (35–37 weeks) and pre-pregnancy weight; it was analyzed both as continuous (kg) and as categorical below, within and above the Institute of Medicine (IOM) 2009 guidelines (7–11.5 kg for overweight and 5–9 kg for obesity) [20].

Food consumption was measured using an FFQ applied twice by trained nutritionists. The first was applied when the participant was enrolled in the study (before 15 weeks of gestation) and the second was applied after the last visit (weeks 35–37 of gestation); both measured food consumption in the previous month.

The semi-quantitative FFQ was designed to evaluate the habitual intake of foods that were a relevant dietary source of total sugars and energy and, according to data derived from R24h mentioned above, in which all foods and culinary preparations eaten by the participant were compiled. One R24h was collected for each participant by a single trained dietitian using the multi-pass method [21] and a photographic atlas of standard portions of usual foods and culinary preparations [22,23]. Contents of energy and total sugars of foods were computed according to the weight/volume consumed and based on nutrient composition databases. In the case of natural foods, we used the Chilean database [24] and the United States Department of Agriculture [25] (USDA) nutrient database; in the case of packaged foods, we used a database previously built by our research group, based on nutrient fact panels of packaged foods collected from supermarkets during 2015 [26]. No specific software was used for dietary assessments. Food items consumed by participants were identified with numbers from 1 to 489 and then grouped into 171 food items according to their similarity in physical characteristics or the content of energy or sugars in 100 g of the foods. We then calculated the percentage that each food contributed to the total intake of energy and total sugars of the diet. These food items were listed in decreasing order according to their contribution and the accumulated percentage calculated. Food groups that provided 95% of the energy and total sugars were selected to design the FFQ. The R24H results were also used to estimate the portions of each food usually consumed; these were selected as the standard portions.

The final FFQ instrument contained 86 food items grouped in 16 food groups: cereals, tubers, bread and cakes, cookies, snacks, sugars and sweets, ice cream, soft drinks and juices, fruits, vegetables, legumes, meats, sausages, whole milk products, Chilean prepared foods, and oils. There were six fruits that were asked about in all months, and some that were asked about only during part of the year: an autumn–winter group from 21 March to 20 September and a spring–summer group from 21 September to 2 March.

To help visualize portions and aid in asking about consumption, there was a specific photographic atlas for this project with photographs of the standard portion of each food item, in the same order in which they were asked. The participant answered how much and how often she consumed this food in the month previous to the application of the FFQ.

2.6. Statistical Analysis

Energy and total sugars consumption were estimated by multiplying the consumption frequency of a food by the mass of the standard portion of the FFQ, giving the monthly intake in grams per food item. This was transformed to total daily intake in grams per item and then to daily energy and total sugars consumption using the nutritional information collected at the International Network for Food and Obesity/Non-Communicable Diseases Research, Monitoring and Action Support Project (INFORMAS) of the Instituto de Nutrición y Tecnología de los Alimentos (INTA) of the Universidad de Chile, and from the nutritional composition of foods of the USA Department of Agriculture (USDA).

The continuous variables were described by means and standard deviations, and the categories by frequencies and proportions.

Differences in the consumption of total sugars and energy between the IG and CG in the baseline and after the intervention were evaluated through the Student's t-test.

The change in energy and total sugars consumption pre-and post-intervention within each group was evaluated using t-test for paired samples.

Finally, the effect of dietary counseling intervention (categorical predictor variable: 0 = no, 1 = yes) on total sugars consumption and energy (outcomes) was tested by multiple linear regression models adjusted for the baseline value and by covariates. We repeated the models considering the change between initial and final values in sugars/energy consumption as outcome.

All analyses were performed on an intention-to-treat basis, according to the treatment group allocated at randomization.

Statistical analyses were performed using the Stata program version 13 (StataCorp LP: College Station, TX, USA). The significance level for all tests was $p < 0.05$.

2.7. Ethical Aspects

The ethics committee of INTA of the Universidad de Chile approved the protocol and informed consent procedure of the study. The women studied accepted to participate by signing an informed consent form.

3. Results

3.1. Characterization of the Participants

The characteristics of the 433 women studied at time of recruitment are shown in Table 1. Mean age was 28 years (SD = 5.8); 63% were obese, no difference between groups. A total of 61% of the participants responded to the FFQ during autumn and winter. A percentage of 74% of the participants had finished secondary school and 55% were married; this percentage was higher in the IG than in the CG ($p < 0.05$). The percentage diagnosis of depression was 27% ($n = 77$), and not different between groups.

Table 1. Characterization of the study sample.

Variables	Control Group N = 161		Intervention Group N = 272		p-Value
	Mean (SD) *	N (%) **	Mean (SD) *	N (%) **	
Age (years)	28.2 (6.1)		27.9 (5.3)		0.629
Nutritional status					
Overweight		67 (42)		94 (35)	0.142
Obese		94 (58)		178 (65)	
Marital state					
Single		82 (51)		100 (37)	0.004
Married		79 (49)		172 (63)	
Level of studies					
Primary school		33 (20)		70 (26)	
High school		119 (74)		188 (69)	0.465
University		9 (6)		14 (5)	
Activity					
Work		80 (50)		129 (47)	
Household labors		54 (33)		106 (39)	0.449
Other		27 (17)		37 (14)	
Number of persons in the home	4.4 (1.9)		4.4 (1.7)		0.651
Monthly household income (US dollars)					
≤ US 717		86 (53)		143(52)	
≥ US 717		64 (40)		99 (36)	0.333
Other (no answer, doesn't know)		11(7)		30 (11)	
Relation to head of household					
Head of household		33 (21)		47 (17)	
Spouse, partner		57 (35)		119 (44)	0.229
Other		71 (44)		106 (39)	
Depression before pregnancy					
Yes		48 (30)		65 (24)	
No		111 (69)		206 (76)	0.211
Doesn't know		2 (1)		1 (0)	
DHA supplementation					
Group 1		82 (51)		138 (51)	0.968
Group 2		79 (49)		134 (49)	
Health center					
Low NSE		79 (49)		120 (44)	0.318
Medium NSE		82 (51)		152 (56)	

* Mean (standard deviation). ** N (%), Number (percentage).

Mean GWG at 35–37 weeks was similar in both groups: 8.6 kg (SD = 4.6) in the IG and 8.7 kg (SD = 4.5) in the CG ($p > 0.05$). The proportion of women who exceeded the IOM recommendations for GWG did not differ between groups (36.2% in the IG and 34.0% in the CG, $p > 0.05$).

3.2. Change in Total Sugars and Energy Consumption between Groups

The total dietary changes in mean intake of total sugars and energy before and after the intervention and the differences between IG and CG are given in Table 2. The mean consumption of total sugars in the group intervened with counseling was 143 g/day before the intervention and 111 g/day after the intervention, with a significant reduction of −33 g/day ($p < 0.05$). There was also a significant reduction in total sugars consumption in the CG, from 137 g/day at the beginning to 124 g/day at the end ($p < 0.05$); however, the reduction in the IG was significantly greater (by 20 g) than in the CG ($p < 0.05$).

Table 2. Change in consumption of total sugars and energy.

	Control Group N = 161				Intervention Group N = 272				Intervention vs. Control					
	<15 Weeks of Pregnancy[1]	35–37 Weeks of Pregnancy[1]	Change[1]	p-Value	15 Weeks of Pregnancy[1]	35–37 Weeks of Pregnancy[1]	Change[1]	p-Value	Difference between Groups at <15 Weeks Pregnancy[2]	p-Value	Difference between Groups at 35–37 Weeks Pregnancy[2]	p-Value	Difference in the Change[2]	p-Value
Total sugars, g/day	136.94 (65.10)	123.94 (53.44)	−13.00 (60.11)	0.006	143.36 (70.33)	110.72 (55.01)	−32.64 (69.74)	<0.001	6.42 (6.80)	0.345	−13.22 (5.41)	0.015	−19.64 (6.59)	0.003
Total kilocalories, kcal/day	2064 (627)	1996 (627)	−68 (658)	0.191	2203 (794)	1924 (652)	−279 (761)	<0.001	139 (73)	0.058	−72 (64)	0.264	−211 (72)	0.003
Kilocalories of energy derived from sugars, kcal/day	548 (260)	496 (214)	−52 (240)	0.007	573 (281)	443 (220)	−130 (279)	<0.001	25 (27)	0.345	−53 (22)	0.015	−78 (26)	0.003
Percentage of energy derived from sugars, %	26 (8)	25 (6)	−1 (8)	0.016	26 (7)	23 (6)	−3 (8)	<0.001	0 (0)	0.519	−2 (1)	0.001	−2 (1)	0.060

Statistical differences were evaluated using Student's *t*-test or *t*-test for paired samples. [1] Mean (standard deviation) [2] Mean difference (standard deviation).

There was also a significant decrease in the total energy consumption in the IG, from 2203 to 1924 kcal/day, a decrease of 279 kcal/day. Total energy consumption also decreased in the CG, from 2064 kcal/day to 1996 kcal/day, but the difference was not significant ($p = 0.191$). The difference between the groups after the intervention (72 kcal/day) was not significant ($p = 0.264$). However, the reduction in energy consumption was significantly greater in the CG ($p < 0.05$).

3.3. Changes in the Consumption of Total Sugars and Energy by Food Group

Tables 3 and 4 show the changes in mean consumption of total sugars and energy, respectively, in the food groups of the FFQ. There were significant decreases in total sugars consumption in the group with dietary counseling in soft drinks and juices, table sugar and sweets, fruits, and milk products, with decreases of 12, 9, 7, and 4 g/day, respectively ($p < 0.05$). Small but significant decreases in sugars consumption were also observed in tubers, ice cream, vegetables, prepared foods, and oils. The CG had significant decreases in sugars and sweets (-4 g/day) and soft drinks and juices (-9 g/day), and very slight decreases in vegetables, prepared foods, and oils. The decrease in consumption was significantly greater in the IG in sugars from fruit (-6 g/day), milk products (-3 g/day), and soft drinks and juices, and it showed a tendency to decrease in ice cream (-1 g/day, $p = 0.08$) and sugars and sweets (-4 g/day, $p = 0.08$). There was a significant reduction of total energy in the IG in the food groups in which the dietary consumption was focused; snacks decreased by 16 kcal/day, sugars and sweets by 44 kcal/day, soft drinks and juices by 50 kcal/day, and whole milk products by 52 kcal/day ($p < 0.05$). There were smaller but significant reductions in ice cream, tubers, fruits, vegetables, prepared meats, and oils ($p < 0.05$).

In the CG, there were significant decreases in the consumption of table sugar and sweets (-22 kcal/day), soft drinks and juices (-39 kcal/day) and a smaller decrease in the energy from vegetables (-3 kcal/day) ($p < 0.05$). Snacks, fruits, and whole milk products that showed greater decrease in the IG after counseling than in the CG, with -14 kcal/day, -30 kcal/day, and -42 kcal/day, respectively ($p < 0.05$).

3.4. Effect of the Nutritional Intervention on Consumption of Sugars and Energy

Table 5 shows the results of the linear and multiple regression models. The multiple model showed that the group with dietary counseling had a significant reduction in total sugars consumption of 15.23 g/day (95% CI -25.01 to -5.46 g/day; $p = 0.002$), in energy by 125 kcal/day (95% CI -239 to -10 kcal/day; $p = 0.032$), and in the percentage of energy derived from sugars by 2% (95% CI -3% to -1%; $p = 0.002$), compared to the control group.

Table 3. Consumption of total sugars (g/day) by food groups.

Food Groups	Control Group N = 161				Intervention Group N = 272				Intervention Group vs. Control Group					
	<15 Weeks of Pregnancy[1]	35-37 Weeks of Pregnancy[1]	Change[1]	p-Value	15 Weeks of Pregnancy[1]	35-37 Weeks of Pregnancy[1]	Change[1]	p-Value	Difference between Groups at <15 Weeks of Pregnancy[2]	p-Value	Difference between Groups at 35-37 Weeks of Pregnancy[2]	p-Value	Difference in the Change[2]	p-Value
Cereals	3.88 (4.77)	3.52 (3.68)	-0.35 (5.26)	0.393	3.35 (3.15)	2.97 (3.43)	-0.38 (3.79)	0.099	-0.53 (0.38)	0.161	-0.55 (0.35)	0.116	-0.03 (0.44)	0.952
Tubers	0.57 (0.42)	0.53 (0.44)	-0.04 (0.53)	0.349	0.62 (0.57)	0.52 (0.39)	-0.10 (0.57)	0.003	0.05 (0.05)	0.307	-0.01 (0.04)	0.768	-0.06 (0.05)	0.259
Bread	2.16 (1.81)	2.29 (2.14)	0.13 (2.38)	0.474	2.44 (2.03)	2.36 (1.76)	-0.8 (2.43)	0.594	0.28 (0.19)	0.139	0.07 (0.18)	0.742	-0.21 (0.24)	0.375
Pies and cakes	9.33 (8.08)	9.86 (9.12)	0.53 (10.63)	0.526	10.36 (10.73)	8.92 (8.69)	-1.44 (13.00)	0.068	1.12 (0.97)	0.250	-0.94 (0.88)	0.283	-1.97 (1.21)	0.103
Cookies	3.81 (6.83)	3.57 (4.90)	-0.24 (7.20)	0.672	3.70 (6.26)	3.06 (7.36)	-0.64 (8.09)	0.192	-0.05 (0.64)	0.931	-0.51 (0.65)	0.430	-0.40 (0.77)	0.602
Snacks	1.97 (2.73)	2.46 (3.64)	0.49 (4.36)	0.155	1.75 (2.79)	1.84 (3.43)	0.09 (4.19)	0.701	-0.23 (0.27)	0.394	-0.62 (0.35)	0.077	-0.39 (0.42)	0.353
Sugars and sweets	23.49 (18.69)	19.19 (18.50)	-4.30 (20.57)	0.008	27.98 (24.68)	19.38 (19.18)	-8.60 (27.37)	<0.001	4.88 (2.23)	0.029	0.19 (1.88)	0.921	-4.30 (2.49)	0.084
Ice cream	2.28 (4.94)	2.65 (7.39)	0.36 (8.65)	0.595	2.71 (5.23)	1.83 (2.66)	-0.88 (5.95)	0.015	0.70 (0.45)	0.125	-0.82 (0.49)	0.096	-1.24 (0.70)	0.077
Soft drinks	29.23 (35.80)	20.50 (23.96)	-8.73 (31.88)	<0.001	27.08 (32.45)	15.05 (20.51)	-12.03 (34.47)	<0.001	-2.15 (3.35)	0.521	-5.45 (2.17)	0.012	-3.30 (3.33)	0.323
Fruits and juices	31.23 (23.16)	30.69 (23.77)	-0.54 (27.23)	0.799	33.03 (25.71)	26.41 (20.97)	-6.61 (25.29)	<0.001	1.89 (2.46)	0.445	-4.28 (2.18)	0.051	-6.07 (2.58)	0.019
Vegetables	1.79 (1.38)	1.49 (1.08)	-0.30 (1.35)	0.005	2.00 (1.73)	1.59 (1.27)	-0.41 (1.75)	<0.001	0.20 (0.16)	0.208	0.10 (0.12)	0.418	-0.11 (0.16)	0.478
Legumes	0.27 (0.30)	0.28 (0.30)	0.01 (0.34)	0.691	0.26 (0.31)	0.27 (0.33)	0.01 (0.40)	0.572	-0.01 (0.03)	0.865	-0.01 (0.03)	0.776	0.00 (0.04)	0.934
Meats	0.29 (0.25)	0.28 (0.19)	-0.01 (0.29)	0.570	0.28 (0.25)	0.26 (0.24)	-0.02 (0.29)	0.295	-0.01 (0.02)	0.839	-0.01 (0.02)	0.579	-0.01 (0.03)	0.845
Sausage	0.20 (0.21)	0.18 (0.19)	-0.02 (0.23)	0.296	0.25 (0.24)	0.20 (0.21)	-0.05 (0.26)	0.002	0.05 (0.02)	0.039	0.02 (0.02)	0.408	-0.03 (0.02)	0.219
Whole milk products	11.67 (14.33)	10.67 (14.58)	-1.00 (15.11)	0.404	12.91 (16.02)	8.51 (13.32)	-4.40 (15.09)	<0.001	1.59 (1.51)	0.294	-2.16 (1.37)	0.115	-3.40 (1.49)	0.023
Low-fat milk products	12.23 (12.19)	14.18 (14.66)	1.95 (15.21)	0.103	10.00 (11.87)	14.19 (15.15)	4.19 (15.74)	<0.001	-2.30 (1.19)	0.054	0.01 (1.48)	0.995	2.24 (1.54)	0.148
Preparations	3.73 (3.68)	3.07 (3.04)	-0.66 (3.49)	0.019	4.54 (8.19)	3.27 (3.63)	-1.27 (7.67)	0.007	0.93 (0.68)	0.171	0.20 (0.34)	0.561	-0.62 (0.64)	0.334
Oils	0.06 (0.06)	0.04 (0.04)	-0.02 (0.06)	0.010	0.06 (0.08)	0.04 (0.04)	-0.03 (0.09)	<0.001	0.01 (0.01)	0.112	0.00 (0.00)	0.665	0.01 (0.01)	0.173

Statistical differences were evaluated using Student's *t*-test or *t*-test for paired samples. [1] Mean (standard deviation) [2] Mean difference (standard deviation).

Table 4. Energy consumption (kcal/day) by food group.

Food Groups	Control Group N = 161				Intervention Group N = 272						Intervention Group vs. Control Group			
	<15 Weeks of Pregnancy [1]	35–37 Weeks of Pregnancy [1]	Change [1]	p-Value	15 Weeks of Pregnancy [1]	35–37 Weeks of Pregnancy [1]	Change [1]	p-Value	Difference between Groups at < 15 Weeks of Pregnancy [2]	p-Value	Difference between Groups at 35–37 Weeks of Pregnancy [2]	p-Value	Difference in the Change [2]	p-Value
Cereals	228 (119)	243 (125)	15 (139)	0.182	238 (132)	231 (143)	−7 (160)	0.527	10 (12)	0.452	−12 (13)	0.391	−21 (15)	0.170
Tubers	48 (35)	43 (34)	−5 (37)	0.070	54 (50)	44 (34)	−10 (50)	<0.001	6 (4)	0.193	1 (3)	0.845	−5 (4)	0.268
Bread	213 (179)	226 (210)	13 (234)	0.472	240 (200)	232 (174)	−8 (240)	0.595	28 (19)	0.139	6 (18)	0.741	−21 (23)	0.375
Pies and cakes	241 (207)	245 (215)	4 (236)	0.813	271 (226)	237 (229)	−34 (286)	0.053	30 (22)	0.172	−8 (22)	0.706	−38 (27)	0.154
Cookies	73 (102)	63 (73)	−10 (110)	0.263	70 (102)	60 (114)	−10 (128)	0.208	−3 (10)	0.898	−2 (10)	0.813	0 (12)	0.999
Snacks	46 (51)	44 (49)	−2 (67)	0.785	51 (77)	35 (47)	−16 (79)	<0.001	5 (7)	0.428	−9 (5)	0.047	−14 (7)	0.047
Sugars and sweets	136 (114)	114 (100)	−22 (121)	0.019	154 (126)	110 (98)	−44 (144)	<0.001	20 (12)	0.087	−3 (9)	0.731	−21 (13)	0.117
Ice cream	22 (47)	25 (72)	3 (84)	0.621	26 (49)	17 (26)	−9 (56)	0.015	7 (4)	0.125	−8 (5)	0.115	−12 (7)	0.086
Soft drinks and juices	123 (151)	84 (104)	−39 (135)	<0.001	112 (132)	62 (84)	−50 (140)	<0.001	−11 (14)	0.407	−22 (9)	0.014	−10 (14)	0.451
Fruits	169 (120)	165 (121)	−4 (136)	0.692	177 (132)	143 (108)	−34 (130)	<0.001	8 (13)	0.516	−22 (11)	0.047	−30 (13)	0.022
Vegetables	16 (12)	13 (9)	−3 (11)	0.001	18 (15)	14 (11)	−4 (15)	<0.001	2 (1)	0.085	1 (1)	0.268	−1 (1)	0.348
Legumes	31 (31)	32 (30)	1 (37)	0.537	30 (32)	30 (33)	0 (39)	0.891	0 (3)	0.975	−2 (3)	0.520	−1 (4)	0.760
Meats	190 (99)	191 (85)	1 (107)	0.923	189 (100)	188 (97)	−1 (112)	0.891	−1 (9)	0.956	−3 (9)	0.763	−2 (11)	0.873
Sausage	42 (35)	39 (35)	−3 (41)	0.459	48 (41)	42 (38)	−6 (47)	0.021	6 (4)	0.089	2 (4)	0.534	−4 (4)	0.345
Whole milk products	143 (172)	133 (165)	−10 (160)	0.403	164 (171)	112 (135)	−52 (164)	<0.001	24 (17)	0.161	−20 (14)	0.162	−41 (16)	0.010
Low-fat milk products	129 (117)	141 (130)	12 (143)	0.287	115 (135)	145 (139)	30 (156)	0.001	−15 (13)	0.246	4 (13)	0.748	18 (15)	0.218
Preparations	152 (83)	143 (96)	−9 (104)	0.288	159 (106)	150 (96)	−9 (120)	0.236	8 (9)	0.379	7 (9)	0.442	0 (11)	0.994
Oils	75 (50)	69 (44)	−6 (55)	0.134	86 (68)	69 (43)	−17 (69)	<0.001	12 (6)	0.060	0 (4)	0.967	−11 (6)	0.094

Statistical differences were evaluated using Student's *t*-test or t-test for paired samples. [1] Mean (standard deviation) [2] Mean difference (standard deviation).

Table 5. Effect of the nutritional intervention on the consumption of total sugars and energy. Linear regression models.

	MODEL 1	p-Value	MODEL 2	p-Value	MODEL 3	p-Value
Total sugars consumption post intervention, g/day	−13.22 (−23.85, −2.57)	0.015	−15.44 (−25.04, −5.84)	0.002	−15.23 (−25.01, −5.46)	0.002
Total kilocalorie consumption post intervention, kcal/day	−71 (−197, 54)	0.265	−127 (−239, −14)	0.028	−125 (−239, −10)	0.032
Total percentage of energy consumption derived from sugars, %	−2 (−3, −1)	0.001	−2 (−3, −1)	0.001	−2 (−3, −1)	0.002

Reference group is the control group. Figures represent β coefficients of the regression models and 95% confidence intervals. MODEL 1: Sugars/kilocalories/% calories from sugars at 35–37 weeks of pregnancy = β0 + β1 (dietary counseling). MODEL 2: Sugars/kilocalories/% calories from sugars at 35–37 weeks of pregnancy = β0 + β1 (dietary counseling) + β2 (sugars/kilocalories/% calories at < 15 weeks of pregnancy). MODEL 3: Sugars/kilocalories/% calories from sugars at 35–37 weeks of pregnancy = β0 + β1 (dietary counseling) + β2 (sugars/kilocalories/% calories at < 15 weeks of pregnancy) + β3 (marital state) + β4 (persons in home) + β5 (age) + β6 (level of studies) + β7 (monthly income) + β8 (relation to head of household) + β9 (activity) + β10 (depression) + β11 (nutritional status) + β12 (gestational weight gain) + β13 (season) + β14 (health center).

In addition to baseline values, significant covariables in the multivariable models were maternal age (β coefficient = −1.1, 95% CI = −2.1, −0.1) and health care center (β coefficient= −14.7, 95% CI = −24.4, −4.9) for the model of sugars consumption, maternal age (β coefficient −16, 95% CI −27, −5), and GWG (β coefficient 18, 95% CI 6, 30) for the model of energy, and health care center (β coefficient −2, 95% CI −3, −1) for the model of percentage of energy derived from sugars.

We repeated the analysis using the change between initial and final consumption as the response variable in a sensitivity analysis. The multiple models had similar results (Table S1).

4. Discussion

The present study showed that an intervention with dietary counseling in overweight or obese pregnant women was effective in decreasing the consumption of foods that most contribute to total sugars, energy, and the percentage of energy from sugars. Compared to participants who only received routine counseling, the IG reduced the mean consumption of total sugars by 15 g/day and energy by 125 kcal/day, and also reduced the percentage of energy from sugars by 2%.

We did not find reported nutritional interventions focused on decreasing sugars comparable to the present study. However, our results are consistent with other interventions of a similar nature. A study in Mexico determined if the nutritional status of overweight and obese pregnant women improved by implementing a personalized diet. These results indicated that the group of sugars had a greater decrease in consumption (15.1%) than the other food groups evaluated [5]. This could be attributed to the fact that the personalized diet emphasized eating foods with low glycemic index. Poston et al. provided diet counseling to obese pregnant women centered on consumption of foods with low glycemic index, including replacement of sugared soft drinks by ones with low glycemic index and reduction of saturated foods; they showed that intervened participants reduced mean intake of saturated fats, carbohydrates, and total energy, and increased intake of protein and fiber [10]. Petrella et al. [16] evaluated a nutritional intervention in overweight and obese pregnant women in which the primary focus of the dietary intervention was decreasing high-glycemic index foods consumption and substituting them with healthier alternatives. They observed that the group who received the intervention decreased the consumption of sugar and increased the consumption of vegetables and fruits [16]. Wolff et al. [27] performed a dietetic intervention restricted to energy consumption, providing a distribution of macronutrients according to Danish recommendations; they found changes in the composition of the diet and a decrease in total energy consumption similar to that found in our study [27]. In contrast, an Australian randomized LIMIT trial performed an intervention in lifestyle focused on maintaining equilibrium of carbohydrates, fats, and proteins to reduce intake of foods rich in refined carbohydrates and saturated fats. They did not find a significant difference in the mean daily energy consumption between the group of pregnant women randomized to the intervention and the control group [28]. Our intervention may have been more effective because it was focused on the foods that contribute most to sugars consumption, thus it may have been easier for the participants to follow the indications. It has been shown that dietary interventions during pregnancy are more effective than a combination of diet and physical activity in terms of weight gained during pregnancy, suggesting that there may be poorer compliance by participants when a large number of indications are given to the participants [29].

As far as we know, ours is the first Chilean study that presents detailed information on the consumption of energy and total sugars in different food groups of the diet of overweight and obese pregnant women. The dietary counseling was effective in differentially reducing some of the foods that most contribute to sugars consumption (the Top 7). Compared to women in the CG, participants from the IG had significant greater reductions of the intake (calories) of snacks and whole fat milk products, which was also the case for total sugars coming from whole fat milk products. Besides this, the intervention also decreased the intake of energy and sugars coming from fruits, even if fruits were not part of the targeted food groups. The intervention also decreased the consumption (calories) of sugars and sweets, and soft drinks and juices, but this was not significantly different from what happened in the CG. The intake of bread and cookies did not decrease. The decrease in more than 15 g of total sugars intake observed in the adjusted models corresponded to a decrease of 2% of energy from total sugars in the studied population. The baseline intake of total sugars was 26% of calories, which is much greater than the recommendations made for the general population (i.e., <20% of daily energy intake from total sugars) [30]. Thus, the reported intervention helped decrease one-third of the excessive intake of total sugars. Total sugars intake decreased mainly from whole fat milk products and fruits, which could be interpreted as an unexpected outcome. In the case of fruits, the intake of total sugars in the follow-up of the IG was 26 g, which corresponds to 2–3 servings of fruits, according to the recommendations (considering that most of them have about 10% of their weight as sugars). The women counseled also showed greater reduction (42 kcal/day) of intake of whole milk foods and increase of 18 kcal/day in low-fat dairy products. These results comply with the nutritional recommendations of the Health ministry for low-fat dairy products for pregnant women, considered essential during pregnancy [31]. Evidence has shown that the absorption of calcium increases during pregnancy, fulfilling the requirements of bone mineralization of the fetus [32]. The study of Osorio et al. [33] suggests that greater intake of low-fat dairy products is significantly associated with lower risk of gestational diabetes mellitus. These findings lend weight to the context of the present research [33].

The decrease in dairy products also implies a decrease in the consumption of both total and added sugars, since the amount of total sugars among all yogurts and most whole fat milks were 2–3 times the natural content of lactose (about 5% among milks and 7% among yogurts). Thus, the intervention also helped in decreasing the gap between the intake and the recommendation of added sugars [34] even when we cannot indicate in which extent given added sugars were not assessed.

As far as we know, it is not possible to anticipate the exact clinical impact of such dietary improvement, but given the current knowledge on the detrimental impact of an excessive intake of sugars [35] and the better understanding of the Developmental Origin of Disease paradigm [36,37], it is plausible that this changes would imply relevant improvements in women and offspring health. Although the evidence of the effect of dietary interventions in mothers with obesity focused on reducing foods with high glycemic index on maternal and offspring cardiometabolic health is not consistent and needs to be further studied, some animal experiments have shown positive results [19,38]. In our study, the intervention was not associated with differences in GWG or in the percentage of women who exceeded the IOM 2009 recommendations for GWG between groups; therefore, any effect of maternal diet on maternal and offspring outcomes could be not mediated by changes in GWG. However, it is possible that sample size of the present study is not large enough to ensure adequate power to detect differences on GWG between groups.

The strengths of this study include its cohort design with close follow-up of the participants and measurement of multiple variables, which helped fit the models. The construction of an FFQ aimed at the consumption of sugars and energy based on an R24H questionnaire in a population with similar characteristics allowed a reliable representation of daily consumption. FFQ is a method of dietary evaluation used in most epidemiological studies [39,40], considered to provide a valid model of long-term habitual consumption as well as being cheap, rapid, and easy to apply [41]. However, since FFQs are self-reported, they may be subject to report bias. The IG may have reported lower intake than the CG, given the fact that our intervention was not blind. However, the results of consumption in the baseline were similar to those reported in the R24h of women who did not participate in the intervention. Both groups showed decreases in the consumption of sugars and total energy, greater in the IG, although evidence shows that pregnant women tend to increase energy consumption during pregnancy [42]. Part of the decrease in the CG group may be due to the routine counseling received in the prenatal attention.

This study showed that focused dietary counseling provided in only three sessions during a 12-week period can be effective in decreasing the consumption of sugars and energy. This is an alternative to the current guidelines of the Chilean National Health Services, which recommend a generally healthy diet [31]. Future studies should evaluate if these changes produce maternal and child results (lower incidence of gestational diabetes and macrosomia, among others). It would also be interesting to know if these maternal dietary changes that may be motivated by the pregnancy are maintained after the women give birth.

5. Conclusions

The implementation of a dietary intervention with nutritional counseling focused on sugars consumption decreased the consumption of energy and total sugars in overweight and obese pregnant women, mainly in the food groups high in free sugars. Future studies should evaluate if these changes produce short- and long-term maternal and child results. Given its focus, this intervention may be replicable in all primary health centers for pregnant women with the highest nutritional risk.

Supplementary Materials: The following are available online at http://www.mdpi.com/2072-6643/11/2/385/s1, Table S1: Effect of the nutritional intervention on the change of consumption of total sugars and energy. Linear regression models.

Author Contributions: M.L.G. helped design the study and provided advice on statistical analysis. E.A. performed the research, inscription of participants, home nutritional counseling, compilation and analysis of data. M.L.G. and E.A. wrote the first version of the manuscript. M.R., M.A.B., M.F., and R.U. read and approved the final manuscript.

Funding: This research was funded by the Chilean National Fund for Scientific and Technological Development, FONDECYT Regular grant number # 1150878.

Acknowledgments: DSM Nutritional Products donated the capsules for the original study, but had no role in the design of the study, in its analysis, interpretation, nor in its dissemination. The South East Health Service and Puente Alto county have provided kind support for the implementation of the study. We thank all the participants in this study for generously helping us in this research.

Conflicts of Interest: The authors declare that there are no conflict of interest. The sponsors had no role in the design, execution, interpretation, or writing of the study.

References

1. Organización de las Naciones Unidas para la Alimentación y la Agricultura. América Latina y el Caribe. Panorama de la Seguridad Alimentaria y Nutricional. 2017. Available online: http://www.fao.org/americas/publicaciones-audio-video/panorama/es/ (accessed on 10 May 2017).

2. Gobierno de Chile Ministerio de Salud. *Encuesta Nacional de Salud ENS Chile 2009–2010*; Ministerio de Salud: Santiago, Chile, 2010.

3. Araya, B.; Padilla, O.; Garmendia, M.L.; Atalah, E.; Uauy, R. Prevalence of obesity among Chilean women in childbearing ages. *Revista Medica de Chile* **2014**, *142*, 1440–1448.

4. Huidobro, A.; Fulford, A.; Carrasco, E. Incidence of gestational diabetes and relationship to obesity in Chilean pregnant women. *Revista Medica de Chile* **2004**, *132*, 931–938. [PubMed]

5. Sandoval, K.V.; Nieves, E.R.; Luna, M.Á. Efecto de una dieta personalizada en mujeres embarazadas con sobrepeso u obesidad. *Revista Chilena de Nutrición* **2016**, *43*, 233–246. [CrossRef]

6. Renault, K.M.; Carlsen, E.M.; Nørgaard, K.; Nilas, L.; Pryds, O.; Secher, N.J.; Olsen, S.F.; Halldorsson, T.I. Intake of sweets, snacks and soft drinks predicts weight gain in obese pregnant women: Detailed analysis of the results of a randomised controlled trial. *PLoS ONE* **2015**, *10*, e0133041. [CrossRef] [PubMed]

7. Aşcı, Ö.; Rathfisch, G. Effect of lifestyle interventions of pregnant women on their dietary habits, lifestyle behaviors, and weight gain: A randomized controlled trial. *J. Health Popul. Nutr.* **2016**, *35*, 7. [CrossRef] [PubMed]

8. Hui, A.L.; Ludwig, S.; Gardiner, P.; Sevenhuysen, G.; Dean, H.J.; Sellers, E.; McGavock, J.; Morris, M.; Shen, G.X.; Jiang, D.; et al. Effects of lifestyle intervention on dietary intake, physical activity level, and gestational weight gain in pregnant women with different pre-pregnancy Body Mass Index in a randomized control trial. *BMC Pregnancy Childbirth* **2014**, *14*, 331. [CrossRef] [PubMed]

9. Shepherd, E.; Gomersall, J.C.; Tieu, J.; Han, S.; Crowther, C.A.; Middleton, P. Combined diet and exercise interventions for preventing gestational diabetes mellitus. *Cochrane Database Syst Rev.* **2017**, *11*, CD01443. [CrossRef] [PubMed]

10. Poston, L.; Bell, R.; Croker, H.; Flynn, A.C.; Godfrey, K.M.; Goff, L.; Hayes, L.; Khazaezadeh, N.; Nelson, S.M.; Oteng-Ntim, E.; et al. Effect of a behavioural intervention in obese pregnant women (the UPBEAT study): A multicentre, randomised controlled trial. *Lancet Diabetes Endocrinol.* **2015**, *3*, 767–777. [CrossRef]

11. Bruno, R.; Petrella, E.; Bertarini, V.; Pedrielli, G.; Neri, I.; Facchinetti, F. Adherence to a lifestyle programme in overweight/obese pregnant women and effect on gestational diabetes mellitus: A randomized controlled trial. *Matern. Child Nutr.* **2017**, *13*, e12333. [CrossRef]

12. Luoto, R.; Kinnunen, T.I.; Aittasalo, M.; Kolu, P.; Raitanen, J.; Ojala, K.; Mansikkamäki, K.; Lamberg, S.; Vasankari, T.; Komulainen, T.; et al. Primary prevention of gestational diabetes mellitus and large-for-gestational-age newborns by lifestyle counseling: A cluster-randomized controlled trial. *PLoS Med.* **2011**, *8*, e1001036. [CrossRef]

13. Garmendia, M.L.; Corvalán, C.; Casanello, P.; Araya, M.; Flores, M.; Bravo, A.; Kusanovic, J.P.; Olmos, P.; Uauy, R. Effectiveness on maternal and offspring metabolic control of a home-based dietary counseling intervention and DHA supplementation in obese/overweight pregnant women (MIGHT study): A randomized controlled trial-Study protocol. *Contemp. Clin. Trials* **2018**, *70*, 35–40. [CrossRef]

14. Governement of Chile, Ministry of Health. *Guía Perinatal*; Ministry of Health: Santiago, Chile, 2015.

15. Poston, L.; Briley, A.L.; Barr, S.; Bell, R.; Croker, H.; Coxon, K.; Essex, H.N.; Hunt, C.; Hayes, L.; Howard, L.M.; et al. Developing a complex intervention for diet and activity behaviour change in obese pregnant women (the UPBEAT trial); assessment of behavioural change and process evaluation in a pilot randomised controlled trial. *BMC Pregnancy Childbirth* **2013**, *13*, 148. [CrossRef]

16. Petrella, E.; Malavolti, M.; Bertarini, V.; Pignatti, L.; Neri, I.; Battistini, N.C.; Facchinetti, F. Gestational weight gain in overweight and obese women enrolled in a healthy lifestyle and eating habits program. *J. Matern. Fetal Neonatal Med.* **2014**, *27*, 1348–1352. [CrossRef] [PubMed]

17. Moses, R.G.; Casey, S.A.; Quinn, E.G.; Cleary, J.M.; Tapsell, L.C.; Milosavljevic, M.; Petocz, P.; Brand-Miller, J.C. Pregnancy and Glycemic Index Outcomes study: Effects of low glycemic index compared with conventional dietary advice on selected pregnancy outcomes. *Am. J. Clin. Nutr.* **2014**, *99*, 517–523. [CrossRef] [PubMed]

18. Walsh, J.M.; McGowan, C.A.; Mahony, R.; Foley, M.E.; McAuliffe, F.M. Low glycaemic index diet in pregnancy to prevent macrosomia (ROLO study): Randomised control trial. *BMJ* **2012**, *345*, e5605. [CrossRef] [PubMed]

19. Zhang, R.; Han, S.; Chen, G.C.; Li, Z.N.; Silva-Zolezzi, I.; Parés, G.V.; Wang, Y.; Qin, L.Q. Effects of low-glycemic-index diets in pregnancy on maternal and newborn outcomes in pregnant women: A meta-analysis of randomized controlled trials. *Eur. J. Nutr.* **2018**, *57*, 167–177. [CrossRef] [PubMed]

20. Institute of Medicine; National Research Council. *Weight Gain During Pregnancy: Reexamining the Guidelines*; National Academies Press: Washington, DC, USA, 2009.

21. Moshfegh, A.J.; Rhodes, D.G.; Baer, D.J.; Murayi, T.; Clemens, J.C.; Rumpler, W.V.; Paul, D.R.; Sebastian, R.S.; Kuczynski, K.J.; Ingwersen, L.A.; et al. The US Department of Agriculture Automated Multiple-Pass Method reduces bias in the collection of energy intakes. *Am. J. Clin. Nutr.* **2008**, *88*, 324–332. [CrossRef]

22. Mendonça, R.D.; Pimenta, A.M.; Gea, A.; de la Fuente-Arrillaga, C.; Martinez-Gonzalez, M.A.; Lopes, A.C.; Bes-Rastrollo, M. Ultraprocessed food consumption and risk of overweight and obesity: The University of Navarra Follow-Up (SUN) cohort study. *Am. J. Clin. Nutr.* **2016**, *104*, 1433–1440. [CrossRef]

23. Gobierno de Chile Ministerio de Salud. Encuesta Nacional de Consumo Alimentario. 2011. Available online: https://www.minsal.cl/enca/ (accessed on 10 November 2016).

24. Schmidt-Hebbel, H.; Pennacchiotti, I.; Masson, L.; Mella, M. *Tabla de Composición Química de Alimentos Chilenos*, 8th ed.; Facultad de Ciencias Químicas y Farmacéuticas, Departamento de Ciencia de los Alimentos y Tecnología Química, Universidad de Chile: Santiago, Chile, 1992.

25. United States Department of Agriculture, Agricultural Research Service, USDA Food Composition Databases. USDA National Nutrient Database for Standard Reference. Available online: https://ndb.nal.usda.gov/ndb/ (accessed on 15 November 2017).

26. Kanter, R.; Reyes, M.; Corvalán, C. Photographic Methods for Measuring Packaged Food and Beverage Products in Supermarkets. *Curr. Dev. Nutr.* **2017**, *1*, e001016. [CrossRef]

27. Wolff, S.; Legarth, J.; Vangsgaard, K.; Toubro, S.; Astrup, A. A randomized trial of the effects of dietary counseling on gestational weight gain and glucose metabolism in obese pregnant women. *Int. J. Obes. (Lond.)* **2008**, *32*, 495–501. [CrossRef]

Nutrients **2019**, *11*, 385

28. Dodd, J.M.; Cramp, C.; Sui, Z.; Yelland, L.N.; Deussen, A.R.; Grivell, R.M.; Moran, L.J.; Crowther, C.A.; Turnbull, D.; McPhee, A.J. The effects of antenatal dietary and lifestyle advice for women who are overweight or obese on maternal diet and physical activity: The LIMIT randomised trial. *BMC Med.* **2014**, *12*, 161. [CrossRef]

29. Thangaratinam, S.; Rogozinska, E.; Jolly, K.; Glinkowski, S.; Roseboom, T.; Tomlinson, J.W.; Kunz, R.; Mol, B.W.; Coomarasamy, A.; Khan, K.S. Effects of interventions in pregnancy on maternal weight and obstetric outcomes: Meta-analysis of randomised evidence. *BMJ* **2012**, *344*, e2088. [CrossRef]

30. French Agency for Food, Environmental and Occupational Health & Safety. *Opinion of the French Agency for Food, Environmental and Occupational Health & Safety on the Establishment of Recommendations on Sugar Intake*; Request No 2012-SA-0186; ANSES: Maisons-Alfort, France, 2016.

31. Gobierno de Chile, Ministerio de Salud, Subsecretaria de Salud Pública. *Guía Perinatal*; Ministerio de Salud: Santiago, Chile, 2014.

32. Hacker, A.N.; Fung, E.B.; King, J.C. Role of calcium during pregnancy: Maternal and fetal needs. *Nutr. Rev.* **2012**, *70*, 397–409. [CrossRef] [PubMed]

33. Osorio-Yáñez, C.; Qiu, C.; Gelaye, B.; Enquobahrie, D.A.; Williams, M.A. Risk of gestational diabetes mellitus in relation to maternal dietary calcium intake. *Public Health Nutr.* **2017**, *20*, 1082–1089. [CrossRef] [PubMed]

34. U.S. Department of Health and Human Services and U.S. Department of Agriculture. 2015–2020 DietaryGuidelines for Americans. 8th Edition. December 2015. Available online: http://health.gov/dietaryguidelines/2015/guidelines/ (accessed on 15 January 2019).

35. Vos, M.B.; Kaar, J.L.; Welsh, J.A.; Van Horn, L.V.; Feig, D.I.; Anderson, C.A.M.; Patel, M.J.; Cruz Munos, J.; Krebs, N.F.; Xanthakos, S.A.; et al. Added Sugars and Cardiovascular Disease Risk in Children: A Scientific Statement from the American Heart Association. *Circulation* **2017**, *135*, e1017–e1034. [CrossRef] [PubMed]

36. Gluckman, P.D.; Hanson, M.A.; Beedle, A.S. Early life events and their consequences for later disease: A life history and evolutionary perspective. *Am. J. Hum. Biol.* **2007**, *19*, 1–19. [CrossRef] [PubMed]

37. Uauy, R.; Kain, J.; Corvalan, C. How can the Developmental Origins of Health and Disease (DOHaD) hypothesis contribute to improving health in developing countries? *Am. J. Clin. Nutr.* **2011**, *94*, 1759S–1764S. [CrossRef] [PubMed]

38. Menting, M.D.; Mintjens, S.; van de Beek, C.; Frick, C.J.; Ozanne, S.E.; Limpens, J.; Roseboom, T.J.; Hooijmans, C.R.; van Deutekom, A.W.; Painter, R.C. Maternal obesity in pregnancy impacts offspring cardiometabolic health: Systematic review and meta-analysis of animal studies. *Obes. Rev.* **2019**. [CrossRef]

39. Trinidad Rodríguez, I.; Fernández Ballart, J.; Cucó Pastor, G.; Biarnés Jordà, E.; Arija Val, V. Validación de un cuestionario de frecuencia de consumo alimentario corto: Reproducibilidad y validez. *Nutrición Hospitalaria* **2008**, *23*, 242–252.

40. Rauh, K.; Kunath, J.; Rosenfeld, E.; Kick, L.; Ulm, K.; Hauner, H. Healthy living in pregnancy: A cluster-randomized controlled trial to prevent excessive gestational weight gain-rationale and design of the GeliS study. *BMC Pregnancy Childbirth* **2014**, *14*, 119. [CrossRef]

41. Barbieri, P.; Nishimura, R.Y.; Crivellenti, L.C.; Sartorelli, D.S. Relative validation of a quantitative FFQ for use in Brazilian pregnant women. *Public Health Nutr.* **2013**, *16*, 1419–1426. [CrossRef] [PubMed]

42. Hillier, S.E.; Olander, E.K. Women's dietary changes before and during pregnancy: A systematic review. *Midwifery* **2017**, *49*, 19–31. [CrossRef] [PubMed]

nutrients

MDPI

Article

Association of Full Breastfeeding Duration with Postpartum Weight Retention in a Cohort of Predominantly Breastfeeding Women

Muna J. Tahir [1,*]**, Jacob L. Haapala** [1]**, Laurie P. Foster** [1]**, Katy M. Duncan** [2]**, April M. Teague** [2]**,
Elyse O. Kharbanda** [3]**, Patricia M. McGovern** [4]**, Kara M. Whitaker** [5,6]**, Kathleen M. Rasmussen** [7]**,
David A. Fields** [2]**, Lisa J. Harnack** [1]**, David R. Jacobs Jr.** [1] **and Ellen W. Demerath** [1]

[1]   Division of Epidemiology and Community Health, University of Minnesota, Minneapolis, MN 55454, USA;
      jacob.l.haapala@healthpartners.com (J.L.H.); fost0112@umn.edu (L.P.F.); harna001@umn.edu (L.J.H.);
      jacob004@umn.edu (D.R.J.J.); ewd@umn.edu (E.W.D.)

[2]   Department of Pediatrics, University of Oklahoma Health Sciences Center, Oklahoma City, OK 73104, USA;
      Katy-Duncan@ouhsc.edu (K.M.D.); April-Teague@ouhsc.edu (A.M.T.); David-Fields@ouhsc.edu (D.A.F.)

[3]   HealthPartners Institute, Minneapolis, MN 55425, USA; elyse.o.kharbanda@healthpartners.com

[4]   Division of Environmental Health Sciences, University of Minnesota, Minneapolis, MN 55455, USA;
      pmcg@umn.edu

[5]   Department of Health and Human Physiology, University of Iowa, Iowa City, IA 52242, USA;
      kara-whitaker@uiowa.edu

[6]   Department of Epidemiology, University of Iowa, Iowa City, IA 52242, USA

[7]   Division of Nutritional Sciences, Cornell University, Ithaca, NY 14853, USA; kathleen.rasmussen@cornell.edu

*   Correspondence: mtahir@umn.edu; Tel.: +1-512-960-5035

Received: 19 March 2019; Accepted: 23 April 2019; Published: 25 April 2019

Abstract: Full breastfeeding (FBF) is promoted as effective for losing pregnancy weight during the postpartum period. This study evaluated whether longer FBF is associated with lower maternal postpartum weight retention (PPWR) as compared to a shorter FBF duration. The MILK (Mothers and Infants Linked for Healthy Growth) study is an ongoing prospective cohort of 370 mother–infant dyads, all of whom fully breastfed their infants for at least 1 month. Breastfeeding status was subsequently self-reported by mothers at 3 and 6 months postpartum. Maternal PPWR was calculated as maternal weight measured at 1, 3, and 6 months postpartum minus maternal prepregnancy weight. Using linear mixed effects models, by 6 months postpartum, adjusted means ± standard errors for weight retention among mothers who fully breastfed for 1–3 (3.40 ± 1.16 kg), 3–6 (1.41 ± 0.69 kg), and ≥6 months (0.97 ± 0.32 kg) were estimated. Compared to mothers who reported FBF for 1–3 months, those who reported FBF for 3–6 months and ≥6 months both had lower PPWR over the period from 1 to 6 months postpartum ($p = 0.04$ and $p < 0.01$, respectively). However, PPWR from 3 to 6 months was not significantly different among those who reported FBF for 3–6 versus ≥6 months ($p > 0.05$). Interventions to promote FBF past 3 months may increase the likelihood of postpartum return to prepregnancy weight.

Keywords: full breastfeeding; postpartum; weight retention; obesity

---

1. Introduction

Pregnancy is a period of rapid weight gain and change in body composition [1] as maternal metabolism accommodates the demands of a growing fetus [2]. Gestational weight gain within the recommended ranges based on prepregnancy body mass index (BMI) is important for optimal fetal development [1,3] and accretion of fat depots necessary to support the energy cost of lactation [2]. Unfortunately, 47% of women in the United States have excessive gestational weight gain [4] and

13–20% fail to return to their prepregnancy weight after delivery, weighing approximately 5 kg more at 6–18 months postpartum compared to prepregnancy [5,6]. Excessive postpartum weight retention (PPWR) may in turn initiate a vicious cycle of obesity among women during their reproductive years and contribute to future unhealthy pregnancies, diabetes, and cardiovascular disease [1].

Many factors influence variation in PPWR, including maternal prepregnancy BMI, gestational weight gain, parity, age, race/ethnicity, education, diet, physical activity, and breastfeeding duration [7–10]. Breastfeeding is of particular interest due to its established physical and psychological benefits for both mother and infant during the postpartum period and beyond [11–13]. Prolonged breastfeeding may theoretically promote postpartum weight loss owing to the energy expenditure requirements of lactation [14] and mobilization of pregnancy-related accumulated fat stores [2].

To date, studies examining the association between breastfeeding and PPWR or postpartum weight loss have yielded inconclusive findings [15–17]. In a recent systematic review, Neville et al. concluded that, compared to other forms of infant feeding, there was insufficient evidence indicating that breastfeeding is directly associated with postpartum weight change [18]. Of particular concern is the issue of confounding, whereby women who breastfeed longer are more likely to have intended to breastfeed [19], have greater social support [11], and have lower prepregnancy BMI [20] and gestational weight gain [21] than women who breastfeed for short durations, never breastfeed or breastfeed only partially. All these factors are directly associated with PPWR [7–10]. Most studies included in this comprehensive review had relatively few women who were exclusively breastfeeding to 6 months as recommended, and/or did not adjust for important potentially confounding variables. They also often relied on self-reported maternal weight and lacked repeated measurements to assess patterns of weight retention over time, both of which decrease precision. Additional studies that can characterize the relationship between longer versus shorter full breastfeeding (FBF) and PPWR, including in women who are strongly committed to and initiate full breastfeeding, are therefore warranted.

In this exploratory study, we aimed to examine the association between FBF duration and maternal PPWR from 1 to 6 months postpartum using repeated measures of FBF and objectively measured maternal anthropometry within a cohort of healthy, predominantly breastfeeding women. We hypothesized that, compared to mothers who fully breastfed for 1–3 months and 3–6 months, those who fully breastfed for ≥6 months would exhibit lower PPWR from 1 to 6 months, and from 3 to 6 months, respectively.

2. Materials and Methods

2.1. Study Population

The Mothers and Infants Linked for Healthy Growth (MILK) study is an ongoing prospective cohort of mother–infant dyads recruited from Minneapolis, MN and Oklahoma City, OK [22]. Women were included in the study if they: were 21–45 years of age at delivery; had a prepregnancy BMI of 18.5–40.0 kg/m^2; had a healthy singleton pregnancy (i.e., spent <3 days in the hospital post-delivery for vaginal deliveries and <5 days for caesarean section deliveries); delivered an infant at-term with a birthweight of ≥2500 g but ≤4500 g; and reported an intention to breastfeed exclusively for at least 3 months. Women were excluded from the study if they consumed tobacco or >1 alcoholic drink per week during pregnancy/lactation, had a history of Type 1 or Type II diabetes or current diagnosis of gestational diabetes, were unable to speak or understand English, or if the infant had a known congenital illness affecting feeding and/or growth.

Of the 370 mother–infant dyads enrolled into the MILK study (all fully breastfeeding at 1 month postpartum), 32 were excluded due to missing information on breastfeeding status at 3 or 6 months or maternal prepregnancy weight. The final analytic sample size included 338 mother–infant dyads with complete information on FBF duration, maternal prepregnancy weight, and maternal weight on at least one of the follow-up time points (i.e., at 1, 3 or 6 months postpartum).

Written informed consent was obtained at baseline, and participants were compensated for completion of each visit. All study protocols were approved by the institutional review boards at the University of Minnesota, HealthPartners Institute, and the University of Oklahoma Health Sciences Center.

2.2. Breastfeeding Status

Maternal breastfeeding status was ascertained at the 1-, 3-, and 6-month study visits. Mothers were asked to report their breastfeeding habits, which were further classified as FBF, mixed-feeding or fully formula feeding. Given our inability to discern "exclusive" (no other solid foods/liquids) from "almost exclusive" breastfeeding (water, vitamins and minerals occasionally provided), we defined FBF as maintaining breastfeeding with <24 oz of formula during the entire time period, and only breast milk for the 2 weeks prior to the visit. Mixed feeding was defined as providing infants with >24 oz of formula for each time period but also some breast milk. Fully formula feeding was defined as providing infants only formula for each time period. We categorized the duration of FBF in terms of the number of completed months of FBF as follows: 1–3 months, 3–6 months, and ≥6 months. The frequencies of various barriers to lactation were queried at the 1-month study visit.

2.3. Maternal Anthropometry

Maternal prepregnancy weight (kg) was measured within 6 weeks of conception and abstracted from electronic medical records. Maternal weight was then measured at the 1-, 3-, and 6-month study visits using calibrated digital scales. Standard procedures were followed after cross-training of staff [23]. PPWR was calculated and defined as maternal weight at the 1-, 3-, and 6-month visits minus maternal prepregnancy weight.

2.4. Covariates

Maternal educational attainment (high school/GED/Associates degree, Bachelor's degree, graduate degree), race/ethnicity (white, other), household income (<$60,000, $60,000–$90,000, >$90,000), physical activity at 3 months postpartum (meets/does not meet moderate-to-vigorous physical activity (MVPA) guidelines of 150 min/week) [24], and frequency of feeds at 1, 3, and 6 months postpartum (≤6, >6 times/day) were self-reported by mothers. Dietary intake was self-reported by mothers at 1 month postpartum using a modified version of the Diet History Questionnaire II to reflect intake during the past month [25]. Maternal age (years), parity (0, 1, ≥2), delivery mode (vaginal, caesarean section), infant birthweight (grams), and infant sex (male, female) were abstracted from the mother's electronic medical records. Maternal gestational weight gain (kg) was calculated by subtracting maternal prepregnancy weight from weight at delivery (measured within 2 weeks of birth and abstracted from medical records).

2.5. Statistical Analyses

Characteristics of the mother–infant dyads were described using raw means ± standard deviations (SD) and raw frequencies stratified by FBF duration. Chi-square tests and one-way ANOVAs were used to compare participant characteristics by FBF duration for categorical and continuous variables, respectively. The purpose of these statistical tests was to identify potential confounding variables. One inclusion criterion for the study included an intention to fully breastfeed for at least 3 months and the presence of breastfeeding support; however, $n = 23$ mothers stopped breastfeeding before 3 months. As such, we examined differences in frequencies of barriers to lactation between women who fully breastfed to 1–3 months and those who fully breastfed for ≥3 months using chi-square tests.

Linear mixed effects models (PROC MIXED) were then used to test the association of FBF duration with repeated within-subject measures of maternal PPWR at 1, 3, and 6 months postpartum using an unstructured covariance matrix. These models can accommodate unbalanced intervals of measurement, time-dependent and independent exposures/covariates, and provide greater statistical power with serial measurements [26]. We first examined the crude association of FBF duration with maternal PPWR. The model was then adjusted for maternal education, race/ethnicity, household income, MVPA at 3 months postpartum, frequency of feeds (time-varying), age, parity, delivery mode, gestational weight gain, and infant birthweight and sex. The main exposure (FBF duration) was included as a main effect and as an interaction with time (time-varying). All covariates were included as main effects, but we only retained an interaction with time if the corresponding estimate for the interaction had a *p*-value < 0.05. Since maternal prepregnancy weight was part of the outcome variable, it was not included as a potential confounder in adjusted analyses. However, in sensitivity analyses, we adjusted for maternal prepregnancy BMI in final models to assess whether any observed associations persisted after accounting for prepregnancy weight status. In another sensitivity analysis, we included maternal Healthy Eating Index-2015 (HEI-2015) total scores and energy intakes at 1 month postpartum as covariates in the final models. Dietary intake was not included in the main analysis owing to the relatively large number of missing data (approximately 13%). Lastly, we tested whether similar findings were observed in main analyses when maternal postpartum weight loss (defined as maternal weight at 1, 3, and 6 months minus maternal weight at delivery) was utilized as an outcome, rather than PPWR.

Because longitudinal mixed effects regression model estimates can be difficult to interpret in the presence of time interactions, we derived and plotted the adjusted means ± SEs for PPWR from the linear mixed model analysis specified above for each FBF group and each time point. Next, a difference-of-differences analysis was conducted, such that the change in adjusted means of weight retention over time (slopes) within each FBF duration category (e.g., PPWR at 1 month minus PPWR at 6 months) were compared across FBF duration categories. Specifically, differences in PPWR from 1 to 6 months were compared for mothers who fully breastfed for 3–6 and ≥6 versus 1–3 months and differences in PPWR from 3 to 6 months were compared for mothers who fully breastfed for 3–6 versus ≥6 months, with corresponding *p*-values derived from *t*-tests presented. Statistical analyses were conducted using SAS, version 9.4 (SAS Institute, Inc., Cary, NC, USA).

3. Results

3.1. Participant Characteristics

Mothers included in the analyses were predominantly white, highly educated, and multiparous; most mothers delivered vaginally, met MVPA guidelines at 3 months postpartum, and fully breastfed for ≥6 months (Table 1). By the 6-month visit, less than half of the women who stopped FBF at 1–3 months were partially breastfeeding (48% were mixed feeding), while 73% of the women who continued to fully breastfeed for 3–6 months were still partially breastfeeding (mixed feeding). Maternal age, race, weight retention at 6 months postpartum, frequency of feeds at 3 and 6 months postpartum, and infant birthweight differed significantly by FBF duration, such that mothers who fully breastfed for longer durations were more likely to be older, white, retain less weight, feed more frequently at 3 and 6 months, and give birth to heavier infants (*p* < 0.05). There were no significant differences in delayed lactogenesis, milk flow, milk supply, sore, cracked or bleeding nipples, breast engorgement, breast yeast infection, clogged milk ducts, infected or abscessed breasts or breast milk leakage between mothers who fully breastfed for 1–3 months versus ≥3 months (all *p* > 0.20; Table S1).

Table 1. Demographic, reproductive, and lifestyle characteristics of mother–infant dyads by full breastfeeding duration (*n* = 338).

Participant Characteristics	Total	Full Breastfeeding Duration			
		1–3 months (*n* = 23)	3–6 months (*n* = 63)	≥6-Months (*n* = 252)	
		n (%) or mean ± SD			*p*-Value
Maternal Age, years	30.8 ± 4.1	28.7 ± 4.9	30.7 ± 4.1	31.0 ± 4.0	0.03 *

Table 1. *Cont.*

Participant Characteristics	Full Breastfeeding Duration				
	Total	1–3 months (*n* = 23)	3–6 months (*n* = 63)	≥6-Months (*n* = 252)	
	n (%) or mean ± SD				*p*-Value
Maternal Race					
White	288 (87)	14 (64)	51 (82)	223 (90)	
Other	44 (13)	8 (36)	11 (18)	25 (10)	<0.01 *
Maternal Education					
High school /GED/Associates degree	74 (23)	10 (44)	11 (19)	53 (22)	
Bachelor's degree	135 (41)	9 (39)	27 (47)	99 (40)	
Graduate degree	117 (36)	4 (17)	19 (33)	94 (38)	0.09
Household income					
<$60,000	99 (30)	11 (48)	22 (39)	66 (27)	
$60,000–90,000	80 (25)	4 (17)	15 (26)	61 (25)	
>90,000	147 (45)	8 (35)	20 (35)	119 (48)	0.11
Parity					
0	141 (42)	13 (59)	21 (35)	107 (43)	
1	128 (39)	4 (18)	29 (48)	95 (38)	
≥2	63 (19)	5 (23)	10 (17)	48 (19)	0.17
Delivery Mode					
Vaginal	264 (80)	17 (77)	50 (81)	197 (79)	0.94
Caesarean section	68 (20)	5 (23)	12 (19)	51 (21)	
Prepregnancy BMI, kg/m^2	26.5 ± 5.6	28.1 ± 5.3	26.9 ± 6.9	26.2 ± 5.2	0.24
Gestational weight gain (kg)	12.3 ± 6.6	12.2 ± 8.8	11.7 ± 7.0	12.5 ± 6.3	0.69
Postpartum weight retention (kg)					
1-month	3.9 ± 5.4	4.2 ± 7.2	3.2 ± 6.5	4.0 ± 5.0	0.56
3-months	2.8 ± 5.5	4.6 ± 7.9	1.9 ± 6.4	2.8 ± 4.9	0.13
6-months	1.4 ± 5.9	4.3 ± 7.8	0.9 ± 6.4	1.2 ± 5.5	0.05 *
HEI-2015 score at 1-month postpartum	65.6 ± 8.9	64.8 ± 9.6	63.7 ± 6.3	66.4 ± 8.9	0.16
Energy intake at 1-month postpartum (kcal)	1952 ± 761	1666 ± 770	1889 ± 633	1975 ± 666	0.14
Meets MVPA guidelines at 3-months postpartum (yes)	178 (54)	13 (62)	25 (40)	140 (57)	0.06
Frequency of feeds (per day)					
1-month	9.7 ± 2.1	9.4 ± 2.3	9.4 ± 2.0	9.8 ± 2.1	0.44
3-months	7.6 ± 2.1	4.6 ± 2.1	7.8 ± 2.0	7.7 ± 1.9	<0.01 *
6-months	6.8 ± 2.1	3.6 ± 1.8	6.0 ± 2.8	7.1 ± 1.7	<0.01 *
Breastfeeding status at 6 months postpartum					
Fully breastfeeding	252 (74)			252 (100)	<0.01 *
Mixed-feeding	57 (17)	11 (48)	46 (73)		
Fully formula feeding	29 (9)	12 (52)	17 (27)		
Infant birthweight, (g)	3534 ± 445	3329 ± 482	3620 ± 425	3530 ± 442	0.03 *
Infant sex					
Male	172 (51)	10 (43)	38 (60)	124 (49)	0.22
Female	166 (49)	13 (57)	25 (40)	128 (51)	

Abbreviations: BMI = body mass index; HEI-2015 = Healthy Eating Index-2015; meeting MVPA guidelines: Moderate to vigorous physical activity >150 min/week. * *p* < 0.05 for tests of differences in participant characteristics by full breastfeeding duration using chi-square or one-way ANOVA for categorical and continuous variables, respectively.

3.2. Full Breastfeeding Duration and Postpartum Weight Retention

A significant interaction between FBF duration and time was observed, suggesting between-group differences in PPWR over time (Table 2). To illustrate the interaction effect of FBF and time on PPWR,

Figure 1 depicts differences in adjusted means ± SE of maternal PPWR from 1 to 6 months postpartum by FBF duration. By 6 months postpartum, mothers who fully breastfed for 1–3, 3–6, and ≥6 months had adjusted means ± SEs for PPWR of 3.40 ± 1.16 kg, 1.41 ± 0.69 kg, and 0.97 ± 0.32 kg. Compared to mothers who fully breastfed for 1–3 months only, those who fully breastfed for 3–6 and ≥6 months retained significantly less weight from 1 to 6 months postpartum ($p = 0.04$ and $p < 0.01$, respectively). Although mothers who fully breastfed for 6 months or longer weighed approximately 0.45 kg less by 6 months postpartum than those who fully breastfed for 3–6 months only, this difference in PPWR at 6 months and slope of change in PPWR from 3 to 6 months was not significant between these two groups ($p = 0.15$). These findings were independent of the statistical effects of numerous potential confounders included in the model. Significant covariate effects were observed for MVPA and maternal age, and time-dependent effects were observed for GWG, maternal education, parity, and mode of delivery (Table S2).

Table 2. Association of full breastfeeding duration with maternal postpartum weight retention measured from 1 to 6 months postpartum.

	Crude Model (*n* = 338)			Covariate-Adjusted Model [a] (*n* = 301)		
	β	95% CI	*p*-Value	β	95% CI	*p*-Value
FBF duration						
1–3 months	Ref	Ref		Ref	Ref	
3–6 months	−1.07	−3.67, 1.54		0.33	−1.55, 2.20	
>6 months	−0.24	−2.57, 2.08	0.18	0.49	−1.18, 2.16	0.53
Time [c]						
1–3 months	Ref.	Ref		Ref	Ref	
3 months	0.39	−0.68, 1.46		0.26	−1.34, 1.86	
6 months	0.04	−1.79, 1.86	<0.01	1.97	−0.66, 4.61	0.20
FBF duration x time [b,c]						
FBF 3–6 months at 3 months	−1.60	−2.86, −0.35		−1.60	−2.95, −0.25	
FBF 3–6 months at 6 months	−2.27	−4.41, −0.13		−2.32	−4.54, −0.09	
FBF ≥6 months at 3 months	−1.55	−2.67, −0.43	0.03	−1.66	−2.86, −0.45	0.04
FBF ≥6 months at 6 months	−2.76	−4.67, −0.85		−2.92	−4.90, −0.94	

Abbreviations: CI = confidence interval; FBF = full breastfeeding. [a] Adjusted for maternal education, race/ethnicity, household income, age, parity, delivery mode, prepregnancy weight, gestational weight gain, frequency of feeds at 1, 3 and 6 months (time-varying), physical activity level at 3 months postpartum, and infant birthweight and sex. [b] Reference category = FBF 1 month. [c] Reference category = time 1 month.

In sensitivity analyses, after including maternal prepregnancy BMI as a potential covariate in the final model examining the association between FBF duration and maternal PPWR, comparable findings were observed for adjusted means and differences in slopes of PPWR over time (data not shown). After additional adjustment for maternal total HEI-2015 scores and energy intake at 1 month postpartum, adjusted means were consistent with those found in the main analyses. In this sensitivity analysis, differences in slopes of change in weight retention from 1 to 6 months for women who fully breastfed to ≥6 versus 1–3 months were similar to those reported above for the main analysis. However, slopes of change in weight retention were no longer significantly greater among those who fully breastfed to 3–6 versus 1–3 months ($p < 0.10$) (data not shown). Lastly, after examining postpartum weight loss (versus PPWR) as an outcome, similar adjusted means and between-group differences in weight loss over time were observed (data not shown).

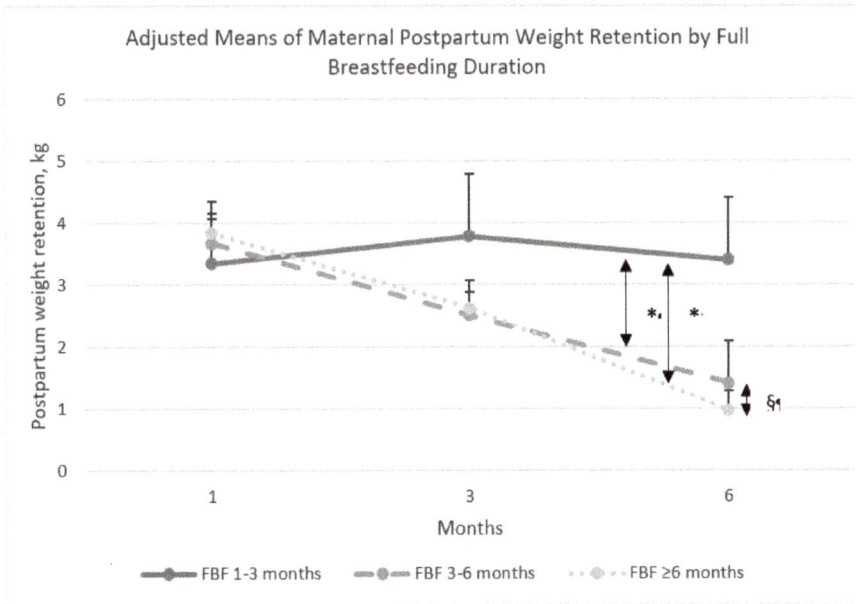

Figure 1. Adjusted means of maternal postpartum weight retention by full breastfeeding duration. Abbreviations: FBF = full breastfeeding; Adjusted for maternal education, race/ethnicity, household income, age, parity, delivery mode, prepregnancy weight, gestational weight gain, frequency of feeds at 1, 3, and 6 months (time-varying), physical activity level at 3 months postpartum, and infant birthweight and sex. * $p < 0.05$ for differences in slopes. § $p > 0.05$ for differences in slopes.

4. Discussion

In this ongoing prospective cohort of mother–infant dyads, we examined the association of FBF duration with maternal weight retention from 1 to 6 months postpartum. We found that, independent of demographic, reproductive, and lifestyle factors, mothers who fully breastfed for longer durations (i.e., greater than 3 months) had significantly greater reductions in PPWR from 1 to 6 months compared to those who fully breastfed to 1–3 months only. Mothers who fully breastfed to 6 months or longer retained approximately 0.45 kg less weight by 6 months postpartum than those who fully breastfed to 3 months only, but differences between these two groups were not statistically significant. The findings from this exploratory analysis suggest that, among healthy women strongly committed and supported to fully breastfeed, encouraging FBF beyond 3 months could reduce PPWR.

The American Academy of Pediatrics recommends exclusive breastfeeding to 6 months, followed by complementary feeding with continued breastfeeding until 1 year and beyond as mutually desired by mothers and their infants [27]. This practice promotes infant health but also has immediate (postpartum) and long-term health benefits for mothers [11–13]. Breastfeeding is hypothesized to mobilize pregnancy-related buildup of visceral and femoral fat stores as a source of energy for milk production [2]. With the concurrent rise in prolactin, lipogenesis is inhibited in peripheral adipose tissue and increases in the mammary glands [28]. Lipoprotein lipase activity in adipocytes (particularly in the femoral region) also surges, resulting in greater weight loss among breastfeeding mothers [29]. Research has demonstrated that maternal total energy expenditure increases by approximately 15–25% during lactation as the production of milk requires an additional 500 kilocalories per day on average [30]. If this excess energy expenditure is not offset by increases in energy intake and decreases in physical activity, the expectation is that mothers will lose weight.

Nonetheless, the 2018 CDC breastfeeding report card suggests that approximately 47% of women in the United States exclusively breastfeed up to 3 months, and only 25% continue to do so up to 6 months with variation between states [31]. A host of social and biological determinants contribute to early cessation of exclusive breastfeeding, including poor family and social support, embarrassment, employment, lactation problems, and perceived lack of sufficient milk supply [32]. Most mothers in the United States do not meet their breastfeeding goals [33], and it is unclear why 23 of the mothers in our highly motivated and supported cohort ceased FBF between 1 and 3 months despite intending to fully breastfeed for at least 3 months. There were no significant differences in early barriers to lactation observed between those with an FBF duration of 1–3 months versus ≥3 months. It is possible that mothers who stopped fully breastfeeding had to return to work, had fewer workplace accommodations for breastfeeding, did not want to pump milk after returning to work or had infants who lost interest in nursing or did not gain enough weight [33]. However, since none of these possible differences are likely to drive postpartum weight loss, they are unlikely to explain the differences we have reported.

Our results on the association between FBF duration and PPWR are in line with those of other studies [15,16,34]. In a cohort of 405 Brazilian women, mean weight retention was approximately 3.6 kg by 9 months postpartum, and longer breastfeeding duration was associated with lower PPWR from 0.5 to 9 months of follow-up [15]. Similarly, Janney et al. reported a mean weight retention of 3.9 kg at 6 months postpartum and lower weight retention from 0.5–18 months postpartum among 110 mothers in the United States who had a longer duration of full and partial breastfeeding [16]. Contrary to our findings, Mullaney et al. reported no association between exclusive breastfeeding and maternal weight change by 4 months postpartum among 470 Irish women enrolled in a prospective cohort study (mean retention at 4 months postpartum was approximately 1.7 kg) [17]. Few studies have had the opportunity to assess the association between prolonged FBF (i.e., up to 6 months) and PPWR due to the limited number of US women who continue to breastfeed beyond 3 months. In one study of over 25,000 mothers in the Danish National Birth Cohort, Baker et al. found that those who breastfed as recommended (i.e., exclusively to 6 months and to any degree for 12 months) had a greater reduction in PPWR at 6 months (regardless of prepregnancy BMI) and at 18 months (among women within the BMI range of 18.5–34.9 kg/m^2) [34].

Discrepancies in findings pertaining to the associations of FBF with PPWR are evident between studies [8,17,35]. Complications in comparisons across research may be attributable to differences in study design, sample sizes (and, accordingly, statistical power), definition and assessment of breastfeeding duration and intensity, time points at which outcomes were assessed and/or periods of collection that were collapsed into one variable, specific outcomes considered and methods used to measure them, quality of exposure and outcome data, duration of follow-up, and confounders included in analyses. It is important to note that our study population is relatively healthy, with 74% of women fully breastfeeding to 6 months (well above the national rates for exclusive breastfeeding) [28] and 55% meeting MVPA guidelines at 3 months postpartum. Among the 63 mothers who breastfed to at least 3 months but not 6 months or more, 73% continued to practice mixed feeding, that is, feeding their infants both breast milk and formula. If only some breastfeeding is required to achieve continued postpartum weight loss, this may explain the similarity in the rate of weight loss from 3 to 6 months in those who breastfed to ≥6 months as compared to only 3–6 months. Further, the mothers in our study are highly educated and had a lower prepregnancy BMI compared to pregnant women in the United States. Our findings of relatively low weight retention by 6 months postpartum even among those with 1–3 months of FBF (3.9 kg) and a nonsignificant difference in PPWR over time between mothers who fully breastfed for 3–6 months versus to 6 months or longer may be attributable to these characteristics.

Our findings can be interpreted within the context of the strengths and limitations of the study. A strength of our study is that the data were drawn from a contemporary, multisite, prospective cohort with serial measurements from 1 to 6 months postpartum, which provides greater statistical power and allows for descriptions of patterns of weight retention over time. Furthermore, we objectively measured weight at 1, 3, and 6 months postpartum, extracted prepregnancy weight from electronic medical records and used rigorous definitions of FBF in real time, which minimizes recall bias. Our study population is well characterized; thus, we were able to adjust for a range of clinical, sociodemographic, and lifestyle potential confounders obtained from detailed questionnaires or medical records, including key elements of energy balance, such as physical activity and dietary intake. A characteristic that distinguishes this cohort from others is that the primary aim of the study was to assess the lactational programming hypothesis. As such, we focused on maternal obesity and breastfeeding as main exposures and were able to adjust for gestational weight gain and frequency of feeds within a day in our primary analyses.

Our study also has important limitations. The MILK study was designed to address the relationship of maternal weight status to breast milk composition in fully breastfeeding women. We were therefore not able to assess the broader questions of whether PPWR differs among women who ever versus never breastfeed, or between women who partially breastfeed for different durations versus those who fully breastfeed. In addition, most of the women were fully breastfeeding for ≥6 months and the samples of women with shorter FBF durations were relatively small. As such, the precision of estimates for the 1–3 and 3–6 months FBF groups were lower than for the ≥6 months FBF group, and we had reduced statistical power to detect PPWR differences between the three groups. Replication of these findings is necessary in larger cohorts of women who initiate FBF but then cease FBF early to characterize the benefit of FBF in reducing PPWR among women who follow breastfeeding guidelines. The mother–infant dyads included in the cohort were primarily non-Hispanic white. Therefore, the generalizability of our results is limited for other race/ethnic groups [35]. We did not collect maternal anthropometry beyond 6 months and consequently could not examine whether the lower weight retention associated with FBF was maintained in the long-term [36]. We also did not have a sufficient sample size to stratify our analysis by maternal obesity [34]. This drawback is important, given the observed differences in initiation and maintenance of FBF [37] and in fat distribution during pregnancy among normal-weight, overweight, and obese women [38]. Maternal prepregnancy weight was measured and abstracted from medical records within 6 weeks of conception, which may lead to variation in assessment of true prepregnancy weight and introduce measurement error. Although we adjusted for numerous theoretical and empirical confounders in our analyses, there may still be residual confounding or confounding by unmeasured variables. Lastly, we only had measures of important energy balance variables (i.e., diet and physical activity) at select time points. These behaviors tend to correlate strongly over time, but this limitation may have resulted in residual confounding that could have biased the estimates somewhat.

5. Conclusions

The perinatal period is one of significant weight and body composition change for women [1]. Consequently, this period is important for preventing excessive gestational weight gain and PPWR, both of which may contribute to future metabolic disease risk. Findings from the present study correspond to some prior research showing that longer FBF duration may be associated with reduced maternal PPWR up to 6 months postpartum [15,16,34]. Coupled with data from other studies, it is evident that, among healthy, predominantly breastfeeding mothers, FBF for prolonged periods (at least 3 months) is associated with significantly less pregnancy weight retention during the postpartum period. FBF for at least 3 months is a potentially modifiable infant feeding choice that not only promotes infant health but also may help to prevent mothers from future development of obesity and cardiometabolic disorders. Studies that identify how to reduce obstacles to prolonged full breastfeeding are critical to advancing the health of both women and their children.

Supplementary Materials: The following are available online at http://www.mdpi.com/2072-6643/11/4/938/s1, Table S1. Differences in barriers to lactation among mothers who fully breastfed to one month only and those who fully breastfed for more than one month (*n* = 338); Table S2: Association of maternal and infant covariates with maternal postpartum weight retention from 1 to 6 months postpartum (*n* = 301).

Author Contributions: Conceptualization, D.A.F. and E.W.D.; Data curation, J.L.H.; Formal analysis, M.J.T., D.R.J.J. and E.W.D.; Funding acquisition, D.A.F.; Investigation, L.P.F., K.M.D., and A.M.T.; Supervision, E.W.D.; Writing—original draft, M.J.T.; Writing—review and editing, J.L.H., L.P.F., K.M.D., A.M.T., E.O.K., P.M.M., K.M.W., K.M.R., D.A.F., D.R.J.J., L.J.H., and E.W.D.

Funding: The research reported in this publication was supported by the National Institute of Child Health and Human Development (NICHD) of the National Institutes of Health under Award Number R01HD080444. MJT was supported by the National Heart, Lung, and Blood Institute of the National Institutes of Health under Award Number T32 HL007779. The content is solely the responsibility of the authors and does not necessarily represent the official views of the National Institutes of Health.

Acknowledgments: We would like to thank the mothers and their families for their interest, time, and dedication to the MILK study.

Conflicts of Interest: The authors declare no conflict of interest.

References

1. Rasmussen, K.M.; Yaktine, A.L. *Institute of Medicine (US) and National Research Council (US) Committee to Reexamine IOM Pregnancy Weight Guidelines*; National Academies Press (US): Washington, DC, USA, 2009.
2. Stuebe, A.M.; Rich-Edwards, J.W. The reset hypothesis: Lactation and maternal metabolism. *Am. J. Perinatol.* **2009**, *26*, 81–88. [CrossRef] [PubMed]
3. Rasmussen, K.M.; Catalano, P.M.; Yaktine, A.L. New guidelines for weight gain during pregnancy: What obstetrician/gynecologists should know. *Curr. Opin. Obstet. Gynecol.* **2009**, *21*, 521–526. [CrossRef] [PubMed]
4. Deputy, N.P.; Sharma, A.J.; Kim, S.Y. Gestational Weight Gain—United States, 2012 and 2013. *Morb. Mortal. Wkly. Rep.* **2015**, *64*, 1215–1220. [CrossRef] [PubMed]
5. Gunderson, E.P.; Abrams, B. Epidemiology of gestational weight gain and body weight changes after pregnancy. *Epidemiol. Rev.* **1999**, *21*, 261–275. [CrossRef] [PubMed]
6. Gunderson, E.P. Childbearing and obesity in women: Weight before, during, and after pregnancy. *Obstet. Gynecol. Clin. N. Am.* **2009**, *36*, 317–332.
7. Hollis, J.L.; Crozier, S.R.; Inskip, H.M.; Cooper, C.; Godfrey, K.M.; Harvey, N.C.; Collins, C.E.; Robinson, S.M. Modifiable risk factors of maternal postpartum weight retention: An analysis of their combined impact and potential opportunities for prevention. *Int. J. Obes.* **2017**, *41*, 1091–1098. [CrossRef] [PubMed]
8. Østbye, T.; Peterson, B.L.; Krause, K.M.; Swamy, G.K.; Lovelady, C.A. Predictors of postpartum weight change among overweight and obese women: Results from the Active Mothers Postpartum study. *J. Womens Health (Larchmt)* **2012**, *21*, 215–222. [CrossRef]
9. Endres, L.K.; Straub, H.; McKinney, C.; Plunkett, B.; Minkovitz, C.S.; Schetter, C.D.; Ramey, S.; Wang, C.; Hobel, C.; Raju, T.; et al. Postpartum weight retention risk factors and relationship to obesity at 1 year. *Obstet. Gynecol.* **2015**, *125*, 144–152.
10. Hill, B.; McPhie, S.; Skouteris, H. The Role of Parity in Gestational Weight Gain and Postpartum Weight Retention. *Womens Health Issues* **2016**, *26*, 123–129. [CrossRef]
11. Dieterich, C.M.; Felice, J.P.; O'Sullivan, E.; Rasmussen, K.M. Breastfeeding and health outcomes for the mother-infant dyad. *Pediatr. Clin. N. Am.* **2013**, *60*, 31–48. [CrossRef] [PubMed]
12. Binns, C.; Lee, M.; Low, W.Y. The Long-Term Public Health Benefits of Breastfeeding. *Asia Pac. J. Public Health* **2016**, *28*, 7–14. [CrossRef]
13. Shamir, R. The Benefits of Breast Feeding. In *Nestle Nutrition Institute Workshop Series*; Karger Publishers: Basel, Switzerland, 2016; Volume 86, pp. 67–76.
14. Institute of Medicine (US) Committee on Nutritional Status During Pregnancy and Lactation. *Nutrition During Lactation*; National Academies Press (US): Washington, DC, USA, 1991.
15. Kac, G.; Benício, M.H.; Velásquez-Meléndez, G.; Valente, J.G.; Struchiner, C.J. Breastfeeding and postpartum weight retention in a cohort of Brazilian women. *Am. J. Clin. Nutr.* **2004**, *79*, 487–493. [CrossRef]
16. Janney, C.A.; Zhang, D.; Sowers, M. Lactation and weight retention. *Am. J. Clin. Nutr.* **1997**, *66*, 1116–1124. [CrossRef] [PubMed]
17. Mullaney, L.; O'Higgins, A.C.; Cawley, S.; Kennedy, R.; McCartney, D.; Turner, M.J. Breast-feeding and postpartum maternal weight trajectories. *Public Health Nutr.* **2016**, *19*, 1397–1404. [CrossRef] [PubMed]

18. Neville, C.E.; McKinley, M.C.; Holmes, V.A.; Spence, D.; Woodside, J.V. The relationship between breastfeeding and postpartum weight change—A systematic review and critical evaluation. *Int. J. Obes.* **2014**, *38*, 577–590. [CrossRef] [PubMed]

19. Colaizy, T.T.; Saftlas, A.F.; Morriss, F.H. Maternal intention to breast-feed and breast-feeding outcomes in term and preterm infants: Pregnancy Risk Assessment Monitoring System (PRAMS), 2000–2003. *Public Health Nutr.* **2012**, *15*, 702–710. [CrossRef] [PubMed]

20. Wojcicki, J.M. Maternal prepregnancy body mass index and initiation and duration of breastfeeding: A review of the literature. *J. Womens Health (Larchmt)* **2011**, *20*, 341–347. [CrossRef] [PubMed]

21. Hilson, J.A.; Rasmussen, K.M.; Kjolhede, C.L. Excessive weight gain during pregnancy is associated with earlier termination of breast-feeding among White women. *J. Nutr.* **2006**, *136*, 140–146. [CrossRef] [PubMed]

22. Whitaker, K.M.; Marino, R.C.; Haapala, J.L.; Foster, L.; Smith, K.D.; Teague, A.M.; Jacobs, D.R.; Fontaine, P.L.; McGovern, P.M.; Schoenfuss, T.C.; et al. Associations of Maternal Weight Status Before, During, and After Pregnancy with Inflammatory Markers in Breast Milk. *Obesity* **2017**, *25*, 2092–2099. [CrossRef] [PubMed]

23. Lohman, T.M.R.; Roche, A.F. *Anthropometric Standardization Reference Manual*; Human Kinetics Books: Champaign, IL, USA, 2011.

24. Hayden-Wade, H.A.; Coleman, K.J.; Sallis, J.F.; Armstrong, C. Validation of the telephone and in-person interview versions of the 7-day PAR. *Med. Sci. Sports Exerc.* **2003**, *35*, 801–809. [CrossRef]

25. Subar, A.F.; Thompson, F.E.; Kipnis, V.; Midthune, D.; Hurwitz, P.; McNutt, S.; McIntosh, A.; Rosenfeld, S. Comparative validation of the Block, Willett, and National Cancer Institute food 489 frequency questionnaires: The Eating at America's Table Study. *Am. J. Epidemiol.* **2001**, *154*, 1089–1099. [CrossRef]

26. Gibbons, R.D.; Hedeker, D.; DuToit, S. Advances in analysis of longitudinal data. *Annu. Rev. Clin. Psychol.* **2010**, *6*, 79–107. [CrossRef] [PubMed]

27. Section on Breastfeeding. Breastfeeding and the use of human milk. *Pediatrics* **2012**, *129*, 827–841. [CrossRef] [PubMed]

28. Brandebourg, T.D.; Bown, J.L.; Ben-Jonathan, N. Prolactin upregulates its receptors and inhibits lipolysis and leptin release in male rat adipose tissue. *Biochem. Biophys. Res. Commun.* **2007**, *357*, 408–413. [CrossRef] [PubMed]

29. Rebuffé-Scrive, M.; Enk, L.; Crona, N.; Lönnroth, P.; Abrahamsson, L.; Smith, U.; Björntorp, P. Fat cell metabolism in different regions in women. *Effect of menstrual cycle, pregnancy, and lactation. J. Clin. Investig.* **1985**, *75*, 1973–1976.

30. Gunderson, E.P. Impact of breastfeeding on maternal metabolism: Implications for women with gestational diabetes. *Curr. Diabetes Rep.* **2014**, *14*, 460. [CrossRef] [PubMed]

31. Centers for Disease Control and Prevention. Breastfeeding Report Card-United States. 2018. Available online: https://www.cdc.gov/breastfeeding/pdf/2018breastfeedingreportcard.pdf (accessed on 8 December 2018).

32. U.S. Department of Health and Human Services. *The Surgeon General's Call to Action to Support Breastfeeding*; U.S. Department of Health and Human Services, Office of the Surgeon General: Washington, DC, USA, 2011.

33. Odom, E.C.; Li, R.; Scanlon, K.S.; Perrine, C.G.; Grummer-Strawn, L. Reasons for Earlier Than Desired Cessation of Breastfeeding. *Pediatrics* **2013**, *131*, e726–e732. [CrossRef]

34. Baker, J.L.; Gamborg, M.; Heitmann, B.L.; Lissner, L.; Sørensen, T.I.; Rasmussen, K.M. Breastfeeding reduces postpartum weight retention. *Am. J. Clin. Nutr.* **2008**, *88*, 1543–1551. [CrossRef]

35. Jones, K.M.; Power, M.L.; Queenan, J.T.; Schulkin, J. Racial and ethnic disparities in breastfeeding. *Breastfeed Med.* **2015**, *10*, 186–196. [CrossRef] [PubMed]

36. Jarlenski, M.P.; Bennett, W.L.; Bleich, S.N.; Barry, C.L.; Stuart, E.A. Effects of breastfeeding on postpartum weight loss among U.S. women. *Prev. Med.* **2014**, *69*, 146–150. [CrossRef]

Nutrients **2019**, *11*, 938

37. Bever Babendure, J.; Reifsnider, E.; Mendias, E.; Moramarco, M.W.; Davila, Y.R. Reduced breastfeeding rates among obese mothers: A review of contributing factors, clinical considerations and future directions. *Int. Breastfeed J.* **2015**, *10*, 21. [CrossRef] [PubMed]
38. Straughen, J.K.; Trudeau, S.; Misra, V.K. Changes in adipose tissue distribution during pregnancy in overweight and obese compared with normal weight women. *Nutr. Diabetes* **2013**, *3*, e84. [CrossRef] [PubMed]

nutrients

MDPI

Article

Differences in Maternal Immunoglobulins within Mother's Own Breast Milk and Donor Breast Milk and across Digestion in Preterm Infants

Veronique Demers-Mathieu [1], Robert K. Huston [2], Andi M. Markell [2], Elizabeth A. McCulley [2], Rachel L. Martin [2], Melinda Spooner [1] and David C. Dallas [1,*]

[1]   Nutrition Program, School of Biological and Population Health Sciences, College of Public Health and Human Sciences, Oregon State University, Corvallis, OR 97331, USA; Veronique.Demers-Mathieu@oregonstate.edu (V.D.-M.); spoonerm@oregonstate.edu (M.S.)

[2]   Department of Pediatrics, Randall Children's Hospital at Legacy Emanuel, Portland, OR 97227, USA; Robert_Huston@mednax.com (R.K.H.); amarkell@lhs.org (A.M.M.); emcculle@lhs.org (E.A.M.); rlmartin@lhs.org (R.L.M.)

*   Correspondence: Dave.Dallas@oregonstate.edu; Tel.: +541-737-1751

Received: 13 March 2019; Accepted: 22 April 2019; Published: 24 April 2019

Abstract: Maternal antibody transfer to the newborn provides essential support for the infant's naïve immune system. Preterm infants normally receive maternal antibodies through mother's own breast milk (MBM) or, when mothers are unable to provide all the milk required, donor breast milk (DBM). DBM is pasteurized and exposed to several freeze–thaw cycles, which could reduce intact antibody concentration and the antibody's resistance to digestion within the infant. Whether concentrations of antibodies in MBM and DBM differ and whether their survival across digestion in preterm infants differs remains unknown. Feed (MBM or DBM), gastric contents (MBM or DBM at 1-h post-ingestion) and stool samples (collected after a mix of MBM and DBM feeding) were collected from 20 preterm (26–36 weeks gestational age) mother–infant pairs at 8–9 and 21–22 days of postnatal age. Samples were analyzed via ELISA for the concentration of secretory IgA (SIgA), total IgA (SIgA/IgA), total IgM (SIgM/IgM) and IgG. Total IgA, SIgA, total IgM and IgG concentrations were 55.0%, 71.6%, 98.4% and 41.1% higher in MBM than in DBM, and were 49.8%, 32.7%, 73.9% and 39.7% higher in gastric contents when infants were fed with MBM than when infants were fed DBM, respectively. All maternal antibody isotypes present in breast milk were detected in the infant stools, of which IgA (not sIgA) was the most abundant.

Keywords: passive immunization; antibodies; lactation; prematurity; proteolysis; breast milk

1. Introduction

Mother's own breast milk (MBM) provides maternal exposure-specific antibodies that provide passive immune protection to the infant. These maternal milk antibodies include IgA, IgG and IgM isotypes, as well as the secretory forms of IgA and IgM [1]. The antibodies in MBM that are ingested by preterm infants are comprised of ~80% total IgA (~73% SIgA/27% IgA), ~15% total IgM and ~5% IgG [1]. The provision of maternal milk antibodies helps compensate for the infant's naïve immune system. Neonates have an immature intestinal immune system, as demonstrated by lower numbers of plasma cells (immune cells that can produce IgA, IgM and IgG) in colonic and rectal biopsies from term infants at 1–12 days postnatal age compared with those at 1 to 6 months postnatal age [2]. No study has demonstrated the specific time when the preterm infants are able to produce their own SIgA in the small intestine. Milk SIgA, which bind to and neutralize pathogens to prevent their adherence to epithelial cells and infection [3–6], provide important immune compensation.

Milk IgM and IgG may also play a role in infant intestinal mucosal defense. IgG can bind to viruses and prevent their attachment to the mucosal surface or trap pathogens when IgG binds to mucus [7]. However, IgG was less efficient than SIgA for altering attachment and trapping pathogens in mucin [7]. Though IgM-secreting cells have been identified in the infant gut, the role of IgM in infant mucosal immune defense remains unknown [2].

Preterm infants likely have an even less developed immune system than term infants. For example, from 1–28 weeks postnatal, preterm neonates (24–28 weeks of gestation) produce less diverse IgG antibodies (based on nucleotide sequences present in the variable region of the antibody gene as detected by RT-PCR) in their blood compared with term infants (36–42 weeks of gestation) [8]. As they are born early, preterm infants also miss some of the placenta–fetal IgG transfer that occurs for term infants [9]. For preterm infants, maternal milk antibodies may help compensate for the potentially lower secretion of antibodies, their loss of placenta–fetal IgG transfer time and perhaps lower immune function compared with term infants.

To neutralize pathogens in the preterm infant gut, maternal milk antibodies must survive digestive protease actions through the gastrointestinal tract to their site of action. Our recent studies demonstrated that milk total IgA concentration decreased by 60% from milk to the preterm infant stomach at 2-h post-ingestion, whereas total IgM and IgG were stable [1]. Two oral supplementation studies (in adults fed bovine colostrum SIgA/IgA, IgM and IgG [10] and in preterm infants fed serum IgA and IgG [11]) demonstrated that IgG and IgM survived intact to the stool, whereas SIgA/IgA did not. On the other hand, some studies have demonstrated that SIgA from MBM can survive to the infant stool and urine [12–14]. These oral supplementation studies did not determine the percentage of survival for SIgA, total IgA, total IgM and IgG to the infant stool. Moreover, measuring Igs in infant stool samples does not accurately represent the biological survival of Igs within the upper GI tract as they can be further degraded by colonic bacteria.

Preterm-delivering mothers often have difficulty making enough milk to feed their infants and often supplement with donor breast milk (DBM). Most mothers (72%) of very preterm infants (<27 week of gestational age (GA) are unable to provide all the MBM required, thus DBM is used to complete their diet [15]. Whether MBM and DBM differ in milk antibody concentrations remains unclear. DBM processing includes pooling milks from different mothers, Holder pasteurization (62.5 °C for 30 min) to inactivate viruses [16,17] and kill bacteria [18], and several freeze–thaw cycles. These factors could result in lower concentrations of maternal milk antibodies.

As the degree to which MBM and DBM antibody concentrations differ and how this affects the survival of antibodies during gastric digestion remains unknown, we examined these questions herein. Evidence of lower antibody concentrations in DBM could determine whether modification of the product or processing techniques are needed to improve infant health outcomes.

2. Materials and Methods

2.1. Participants and Sample Collection

2.1.1. Participants and Enrollment

This study was approved by the Institutional Review Boards of Legacy Health Systems and Oregon State University. Samples were collected from twenty premature-delivering mother–infant pairs ranging in GA at birth from 26 to 36 weeks (Table 1) in the NICU. Eligibility criteria included having an indwelling naso/orogastric feeding tube, bolus feeding (<60 min infusion tolerated), feeding volumes of at least 4 mL and mothers who could produce a volume of MBM adequate for one full-volume feed per day. Exclusion criteria included neonates with diagnoses that are incompatible with life, gastrointestinal system anomalies, major gastrointestinal surgery, severe genitourinary anomalies and significant metabolic or endocrine diseases.

Table 1. Demographics of preterm-delivering mother–infant pairs sampled for mother's own breast milk, gastric contents (1-h postprandial time) and stools (24-h post-feeding).

Demographics	Preterm-Delivering Mother Infant Pairs [1,2]
GA, week	30 ± 3 (26–36)
Postnatal age at mother's milk feeding, days	8.6 ± 0.1 and 21.4 ± 0.1
Postnatal age at donor breast milk feeding, days	8.4 ± 0.1 and 21.6 ± 0.1
Birth weight at birth, kg	1.5 ± 0.7 (0.7–3.6)
Weight gain velocity, g/kg/day [3]	11 ± 4
Length gain velocity, cm/week [3]	1.0 ± 0.3
Head circumference gain velocity, cm/week [3]	0.6 ± 0.3
Volume of feeding at 8–9 days of postnatal age, mL	16 ± 12 (2.5–38)
Volume of feeding at 21–22 days, of postnatal age, mL	28 ± 10 (14–49)
Infant sex	10 females; 10 males
Infant of a diabetic mother, *n*	3
SGA10, less than the 10th percentile [4], *n*	2
SGA3, less than the 3rd percentile [4], *n*	1
Intrauterine illicit drug exposure, *n*	2
C-section, *n*	16
Retinopathy of prematurity, *n*	0
Chronic lung disease, *n*	2
Gram-negative sepsis, *n*	1
Gram-positive sepsis or fungal sepsis	0
Late-onset sepsis	1
Antibiotics[5] received postnatally, *n*	5
Necrotizing enterocolitis	0
Gastrointestinal bleeding	0
Cardiopulmonary resuscitation	0
Infant death	0
Length of stay in NICU	47 ± 29 (10–109)

[1] Values are mean ± SD (range); [2] Number of paired milk and gastric and stool samples from 20 preterm infants was $n = 20$ at 8–9 days and $n = 16$ at 21–22 days of postnatal age; [3] Birth to discharge; [4] Small for gestational age (SGA) on the Fenton 2013 Growth chart; [5] Antibiotics were ampicillin/cefdinir.

2.1.2. Feeding and Sampling

In order to compare the concentration of immunoglobulins in DBM and MBM during preterm infant digestion, we gave two separate feedings of DBM and MBM without fortification rather than the typical feed consisting of a mixture of DBM and MBM with fortifier on days in which gastric sampling was accomplished. Milk and gastric samples (1–2 mL) were collected on 8–9 and 21–22 days of life. A separate sample of the donor milk (even though it was from 2 original pools) used to feed each infant was collected and used as biological replicates for all the comparisons between MBM and DBM. At both sample time periods, each infant received 2 of the normal 8 daily feedings as unfortified MBM or DBM on alternate days (randomized order). We randomized the order of feeding MBM and DBM to control for any potential effect of infant day of life on antibody digestion. The pool of DBM was acquired from two batches at Northwest Mother's Milk Bank. Three-liter batches were pasteurized and frozen in 50-mL doses so that only a small fraction was thawed for each infant feeding. The power analysis based on detection of differences in antibody concentrations between MBM and gastric samples from preterm infants in our previous study [1] indicated that at least 15 infants were required to compare DBM and MBM-fed infants, as, in the previous study, most antibody concentrations differed significantly between milk and stomach with this sample size.

Prior to feeding, any gastric residuals were removed by syringe via the feeding tube to remove contamination from the previous feeding. Feedings were prepared at the Randall Children's Hospital at Legacy Emanuel NICU using aseptic technique. Frozen MBM and DBM were thawed in Ameda Penguin warmers at 37 °C. Milk (either MBM or DBM) was fed to the infant via the nasogastric tube with a feeding pump set to deliver the entire bolus over 30–60 min. A 2-mL sample of the gastric fluid was collected 30 min after the completion of feed infusion. This sample collection timing was selected to match the gastric half-emptying time of premature infants to maximize time in the stomach as well as our ability to collect remaining contents [19]. As the mouth and esophagus do not contribute to proteolytic digestion, the use of nasogastric tubes (bypassing this) will likely not alter the results from an enteral feed taking orally. After collecting the 2-mL gastric sample, 2 mL of additional feed plus the additional volume recorded of gastric residue that was removed prior to the feed were provided to avoid any nutritional interruption. Stool (1 g) was collected within 48 h of the gastric sampling time point and was recovered from the diaper and scraped into a sterile jar. Stool sample collection was not specific to DBM/MBM and thus represents stools deriving from a mixture of DBM and MBM feeding. After collection of each sample type (feed, gastric and stool), samples were placed immediately on ice and stored at −80 °C in the NICU. Samples were then be transported on dry ice to Oregon State University for sample analysis.

2.1.3. Clinical Data Collection

Infant GA and postnatal age at mother's milk feeding and at donor breast milk feeding were collected (see Table 1).

2.2. Sample Preparation and ELISAs

Feed (MBM and DBM) and gastric samples were thawed at 4 °C, pH was determined and samples (1 mL) were centrifuged at 4000× *g* for 20 min at 4 °C. The infranate was collected, separated into aliquots (100 µL) and stored at −80 °C. Frozen stool samples (0.1 g) were diluted in 700 µL of phosphate-buffered saline pH 7.4 (Thermo Fisher Scientific, Waltham, MA, USA) with 0.05% Tween-20 (Bio-Rad Laboratories, Irvine, CA, USA) (PBST) and 3% fraction V bovine serum albumin solution (Innovative Research, Novi, MI, USA). Diluted stool samples were mixed by vortex for 2 min and then the vials were centrifuged at 4000× *g* for 20 min at 4 °C. The supernatant was collected, separated into aliquots (100 µL) and stored at −80 °C. Sample pH measurements were performed with an S220 SevenCompact pH/Ion meter (Mettler-Toledo) equipped with a combined sealed glass electrode.

The spectrophotometric ELISAs were measured with a microplate reader (Spectramax M2, Molecular Devices, Sunnyvale, CA, USA) with two replicates of blanks, standards and samples. SoftMax Pro 7.0 Microplate Data Analysis Software (Molecular Devices) was used to create a standard curve with a Four-Parameter Logistic curve fit. Clear flat-bottom Immuno 96-well plates MaxiSorp (Thermo Fisher Scientific) were coated with 100 µL of 1 µg/mL of capture antibodies: goat anti-human IgA alpha-chain for SIgA or total IgA (SIgA/IgA); rat anti-human IgM mu-chain for total IgM (SIgM/IgM) and goat anti-human IgG gamma-chain for IgG (Bio-Rad Laboratories). Plates were incubated overnight at 4 °C. After incubation, plates were washed 3 times with PBST and then 100 µL of blocking buffer (PBST with 3% of fraction V bovine serum albumin solution) was added in all wells for 1 h at room temperature. Standard samples were prepared using purified IgA from human colostrum (Sigma-Aldrich, St. Louis, MO, USA) for SIgA and total IgA, and purified IgM and IgG from human serum (Sigma-Aldrich). The standard curves were prepared using a dilution series of standard antibody in blocking buffer and the final concentration covered a range from 1 to 5000 ng/mL. Feed (MBM or DBM) and gastric samples were diluted 250× with blocking buffer for total IgM and IgG measurements and 500× for SIgA and total IgA measurements. Prediluted stool samples were diluted in 250× for total IgA, total IgM and IgG, and 20× for SIgA. For each step (addition of standards/samples and secondary antibodies at 1 µg/mL), washing and incubation for 1 h at room temperature were performed. To determine SIgA concentration, mouse anti-human IgA secretory-chain was added, the plate was washed and anti-mouse IgG:horseradish peroxidase (HRP) was added. To determine total IgA concentration, goat anti-human IgA alpha-chain:HRP was used. For total IgM, goat anti-human IgM mu-chain:HRP was used. For IgG, goat anti-human IgG gamma-chain: HRP was used (all antibodies from Bio-Rad). The substrate (1×, 100 µL), 3,3′,5,5′-tetramethylbenzidine (Thermo Fisher

Scientific), was added for 5 min at room temperature followed by addition of 50 µL of 2 N sulfuric acid to stop the coloration reaction. Optical density was measured at 450 nm.

The protein concentration of all samples was measured with the BCA protein assay (Thermo Fisher Scientific). Feed (MBM or DBM) and gastric samples were diluted 10× in the diluent provided in the kit, whereas dissolved stool samples were diluted 250×.

2.3. Pasteurization and Freezing/Thawing Effects on Mother's Milk Antibodies

To directly test the effect of pasteurization and freeze–thaw cycles on mother's milk antibodies, 500 mL of milk from one mother who delivered a term infant at 38 weeks of gestational age was pumped and collected in a sterile plastic bag at 12 days of postnatal age. Six 1-mL aliquots were centrifuged at 4000× g for 30 min at 4 °C and the infranates were collected (skim milk). Three of these skim milk samples were used for the control raw human milk (RHM) and 3 others were pasteurized at 62.5 °C for 30 min for pasteurized skimmed human milk (PSHM). Three other 1-mL aliquots of whole human milk were pasteurized without skimming (pasteurized whole human milk, PWHM). After pasteurization, the PWHM samples were centrifuged at 4000× g for 30 min at 4 °C to remove the lipid layer. Each sample was analyzed via ELISAs for SIgA, total IgA, total IgM and IgG.

To test the effect of freeze–thaw cycles, three 1-mL aliquots of whole raw milk were frozen at –20 °C, 3 aliquots were frozen at –80 °C and 3 aliquots were stored on ice for 1 h. After 1 h in the freezer, the samples were thawed rapidly at 37 °C. All samples were centrifuged at 4000× g for 30 min at 4 °C to remove the lipid layer and analyzed via ELISAs for SIgA, total IgA, total IgM and IgG.

2.4. Statistical Analyses

Student's t-tests were used to determine whether the measurements in samples differed between 8–9 and 21–22 days of postnatal age. As the concentrations of total IgA, SIgA, total IgM, IgG and pH did not differ between 8–9 and 21–22 days of postnatal age (see Table S1 for statistical analyses), these samples were combined to compare groups by type of feeding (MBM versus DBM) in milk and gastric samples within the same mother–infant pairs using Wilcoxon matched-pairs signed-rank test in GraphPad Prism software (version 7.03). All tests were nonparametric as some of the values did not pass the D'Agostino and Pearson normality test. Wilcoxon matched-pairs signed-rank test was also used to compare total protein concentration in samples at 8–9 and 21–22 days of postnatal age. Though Student's t-tests did not reveal differences between days 8–9 and 21–22 days of postnatal age, we performed Wilcoxon matched-pairs signed-rank test to compare antibody concentrations in MBM and DBM in milk and gastric samples at 8–9 days and 21–22 days separately, which are shown in the supplemental data (Figure S1). Though our study was not designed to have statistical power to compare infants by GA groups, we examined whether antibody concentrations in the samples differed by GA groups (26–27 week, 30–31 week and 35–36 week) using Wilcoxon matched-pairs signed-rank test, which is shown in the supplemental data (Figure S2). We also evaluated the effect of the antibiotics that infants received or not in gastric and stool samples from both feedings (MBM and DBM) using Student's t-tests, which is shown in the supplemental data (Table S2). One-way ANOVA followed by Dunnett's multiple comparisons test were performed to compare RHM with PWHM and PSHM as well as raw milk to frozen milks. Differences were designated significant at $p < 0.05$.

3. Results

3.1. Infant Demographics

The demographic details for the preterm-delivering mother–infant pairs are presented in Table 1.

3.2. pH

The pH values of MBM (average 7.27 ± 0.05) were significantly higher than of DBM (average 6.57 ± 0.02, $p < 0.001$, Figure 1A), but they did not differ in gastric contents when infants were fed MBM or DBM (average 5.2 ± 0.1, Figure 1A). Both MBM and DBM had higher pH values than their respective gastric samples ($p < 0.001$) (Figure 1A).

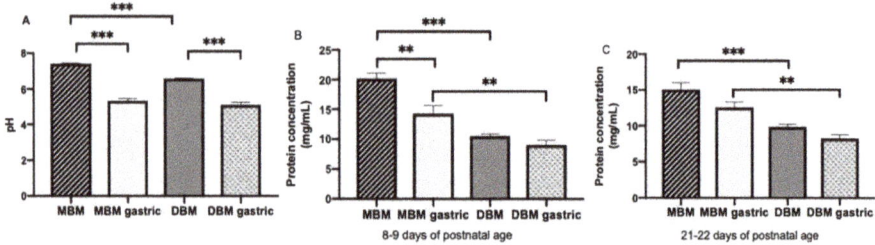

Figure 1. (**A**) pH values of milk and gastric contents at 1-h postprandial time from 20 preterm infants (26–36 weeks of gestational age (GA)) fed mother's own breast milk (MBM) and donor breast milk (DBM). Values are mean ± SEM, $n = 36$ for each group ($n = 20$ for 8–9 days and $n = 16$ for 21–22 days of postnatal age). (**B**) Total protein concentration in milk and gastric contents from infant fed MBM and DBM at 8–9 days ($n = 20$) (**C**) Total protein concentration in milk and gastric contents from infant fed MBM and DBM at 21–22 days ($n = 16$). Asterisks show statistically significant differences between variables (*** $p < 0.001$; ** $p < 0.01$) using the Wilcoxon matched-pairs signed-rank test.

3.3. Protein Concentration

Protein concentration in MBM collected at 8–9 days of life (average 20.2 ± 0.9 mg/mL) was 1.3-fold higher than that in MBM collected at 21–22 days of life (average 15.6 ± 0.9 mg/mL, $p < 0.001$). Protein concentration in the stomach (average 11.2 ± 0.6 mg/mL) did not differ between 8–9 and 21–22 days for MBM or DBM. Protein concentration in MBM was 1.9- and 1.6-fold higher than in DBM at 8–9 (average 10.2 ± 0.4 mg/mL) and 21–22 days (average 9.4 ± 0.6), respectively ($p < 0.001$, Figure 1B). Protein concentration in the stomach from preterm infants fed MBM (average 13.2 ± 0.7 mg/mL) was 1.6-fold higher than those fed DBM (average 9.4 ± 0.8 mg/mL, $p < 0.01$, Figure 1B). Protein concentration decreased 1.3-fold from MBM to gastric contents from infants fed MBM at 8–9 days ($p < 0.01$, Figure 1B) but did not differ at 21–22 days of postnatal age (Figure 1B). Protein concentration decreased 1.2-fold from DBM to gastric contents from infants fed DBM at 8–9 and 21–22 days postnatal age (combined across days, $p = 0.007$, Figure 1B).

3.4. Maternal Milk Antibody Concentrations

3.4.1. Antibody Concentrations in MBM and DBM

Total IgA, SIgA, total IgM and IgG concentrations were respectively 55.0, 71.6, 98.4 and 41.1% higher in MBM compared with DBM ($p < 0.001$, Figure 2A–D).

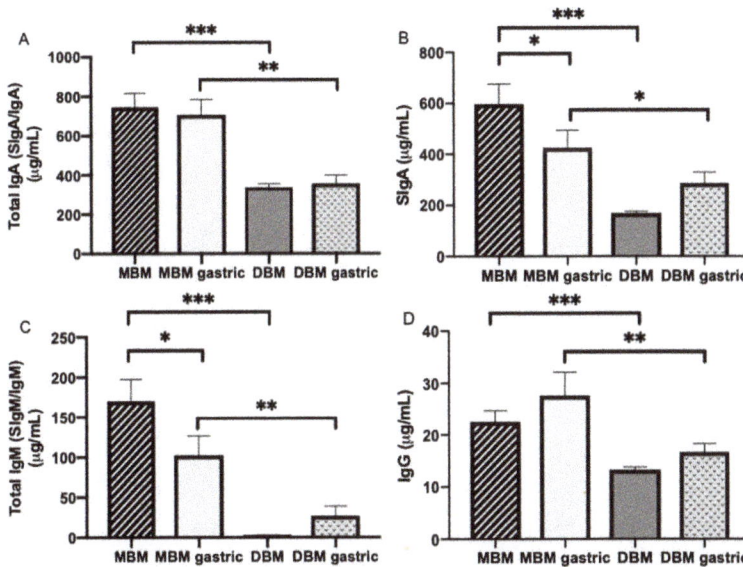

Figure 2. Immunoglobulin concentrations in milk and gastric contents at 1-h postprandial time from 20 preterm infants (26–36 weeks of gestational age (GA)) fed mother's own breast milk (MBM) and donor breast milk (DBM). Concentration of (**A**) total IgA (SIgA/IgA), (**B**) secretory IgA (SIgA), (**C**) total IgM (SIgM/IgM) and (**D**) IgG in milk and gastric samples. Values are mean ± SEM, $n = 36$ for MBM and DBM ($n = 20$ for 8–9 days and $n = 16$ for 21–22 days of postnatal age). Asterisks show statistically significant differences between variables (*** $p < 0.001$; ** $p < 0.01$; * $p < 0.05$) using the Wilcoxon matched-pairs signed-rank test.

3.4.2. Maternal Antibody Digestion

Total IgA, SIgA, total IgM and IgG concentrations were respectively 49.8, 32.7, 73.9% and 39.7% higher in gastric contents from infants fed MBM than infants fed DBM (Figure 2A–D).

SIgA and total IgM concentration significantly decreased (Figure 2B,C) 28.8 and 39.8%, respectively, from MBM to the preterm infant stomach but did not change for total IgA and IgG ($p > 0.05$, Figure 2A,D). Total IgA, SIgA, total IgM and IgG concentrations did not change from DBM to the preterm stomach (Figure 2A–D).

In MBM, the proportions of total IgA, total IgM and IgG were respectively 79.4, 18.1 and 2.4%. In the gastric contents after feeding with MBM, the proportions of total IgA, total IgM and IgG were respectively 84.4, 12.2 and 3.3% (Figure 3A). The proportion of total IgA that was SIgA decreased from 80% in MBM to 60% in the gastric samples, whereas this proportion increased from 51% in DBM to 81% in the gastric samples.

Total IgA, SIgA, total IgM and IgG were detected in stools from preterm infants (Figure 3). These antibodies could derive from the MBM and/or DBM feedings. Antibody concentration did not differ in stool between 8–9 and 21–22 days of postnatal age. The proportion of total IgA, total IgM and IgG was respectively 85.2, 12.7 and 2.1%, whereas SIgA was only 0.7% of total IgA (Figure 3). The proportion of SIgA in stool was much lower than that present in milk and the gastric contents.

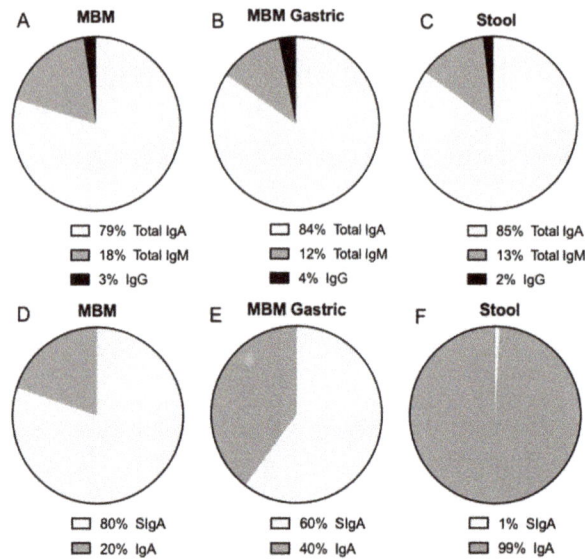

Figure 3. Proportion of total IgA, total IgM and IgG (**A**) in mother's own breast milk (MBM), (**B**) in gastric contents at 1-h postprandial time from preterm infants fed MBM and (**C**) in infant stools (from MBM, DBM and/or infant self). Proportion of SIgA and IgA from total IgA (**D**) in mother's own breast milk (MBM), (**E**) in gastric contents at 1-h postprandial time from preterm infants fed MBM and (**F**) in infant stools. Values are mean, $n = 36$ for MBM ($n = 20$ for 8–9 days and $n = 16$ for 21–22 days of postnatal age).

3.4.3. Pasteurization and Freeze–Thaw Effects

The concentration of total IgM decreased 62% from RHM to PWHM but those of SIgA, total IgA and IgG did not differ (Figure S3). Concentrations of total IgA, SIgA, total IgM and IgG decreased 37, 23, 87 and 54%, respectively from RHM to PSHM.

For all isotypes, the concentration did not differ between fresh milk and that exposed to freeze–thaw cycles at −20 or −80 °C ($p > 0.05$).

4. Discussion

The present study examined whether the concentration of maternal milk antibodies differed between MBM and DBM and across their digestion in the stomach of preterm infants. We also measured the survival of milk antibodies to the infant stool. For the first time, we demonstrated that the concentrations of all the Ig isotypes (total IgA, SIgA, total IgM and IgG) were higher in MBM than in DBM. The apparent lower antibody concentrations in DBM could be due to milk processing (including Holder pasteurization), the different gestational age at which mothers delivered and different postnatal day at which milks were expressed. Total IgM concentration was strongly reduced in both pasteurized whole milk and skim milk, whereas total IgA, SIgA and IgG concentrations were reduced in pasteurized skim milk but not in pasteurized whole milk. Our results that IgM was more sensitive to Holder pasteurization than SIgA agree with those of Ford et al. [20] who reported that SIgA from centrifuged MBM (pooled mature milk from term-delivering mothers) decreased 22% after Holder pasteurization, whereas IgM concentration had 100% loss (IgG was not measured). This observation was in accordance with our results in that the difference between MBM and DBM was the highest for IgM.

A previous study demonstrated that IgA and IgG concentrations were stable after freezing MBM at −20 °C for 3 months [21]. In contrast, Pardou et al. [22] showed that total IgA concentration in MBM from nursing mothers in the maternity ward and neonatal unit decreased 17.6% after freezing MBM at −20 °C for 4 days followed by thawing when contaminating bacteria were detected in milk via culture methods but not when those bacteria were absent. IgA proteases produced by some microbes (e.g., viridans streptococci and *Bacteroides* spp.) can cleave peptide bonds in the hinge region of the IgA1 heavy chain and then reduce their concentration in contaminated milk [22]. Our investigation showed that IgA, SIgA, IgM or IgG concentrations were maintained after freeze–thaw cycles at −20 and −80 °C in human milk. Therefore, the effect of freeze–thaw cycles is likely minimal if the milk is not contaminated but could reduce IgA1 concentration if the milk is contaminated. Before pasteurization, MBM is thawed at least twice and it is manipulated by the donors (pumping milk) when making DBM, which could lead to contamination with bacteria that produce these extracellular peptidases that could degrade proteins.

As a baseline validation, protein concentration was lower in DBM than in preterm milk, which validates the results of a previous study [23]. DBM is often donated from term-delivering mothers at late lactation time, whereas the MBM examined in this study was from preterm mothers at early lactation time, which may have higher antibody concentrations. We observed that the concentration of total IgA or total secretory component (free secretory component (SC)/SIgA/SIgM) did not differ between preterm milk and term milk from 6 to 28 days of postnatal age [1], and this observation was also demonstrated by another study [24]. However, Chandra et al. [25] found that IgA concentration in preterm milk was higher than in term milk from 3 to 15 days of postnatal age. Total IgA in colostrum (1–3 days of postnatal age) from preterm-delivering mothers was higher than in colostrum from term-delivering mothers in several studies [24–26]. Total IgA concentration in preterm milk decreased from 1 week to 2 months of postnatal age [1] and from colostrum to mature milk [24] but did not differ across time in term milk [1]. IgM and IgG did not differ across time postpartum in preterm and term milk [1]. IgM and IgG concentrations in preterm-delivering mothers were lower than [1], higher than [25] or similar to [27,28] that in term-delivering mothers. Therefore, the observed higher total IgA concentration in MBM compared with DBM could be due to the early lactation time of the preterm-delivering mothers. We observed a decrease in total IgM concentration with increasing GA, but no correlation between GA and total IgA, SIgA or IgG concentrations. The present study found a high variation (SD) in antibody concentration for all isotypes among mothers (possibly due to maternal background factors, such as vaccination schedule, diets, etc.). These variations could be responsible for the differing results between studies.

The proportion of SIgA and IgA (without SC) in milk from premature-delivering mothers as well as the slightly higher proportion of IgG compared with IgM are in agreement with our previous study where the proportion of antibody in preterm-delivering mothers was 82% for total IgA, 60% for total secretory component (SC/SIgA/SIgM), 6.9% for total IgM and 11.0% for IgG [1]. Percentages of SIgA in total IgA also matched the observations reported by Goldman et al. [29] that total SC (called "SIgA" by the authors but actually representing total SC, as an anti-SC primary antibody was used) concentration represented 90% of the total IgA in milk from term-delivering mothers. We also found a similar proportion of these isotypes in the gastric samples from infants fed MBM compared with the MBM samples (feeds).

For the first time, this investigation reported that total IgA, SIgA, total IgM and IgG concentrations were higher in the stomach from preterm infants fed MBM than those fed DBM. This higher concentration could be due to the initial higher concentration of maternal antibodies in MBM compared with DBM. SIgA and total IgM in MBM were partially digested in the stomach, total IgA and IgG in MBM were stable in gastric contents and none of the isotypes in DBM were digested. The lower apparent digestibility of antibodies in DBM could be due to the changes in the Ig structure after pasteurization. A previous study observed that pasteurization of breast milk enhanced the gastric digestion of lactoferrin and reduced that of alpha-lactalbumin [30]. Protein susceptibility to protease (pepsin or milk proteases) is influenced by the specific protein structure in the milk emulsion [31]. Pasteurization could change the organization of Igs and reduce the accessibility of cleavage sites to proteases (i.e., milk proteases or pepsin).

Gastric digestion reduced the amount of SIgA and total IgM from MBM but total IgA and IgG from MBM were not affected. No milk antibody isotype decreased significantly from DBM to the gastric contents. SIgA, IgM and IgG after DBM feeding appeared to be slightly, but not significantly, higher in the gastric samples. This observation could be due to contamination of the gastric contents with residual from previous feeds based on MBM.

A small degradation of SIgA from MBM in the preterm infant stomach was observed, which is similar to findings of our previous study where the decrease of total IgA was likely derived from preterm gastric digestion of partly IgA and partly SIgA [1]. IgM was digested but total IgA and IgG were stable in the present study, which differed from our previous findings that preterm infants partially degraded total IgA but not IgG and IgM in the stomach [32]. These differences could be due to the longer postprandial time (2-h post feed initiation) in the previous study [32] compared with the present study (30 min post feed completion). The reduction of SIgA and total IgM in gastric contents is due to the degradation by proteases (pepsin and/or milk proteases [33]) and not acid-induced denaturation in the stomach, as we previously demonstrated that standard IgA and IgM did not decrease in concentration in gastric acid conditions [1].

Antibodies detected in stool samples could derive from the MBM and/or DBM feeding or be potentially generated by the infant. The proportions of Igs from MBM and DBM in infant stool differed significantly from those in feed and gastric contents, especially the proportion of total IgA made up by SIgA. Unlike MBM and DBM samples in which most of the total IgA was SIgA, most of the total IgA in stool was IgA without the SC. IgM and IgG were detected in the stools at concentrations lower than total IgA. The lower proportion of SIgA compared to total IgA in the stool was likely due to release of the SC portion during gastrointestinal proteolytic digestion. A few studies have measured the survival of milk Igs to infant stools [11,34]. A greater reduction of IgA compared with IgG or IgM in preterm infant stools was previously observed [11]. These investigators fed preterm infants (1–28 days of postnatal age (GA unknown) 0.8–2 kg BW) infant formula plus pasteurized pooled breast milk supplemented with 600 mg daily of serum-derived human IgA (non-secretory form) (73%) and IgG (26%). The stool samples collected contained 1–10 mg IgG per g of dried feces (percentage reduction not calculated) and no IgA [11].

Five infants among the total of twenty received antibiotics (ampicillin/cefdinir) (Table 1). Provision of antibiotics could change the microbiome and alter the survival of milk antibodies to the stool. However, no differences were detected in the survival of Ig in the stomach and stool between infants with and without antibiotics (Table S2). We did observe that SIgA concentration tended to be higher in the gastric contents (after MBM feeding) from infants that did not receive antibiotics compared with those that received antibiotics.

The gastric pH was similar between infants fed MBM and those fed DBM even though the initial pH of MBM was higher than that of DBM. A previous study demonstrated that pasteurization of breast milk did not affect its pH [30], whereas storage at −20 °C and thawing reduced its pH [35], likely because of lipolytic release of free fatty acids [36]. The lower pH in DBM compared with MBM is likely due to repeated freeze–thaw cycles during milk processing.

One limitation of this study is the collection of stool samples within a 48-h window after collecting the milk and gastric samples. This window was selected for convenience to collect a stool sample around the time of milk and gastric sample collections. We could not determine whether stool samples derived from the MBM or DBM feeding, thus they represent a mixture of the total diet. Future studies could attempt to separate DBM and MBM-derived stools using indigestible markers, such as food colorants. Moreover, the immunoglobulins detected in the stool could also derive from the infant's own production. Though IgA-, IgG- and IgM-positive plasma cells are low in the lamina propria of the rectum and colon of term infants in the first month, they are indeed present and could contribute to the appearance of antibodies in the stool. In order to confirm that immunoglobulins in stool derive from feed (MBM or DBM) rather than the infant would require a protein labeling approach. The use of a single human milk sample to examine the effect of pasteurization and freeze–thaw cycles on the retention of antibodies could also represent a limitation. We used a single sample in order to focus on the effect of treatment; however, variations in biological samples could modify the treatment effects. Another limitation is that we did not have enough subjects to detect many differences between MBM and DBM feeds and gastric contents when grouped by GA (26–27 weeks, 30–31 weeks and 35–36 weeks) (Figure S2). Moreover, we did not have enough subjects to detect differences between days 8–9 and 21–22, though a larger sample size could have identified differences as many p-values for MBM were close to being significantly different (Table S1, Figure S1).

These findings indicate that the concentration of maternal antibodies in DBM may be lower than in MBM. If that is the case, preterm infants fed DBM may receive lower amounts of antibodies than infants fed MBM. Lower ingestion of milk antibodies could result in lower protection against infections in these vulnerable infants. Indeed, several studies may provide some evidence of lowered protection. A clinical study of 226 infants observed that preterm infants fed raw MBM tended to have lower infection rates than those fed pasteurized MBM (10.5% versus 14.3%) [37]. Another study of 243 preterm infants showed that infants fed MBM had fewer episodes of late-onset sepsis, necrotizing enterocolitis, total infection-related events, shorter hospital stay duration and had a lower percentage of blood samples that tested positive for Gram-negative bacteria than infants fed DBM [38].

5. Conclusions

The present study revealed that antibody concentrations (total IgA, SIgA, total IgM and IgG) were higher in MBM from preterm-delivering mothers than in DBM. Their concentrations in gastric contents were higher from infants fed MBM than those fed DBM, but Igs from DBM were less digested than Igs from MBM. All isotypes were detected in stools from preterm infants, but the proportion of Ig made up by SIgA decreased dramatically; most surviving Ig were IgA (without SC). The higher concentration of antibodies in MBM than in DBM may make MBM more effective in preventing enteric pathogens adhesion and invasion in newborns compared with DBM. This information could lead to changes in processing and collection practices to preserve these immune components in DBM.

Supplementary Materials: The following are available online at http://www.mdpi.com/2072-6643/11/4/920/s1, Table S1: Statistical results (*p*-values) for Student's *t*-tests to compare pH, protein concentration and antibody concentration between samples from 8–9 days (*n* = 20) and 21–22 days (*n* = 16) of postnatal age for each sample type (feed, gastric and stools) (separated days of postnatal age). MBM, mother's own breast milk; DBM, donor breast milk, Table S2: Statistical results (*p*-values) for Student's *t*-tests comparing antibody concentration between infants that received antibiotics (*n* = 5 at 8–9 days and *n* = 5 at 21–22 days of postnatal age) and infants that did not receive antibiotics (*n* = 15 at 8–9 days and *n* = 8 at 21–22 days) in gastric and stool samples (combined time of postnatal age). MBM, mother's own breast milk; DBM, donor breast milk, Figure S1: Immunoglobulin concentrations in milk and gastric contents at 1-h postprandial time from 20 preterm infants (26–36 weeks of gestational age (GA)) fed mother's own breast milk (MBM) and donor breast milk (DBM) at 8–9 days and 21–22 days of postnatal age. Concentration of (A) total IgA (SIgA/IgA), (B) secretory IgA (SIgA), (C) total IgM (SIgM/IgM) and (D) IgG in milk and gastric samples. Values are mean ± SEM, *n* = 20 for 8–9 days for MBM and DBM and *n* = 16 for 21–22 days of postnatal age for MBM and DBM. Asterisks show statistically significant differences between variables (***, *p* < 0.001; **, *p* < 0.01; *, *p* < 0.05) using the Wilcoxon matched-pairs signed-rank test, Figure S2: Immunoglobulin concentrations in milk and gastric contents at 1-h postprandial time from 3 gestational age (GA) groups of preterm infants (26–27 weeks of GA, 30–31 weeks of GA, 35–36 weeks of GA) fed mother's own breast milk (MBM) and donor breast milk (DBM). Concentration of (A) total IgA (SIgA/IgA), (B) secretory IgA (SIgA), (C) total IgM (SIgM/IgM) and (D) IgG in milk and gastric samples. Values are mean ± SEM, *n* = 8 for G 26–27 (*n* = 4 for 8–9 and *n* = 4 for 21–22 days of postnatal age for MBM and DBM); *n* = 8 for G 30–31 (*n* = 4 for 8–9 days and *n* = 4 for 21–22 days of postnatal age for MBM and DBM), *n* = 4 for G 35–36 (8–9 days of postnatal age). Asterisks show statistically significant differences between variables (*** *p* < 0.001; ** *p* < 0.01, * *p* < 0.05) using the Wilcoxon matched-pairs signed-rank test, Figure S3: Comparison of immunoglobulin concentration between raw human milk (RHM) and pasteurized whole human milk (PWHM) or pasteurized skimmed human milk (PSHM) (*n* = 3). Concentration of (A) total IgA (SIgA/IgA), (B) SIgA, (C) total IgM (SIgM/IgM) and (D) IgG in mother's milk. Milk samples were from one mother who delivered one term infants with 38 weeks of gestational age and 12 days of postnatal age. Asterisks show statistically significant differences between variables (***p < 0.001; *p < 0.05) using One-way ANOVA followed by Dunnett's multiple comparisons test.

Nutrients **2019**, *11*, 920

Author Contributions: V.D.-M. conducted the ELISA analyses, analyzed data and conducted the statistical analysis. R.K.H., A.M.M., E.A.M. and R.L.M. provided milk and gastric samples. D.C.D., V.D.-M., R.K.H., M.S. and A.M.M. designed the clinical study. V.D.-M. and D.C.D. designed the study and drafted the manuscript. V.D.M. and D.C.D. have primary responsibility for the final content.

Acknowledgments: This study was supported by the K99/R00 Pathway to Independence Career Award, Eunice Kennedy Shriver Institute of Child Health & Development of the National Institutes of Health (R00HD079561) (D.C.D) and The Gerber Foundation (2017-1586). We thank the Northwest Mother's Milk Bank for providing DBM for this study.

Conflicts of Interest: The authors declare no conflict of interest.

References

1. Demers-Mathieu, V.; Underwood, M.A.; Beverly, R.L.; Nielsen, S.D.; Dallas, D.C. Comparison of human milk immunoglobulin survival during gastric digestion between preterm and term infants. *Nutrients* **2018**, *10*, 631. [CrossRef]

2. Perkkiö, M.; Savilahti, E. Time of appearance of immunoglobulin-containing cells in the mucosa of the neonatal intestine. *Pediatr. Res.* **1980**, *14*, 953. [CrossRef]

3. Mantis, N.J.; Rol, N.; Corthésy, B. Secretory IgA's complex roles in immunity and mucosal homeostasis in the gut. *Mucosal Immunol.* **2011**, *4*, 603. [CrossRef]

4. Apter, F.M.; Lencer, W.; Finkelstein, R.A.; Mekalanos, J.J.; Neutra, M.R. Monoclonal immunoglobulin A antibodies directed against cholera toxin prevent the toxin-induced chloride secretory response and block toxin binding to intestinal epithelial cells in vitro. *Infect. Immun.* **1993**, *61*, 5271–5278.

5. Mantis, N.J.; McGuinness, C.R.; Sonuyi, O.; Edwards, G.; Farrant, S.A. Immunoglobulin A antibodies against ricin A and B subunits protect epithelial cells from ricin intoxication. *Infect. Immun.* **2006**, *74*, 3455–3462. [CrossRef] [PubMed]

6. Stubbe, H.; Berdoz, J.; Kraehenbuhl, J.P.; Corthésy, B. Polymeric IgA is superior to monomeric IgA and IgG carrying the same variable domain in preventing Clostridium difficile toxin A damaging of T84 monolayers. *J. Immunol.* **2000**, *164*, 1952–1960. [CrossRef]

7. Robert-Guroff, M. IgG surfaces as an important component in mucosal protection. *Nat. Med.* **2000**, *6*, 129. [CrossRef]

8. Zemlin, M.; Hoersch, G.; Zemlin, C.; Pohl-Schickinger, A.; Hummel, M.; Berek, C.; Maier, R.F.; Bauer, K. The postnatal maturation of the immunoglobulin heavy chain IgG repertoire in human preterm neonates is slower than in term neonates. *J. Immunol.* **2007**, *178*, 1180–1188. [CrossRef] [PubMed]

9. Malek, A.; Sager, R.; Kuhn, P.; Nicolaides, K.H.; Schneider, H. Evolution of maternofetal transport of immunoglobulins during human pregnancy. *Am. J. Reprod. Immunol.* **1996**, *36*, 248–255. [CrossRef] [PubMed]

10. Roos, N.; Mahe, S.; Benamouzig, R.; Sick, H.; Rautureau, J.; Tome, D. 15N-labeled immunoglobulins from bovine colostrum are partially resistant to digestion in human intestine. *J. Nutr.* **1995**, *125*, 1238–1244. [PubMed]

11. Eibl, M.M.; Wolf, H.M.; Fürnkranz, H.; Rosenkranz, A. Prevention of necrotizing enterocolitis in low-birth-weight infants by IgA–IgG feeding. *N. Engl. J. Med.* **1988**, *319*, 1–7. [CrossRef]

12. Bakker-Zierikzee, A.M.; Van Tol, E.A.F.; Kroes, H.; Alles, M.S.; Kok, F.J.; Bindels, J.G. Faecal SIgA secretion in infants fed on pre-or probiotic infant formula. *Pediatr. Allergy Immunol.* **2006**, *17*, 134–140. [CrossRef]

13. Schanler, R.J.; Goldblum, R.M.; Garza, C.; Goldman, A.S. Enhanced fecal excretion of selected immune factors in very low birth weight infants fed fortified human milk. *Pediatr. Res.* **1986**, *20*, 711. [CrossRef]

14. Goldblum, R.M.; Schanler, R.J.; Garza, C.; Goldman, A.S. Human milk feeding enhances the urinary excretion of immunologic factors in low birth weight infants. *Pediatr. Res.* **1989**, *25*, 184. [CrossRef] [PubMed]

15. Carroll, K.; Herrmann, K.R. The cost of using donor human milk in the NICU to achieve exclusively human milk feeding through 32 weeks postmenstrual age. *Breastfeed. Med.* **2013**, *8*, 286–290. [CrossRef]

16. Orloff, S.L.; Wallingford, J.C.; McDougal, J.S. Inactivation of human immunodeficiency virus type I in human milk: Effects of intrinsic factors in human milk and of pasteurization. *J. Hum. Lact.* **1993**, *9*, 13–17. [CrossRef] [PubMed]

17. Hamprecht, K.; Maschmann, J.; Müller, D.; Dietz, K.; Besenthal, I.; Goelz, R.; Middeldorp, J.M.; Speer, C.P.; Jahn, G. Cytomegalovirus (CMV) inactivation in breast milk: Reassessment of pasteurization and freeze-thawing. *Pediatr. Res.* **2004**, *56*, 529. [CrossRef]

18. Franklin, J.G. A comparison of the bactericidal efficiencies of laboratory Holder and HTST methods of milk pasteurization and the keeping qualities of the processed milks. *Int. J. Dairy Technol.* **1965**, *18*, 115–118. [CrossRef]

19. Bourlieu, C.; Ménard, O.; Bouzerzour, K.; Mandalari, G.; Macierzanka, A.; Mackie, A.R.; Dupont, D. Specificity of infant digestive conditions: Some clues for developing relevant in vitro models. *Cri. Rev. Food Sci. Nutr.* **2014**, *54*, 1427–1457. [CrossRef] [PubMed]

20. Ford, J.; Law, B.; Marshall, V.M.; Reiter, B. Influence of the heat treatment of human milk on some of its protective constituents. *J. Pediatr* **1977**, *90*, 29–35. [CrossRef]

21. Evans, T.J.; Ryley, H.; Neale, L.; Dodge, J.A.; Lewarne, V.M. Effect of storage and heat on antimicrobial proteins in human milk. *Arch. Dis. Child.* **1978**, *53*, 239–241. [CrossRef]

22. Pardou, A.; Serruys, E.; Mascart-Lemone, F.; Dramaix, M.; Vis, H.L. Human milk banking: Influence of storage processes and of bacterial contamination on some milk constituents. *Neonatology* **2018**, *65*, 302–309. [CrossRef]

23. John, A.; Sun, R.; Maillart, L.; Schaefer, A.; Spence, E.H.; Perrin, M.T. Macronutrient variability in human milk from donors to a milk bank: Implications for feeding preterm infants. *PLoS ONE* **2019**, *14*, e0210610. [CrossRef]

24. Ballabio, C.; Bertino, E.; Coscia, A.; Fabris, C.; Fuggetta, D.; Molfino, S.; Testa, T.; Sgarrella, M.; Sabatino, G.; Restani, P. Immunoglobulin-A profile in breast milk from mothers delivering full term and preterm infants. *Int. J. Immunopathol. Pharmacol.* **2007**, *20*, 119–128. [CrossRef]

25. Chandra, R.K. Immunoglobulin and protein levels in breast milk produced by mothers of preterm infants. *Nutr. Res.* **1982**, *2*, 27–30. [CrossRef]

26. Montagne, P.; Cuillière, M.L.; Molé, C.; Béné, M.C.; Faure, G. Immunological and nutritional composition of human milk in relation to prematurity and mothers' parity during the first 2 weeks of lactation. *J. Pediatr. Gastroenterol. Nutr.* **1999**, *29*, 75–80. [CrossRef]

27. Gross, S.J.; Buckley, R.H.; Wakil, S.S.; McAllister, D.C.; David, R.J.; Faix, R.G. Elevated IgA concentration in milk produced by mothers delivered of preterm infants. *J. Pediatr.* **1981**, *99*, 389–393. [CrossRef]

28. Koenig, Á.; Diniz, E.M.D.A.; Barbosa, S.F.C.; Vaz, F.A.C. Immunologic factors in human milk: The effects of gestational age and pasteurization. *J. Hum. Lact.* **2005**, *21*, 439–443. [CrossRef]

29. Goldman, A.S.; Garza, C.; Nichols, B.L.; Goldblum, R.M. Immunologic factors in human milk during the first year of lactation. *J. Pediatr.* **1982**, *100*, 563–567. [CrossRef]

30. De Oliveira, S.C.; Bellanger, A.; Ménard, O.; Pladys, P.; Le Gouar, Y.; Dirson, E.; Kroell, F.; Dupont, D.; Deglaire, A.; Bourlieu, C. Impact of human milk pasteurization on gastric digestion in preterm infants: A randomized controlled trial. *Am. J. Clin. Nutr.* **2017**, *105*, 379–390. [CrossRef]

31. Zhang, Q.; Cundiff, J.K.; Maria, S.D.; McMahon, R.J.; Wickham, M.S.J.; Faulks, R.M.; van Tol, E.A. Differential digestion of human milk proteins in a simulated stomach model. *J. Proteome Res.* **2013**, *13*, 1055–1064. [CrossRef] [PubMed]

32. Demers-Mathieu, V.; Underwood, M.A.; Beverly, R.L.; David, D.C. Survival of immunoglobulins from human milk to preterm infant gastric samples at 1, 2, and 3 hours postprandial. *Neonatology* **2018**, *114*, 242–250. [CrossRef]

33. Demers-Mathieu, V.; Nielsen, S.D.; Underwood, M.A.; Borghese, R.; Dallas, D.C. Changes in proteases, antiproteases and bioactive proteins from mother's breast milk to the premature infant stomach. *J. Pediatr. Gastroenterol. Nutr.* **2018**, *66*, 318–324. [CrossRef]

34. Blum, P.M.; Phelps, D.L.; Ank, B.J.; Krantman, H.J.; Stiehm, E.R. Survival of oral human immune serum globulin in the gastrointestinal tract of low birth weight infants. *Pediatr. Res.* **1981**, *15*, 1256–1260. [CrossRef]

35. Ogundele, M.O. Effects of storage on the physicochemical and antibacterial properties of human milk. *Brit. J. Biomed. Sci.* **2002**, *59*, 205–211. [CrossRef]

36. Bitman, J.; Wood, D.L.; Mehta, N.R.; Hamosh, P.; Hamosh, M. Lipolysis of triglycerides of human milk during storage at low temperatures: A note of caution. *J. Pediatr. Gastroenterol. Nutr.* **1983**, *2*, 521–524. [CrossRef] [PubMed]

37. Narayanan, I.; Murthy, N.S.; Prakash, K.; Gujral, V.V. Randomised controlled trial of effect of raw and Holder pasteurised human milk and of formula supplements on incidence of neonatal infection. *Lancet* **1984**, *2*, 1111–1113. [CrossRef]
38. Schanler, R.J.; Lau, C.; Hurst, N.M.; Smith, E.O.B. Randomized trial of donor human milk versus preterm formula as substitutes for mothers' own milk in the feeding of extremely premature infants. *Pediatrics* **2005**, *116*, 400–406. [CrossRef] [PubMed]

nutrients

MDPI

Article

Impact of Maternal Nutrition and Perinatal Factors on Breast Milk Composition after Premature Delivery

Jean-Michel Hascoët [1,2,*], Martine Chauvin [3], Christine Pierret [3], Sébastien Skweres [1], Louis-Dominique Van Egroo [4], Carole Rougé [5] and Patricia Franck [6]

[1]   Department of Neonatology, Maternite Regionale, CHRU Nancy, 54035 Nancy, France; sebastienskweres@hotmail.fr
[2]   DevAH, Lorraine University, 54500 Vandoeuvre les Nancy, France
[3]   Dietetic and Nutrition Unit, CHRU Nancy, 54035 Nancy, France; m.chauvin@chru-nancy.fr (M.C.); c.pierret@chru-nancy.fr (C.P.)
[4]   Saint Cloud Hospital, 92210 Paris, France; louisdo.vanegroo@club-internet.fr
[5]   Bledina Limonest, 69410 Champagne-au-Mont-d'Or, France; carole.rouge@danone.com
[6]   Biology Laboratory, CHRU Nancy, 54035 Nancy, France; p.franck@chru-nancy.fr
*   Correspondence: jean-michel.hascoet@univ-lorraine.fr; Tel.: +33-383-342-934

Received: 20 January 2019; Accepted: 6 February 2019; Published: 10 February 2019

Abstract: (1) Background: Premature infants require mothers' milk fortification to meet nutrition needs, but breast milk composition may be variable, leading to the risk of inadequate nutrition. We aimed at determining the factors influencing mothers' milk macronutrients. (2) Methods: Milk samples were analyzed for the first five weeks after premature delivery by infrared spectroscopy. Mothers' nutritional intake data were obtained during standardized interviews with dieticians, and then analyzed with reference software. (3) Results: The composition of 367 milk samples from 81 mothers was (median (range) g/100 mL): carbohydrates 6.8 (4.4–7.3), lipids 3.4 (1.3–6.4), proteins 1.3 (0.1–3.1). There was a relationship between milk composition and mothers' carbohydrates intake only ($r = 0.164$; $p < 0.01$). Postnatal age was correlated with milk proteins ($r = -0.505$; $p < 0.001$) and carbohydrates ($r = +0.202$, $p < 0.001$). Multiple linear regression analyses showed (coefficient) a relationship between milk proteins $r = 0.547$ and postnatal age (-0.028), carbohydrate intake ($+0.449$), and the absence of maturation (-0.066); associations were also found among milk lipids $r = 0.295$, carbohydrate intake ($+1.279$), and smoking (-0.557). Finally, there was a relationship among the concentration of milk carbohydrates $r = 0.266$, postnatal age ($+0.012$), and smoking (-0.167). (4) Conclusions: The variability of mothers' milk composition is differentially associated for each macronutrient with maternal carbohydrate intake, antenatal steroids, smoking, and postnatal age. Improvement in milk composition could be achieved by the modification of these related factors.

Keywords: maternal nutrition; breast milk; premature delivery; milk composition

1. Introduction

Premature infants require mothers' milk fortification to meet their nutrition needs [1]. This fortification is usually standardized using an assumed macronutrient milk composition [2]. However, studies have suggested that breast milk composition variability may be much wider than expected [3,4], leading to inadequate newborn nutrition. McLeod et al. performed a survey of protein and energy intake by milk analysis within the first 28 days of life in 63 infants born before 33 weeks' gestation to assess their effect on growth [3]. Their results show that breast milk composition vary for all of the macronutrients with median protein concentrations of 16.6 g/L ranging from 13.4 g/L to 27.6 g/L, and median caloric intake of 73.3 Kcal/100 mL ranging from 63 Kcal/100 mL to 93 Kcal/100 mL. Of note, actual protein intake was correlated with infants' growth. The authors concluded that preterm milk composition is very variable, and routine fortification using assumed averaged composition may result in inadequate nutrition with slower weight gain, as observed in their study [3].

Inadequate nutrition could indeed explain in part the postnatal growth restriction observed in many premature infants [5,6]. We aimed at determining factors, including mothers' nutritional intake, which may be associated and explain breast milk macronutrient variability after premature delivery before 34 weeks' gestational age.

2. Materials and Methods

This is an observational study using milk bank data. In our level III Maternity Hospital, mothers' own milk is pasteurized for premature infants. After pasteurization, we routinely analyze breast milk composition to verify the appropriateness of standardized fortification. For the purpose of the study, we collected data for all of the milk batches that each mother provided to the milk bank regardless of the time of the milk collection to evaluate the milk composition variation throughout the first five weeks of lactation. Each batch is a pool of one to three days, depending on the volume collected and frozen by each mother at home. Once a volume of at least 500 mL was collected, the mother would hand over the frozen batch to the milk bank for pasteurization. Hand-over occurred daily to twice a week.

Mothers who delivered a premature infant before 34 weeks' gestation at our unit were enrolled within five days after delivery when they announced their choice for breastfeeding. Mothers' dietary preference and nutritional intake data were obtained during personalized interviews with experienced dieticians. They used standardized questionnaires based on validated documents in the National French survey. The dieticians recalled dietary intake information from the two weeks prior to the interview. Perinatal data were prospectively collected at the time of the interview and from the mothers' file. We analyzed maternal diet and macronutrient intake from the recall data averaged per day with appropriate software (Geni® V7.0, Micro6, Villers les Nancy-F, France).

For milk analysis, we used the same protocol as described in the literature [7,8] and advised by the manufacturer. In short, mothers milk pools over one to three days were delivered frozen ($-20\,^{\circ}$C) to the milk bank, thawed, pasteurized, aliquoted in bottles of 50 mL, and stored at $-80\,^{\circ}$C. Two samples of one mL from two different bottles of each batch were analyzed after homogenization by ultrasound to compensate for milk thawing and verify the quality of the homogenization. We used infrared spectroscopy pre-calibrated against a chemical reference with analytical accuracy <0.1 g/100 mL (Miris AB® V3, Uppsala, Sweden). Since it has been shown that mid-infrared analyzers may require calibration adjustment, we verified the calibration daily to ensure appropriate measures [9]. The samples were rejected if variability was over 10%; otherwise, the average of the two values was kept for further analysis.

We started our statistical analyses by power calculation relying upon McLeod [3]. We calculated that to demonstrate a significant relationship with a coefficient of at least $r = 0.550$, which is considered clinically significant, with an alpha risk of 0.016 (Bonferroni correction for the three nutrients) and a power of 0.90, 78 mothers with at least two samples would be needed (Power and Precision™ V4, Biostat Inc., Englewood, NJ, USA, 2001). Normally distributed data, assessed by a Shapiro–Wilk test of normality, are presented as mean values with standard deviation (SD), the median, and the interquartile range (IQR); non-normally distributed data are presented as medians with IQR only. To evaluate the differences between groups, we used the Student *t*-test for continuous variables and the Chi-squared test or Fisher exact test when appropriate for categorical variables. For continuous variables not normally distributed, we used the Mann–Whitney U test. To determine which variables were associated with milk composition, we performed a bivariate analysis, and then a stepwise multiple linear regression, including all of the variables associated with milk composition in bivariate analysis, with a tolerance set at 10^{-5} with the probability to remove the set at 0.15 and a confidence interval of 0.95. Observed differences were considered statistically significant if $p < 0.05$. Statistical analysis was performed with SYSTAT 12 software (2007, Systat Software Inc., San Jose, CA, USA).

The study has been approved by our Institutional Ethics and Review Board on 9 March 2013 (Number: MRU13-02).

3. Results

3.1. Description of the Studied Population

From August 2013 to January 2014, 367 milk samples were obtained from 81 mothers (Figure 1). Two to 10 batches were collected from the mothers (median (IQR): three (two to six)).

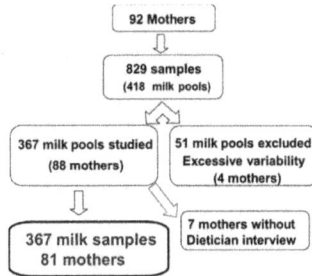

Figure 1. flow chart.

The mothers involved in the study were 29 years old (19–42) (median (range)). Their average height was 1.64 m (1.53–1.85) for a weight of 63 kg (42–110) and a body mass index (BMI) of 23.2 (16.4–43). Weight gain during pregnancy ranged from 0 to 30 kg (mean 10.2 kg). The age of delivery was 31 weeks' gestational age (24–34). Twenty (25%) mothers smoked during pregnancy; 63 (78%) were single pregnancies, and 19 (23%) presented with toxemia. Prenatal maturation with corticosteroids was achieved in 37 (46%) mothers and partial maturation in 33 (41%). Vaginal delivery occurred in 42 mothers (52%) and cesarean section occurred in 39 (48%).

Forty-nine (60%) newborns were males; neonatal adaptation was good with an Apgar score above six at one and five minutes for all infants. Their mean birth weight was 1523 ± 512 g (median (range)) = 1460 (600–2500) g).

3.2. Maternal Nutritional Intake and Milk Composition

Maternal nutritional intake was 2169 ± 562 Kcal/day (2146 (1197–3628)) with 88 ± 28 g/day (88 (40–213)) of fat intake, 86 ± 20 g/day (87 (40–160)) of protein intake, and 257 ± 81 g/day (247 (103–533)) of overall carbohydrate intake.

The global milk sample composition was (median [range]/100 mL): carbohydrates 6.8 g (4.4–7.3), lipids 3.4 g (1.3–6.4), proteins 1.3 g (0.1–3.1). The correlation between mothers' food intake and milk composition is shown in Table 1.

Table 1. Linear regression between mothers' food intake and milk composition (coefficient *r*).

Nutrients Intake	Milk Calories	Milk Proteins	Milk Lipids	Milk Carbohydrates
Log Energy	0.110 *	0.094	0.106 *	0.034
Protein	0.01	0.03	0.01	0.04
Fat	0.03	0.04	0.03	0.03
Carbohydrates	0.131 **	0.109 *	0.127 *	0.035

$* p < 0.05;\ ** p < 0.01.$

3.3. Perinatal Factors' Effect on Milk Composition

3.3.1. Postnatal Age Effect on Milk Composition

Weekly mean protein content significantly decreased for the first four weeks post-delivery, and then remained stable at week five. We observed a comparable but inverse evolution for carbohydrates, and there was no significant difference over the first five weeks after delivery for the fat content of mothers' milk (Table 2).

Table 2. Average milk composition per postnatal week.

Week	1 (n = 52)	2 (n = 125)	3 (n = 91)	4 (n = 68)	5 (n = 31)
Protein	1.78 ± 0.39 *	1.40 ± 0.40 *	1.26 ± 0.34 *	1.08 ± 0.36 *	1.05 ± 0.40
Lipids	3.23 ± 0.80	3.58 ± 0.98	3.59 ± 0.97	3.41 ± 0.96	3.40 ± 1.06
CHO	6.50 ± 0.43 *	6.66 ± 0.38 *	6.70 ± 0.46 *	6.81 ± 0.44 *	6.75 ± 0.44

* $p < 0.01$.

We observed a significant linear regression for carbohydrate and an inverse correlation for protein contents, as shown in Figure 2.

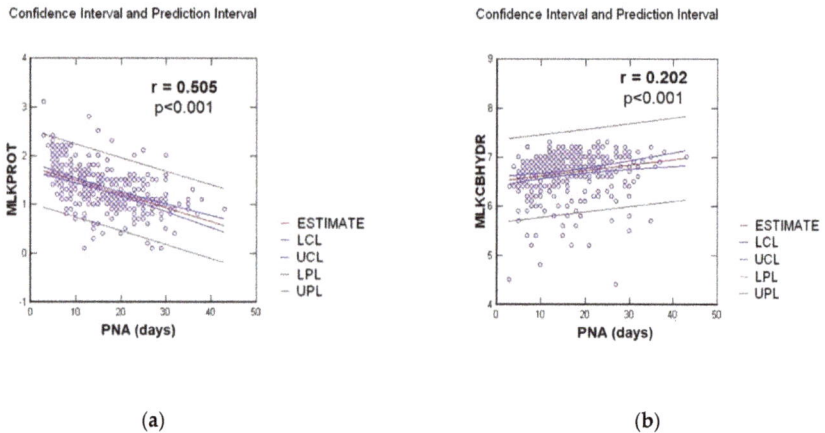

Figure 2. Linear regression between milk composition in g/100 mL and postnatal age (PNA) in days, for: (**a**) Protein milk content (MLKPROT); (**b**) Carbohydrate milk content (MLKCBHYD).

3.3.2. Perinatal Factors' Impact on Milk Composition in Bivariate Analysis

There was no relationship between the milk composition and the mothers' age, weight, height, or BMI before pregnancy. There was no relationship either for the mode of delivery, multiple pregnancies, toxemia, or gestational age at delivery. We observed a weak association between weight gain during pregnancy and milk lipid content (r = 0.117, p = 0.026). Finally, we observed a significant correlation between milk composition and antenatal steroid maturation, smoking during pregnancy, and an inverse relationship with the infants' birth weight. Detailed data are shown in Table 3.

Table 3. Confounding perinatal factors for milk composition.

| Milk Content (Mean, g/100 mL) | Smoking | | Antenatal Steroids | | Birth Weight (Linear Regression: r) |
	Yes	No	Yes	No	
Lipids	3.10	3.59 *	3.6 *	3.1	0.082
Carbohydrates	6.57	6.71 *	6.7	6.6	−0.259 *
Protein	1.34	1.33	1.3	1.2	−0.318 *
Calories	62.8	67.8 *	67.6 *	61.9	−0.106

* $p < 0.05$.

3.3.3. Stepwise Multivariate Regression Analysis of Factors Associated with Milk Composition in Bivariate Analysis

All of the factors showing a significant association with mothers' milk content in bivariate analysis were included in the model presented in Table 4.

Table 4. Multivariate analysis of factors associated with mothers' milk content.

Milk Content	Postnatal Age	Carbohydrate Intake	Smoking	No Steroids	Weight Gain
Lipids ($r^2 = 0.087$)	NS	1.279 *	−0.557 *	NS	NS
Carbohydrates ($r^2 = 0.071$)	0.012 *	NS	−0.167 *	NS	NS
Protein ($r^2 = 0.299$)	−0.028 *	0.449 *	NS	−0.066 *	NS
Calories ($r^2 = 0.101$)	NS	14.053 *	−5.901 *	NS	NS

* $p < 0.05$; NS: non-significant.

4. Discussion

Our study confirms that breast milk composition in macronutrients has a large variability, as suggested in previous studies [3,4]. Therefore, fortification of the assumed averaged macronutrient in milk leads to inadequate nutritional intake for most premature infants. Our data show that indeed, maternal nutrition may influence breast milk macronutrient composition in mothers who delivered prematurely. Our data also show that only overall carbohydrate is positively correlated with protein, fat, and caloric density. To our knowledge, this observation has not been yet reported. In their study comparing the breast milk nitrogen content of mothers from two different Chinese areas, Zhao et al. found no significant difference in 18 studied amino acids, despite significant lower protein intake in one of the studied areas [10]. This is consistent with our results. Likewise, in their study on maternal supplementation with omega-3 precursors, Mazurier et al. showed a qualitative alteration with increased alpha-linoleic acid content, but no significant modification of the overall lipid concentration [11]. This observation also recalls that a true difference in the secretory activity of the mammary gland is not merely a difference in concentration. In a recent survey on food and nutrient intake of women in France, Hebel et al. showed that few lactating mothers met the nutritional guidelines, and may therefore be at risk of food and nutrient inadequacies [12]. One could speculate that improving the maternal nutritional balance, especially an appropriate overall carbohydrate intake, might contribute to improved milk composition for preterm infant feeding.

Longitudinal analysis of milk composition showed a significant decrease in milk protein content with an inverse correlation for carbohydrate and a stable lipid concentration over the first four weeks after delivery. These data are consistent with the publication of Maly et al. [8], who found the same evolution for the three macronutrients and Mahakan et al. [4], who found a 50% decrease in protein concentrations with a 30% increase in carbohydrate concentrations over the first 28 days after delivery. From his review analysis on human milk in premature infants [13], which included the studies Charpak et al. [14] and Bauer et al. [15], Underwood evaluated the changes in milk composition over eight weeks from delivery, and found the same results as in our study over this longer assessment period. Yoneyama et al. suggested that there may be a negative correlation between protein content and milk volume and between early and mature milk [16]. This observation may explain why breastfeeding infants adjust their volume intake in relation with milk caloric density and protein content. However, when feeding premature neonates, the amount of milk is usually fixed to a target volume [17]. Therefore, protein content evolution is important to take into account when considering the fortification of human milk for premature infants.

Few data are available for the impact of perinatal factors on milk composition. In our study, we did not find any significant association among the mothers' age, weight, height, or BMI before pregnancy. In their study on maternal nutrition and body composition during breast feeding, Bzikowska-Jura et al. [7] found a variance in milk fat content related to BMI. However, they studied the actual BMI at three time points, while we differentiated between BMI before pregnancy and weight gain during pregnancy, which was indeed associated with milk fat content. Thus, our results are consistent with published data, but suggest that it is not BMI per se, but rather weight gain during pregnancy that may be associated with milk composition. Also, there was no relationship for the mode of delivery, multiple pregnancies, or toxemia. As in our study, Maly et al. [8] showed no effect of the degree of prematurity at delivery on milk composition. However, we found a moderate effect of birth weight in bivariate analysis, but not after adjusting for other confounding factors in multivariate analysis (data not shown). Finally, we showed that breast milk protein concentrations were positively correlated with mothers' overall carbohydrate intake, and negatively correlated with the duration of lactation from birth onwards, and the absence of steroid maturation; breast milk carbohydrate concentrations were positively correlated with the duration of lactation, and negatively correlated with smoking, as shown in the study of Bachour et al. [18], whereas breast milk lipid concentrations were positively correlated with mothers' carbohydrate intake and negatively correlated with smoking.

Our study has strength and limitations. Our strength relies upon the number of samples, the blinded collection of the data, and the reproducibility of milk content measurements with a rejection when the control variability was above 0.10. Also, the dietary questionnaire over two weeks recall, which was given by experienced dieticians with an illustrated standardized catalog for the amount of food, allowed an appropriate evaluation of the averaged usual diet of the mothers. However, this questionnaire was given two weeks from the interview, and may not strictly reflect the diet of the lactating mother at the time of milk collection. Anyhow, one may speculate that the level of macronutrients intake would be only slightly modified from the routine diet of the patients. We did not find a significant difference from the results with sugar, fibers, or overall carbohydrate; therefore, we presented only the results with overall carbohydrate, but this would be confirmed by further targeting the study. Finally, because this is an observational study, we were not able to standardize the time that the mothers would hand over their milk collected at home to the milk bank, nor the number of collection days within each pool. However, this would only vary from one to three days, and we longitudinally evaluated the results over five weeks. Thus, even though this study does not really present an individual longitudinal analysis, the linear regressions that were observed for the overall population and the multivariate analysis, taking time as a confounding factor, are likely to allow a good reliability for our findings.

In conclusion, our study confirms that human milk macronutrients composition has a wide variability. This variability is differentially associated for each macronutrient, and associated with maternal carbohydrates intake, antenatal steroids, smoking, and the delay from delivery. To improve mothers' milk composition to values closer to the one needed for achieving infants' nutrition needs [19], one could aim for improving the prenatal nutritional balance in pregnant women, particularly for carbohydrate intake, supporting an appropriate weight gain during pregnancy, and advising stopping smoking. In the case of expected prematurity, steroid maturation is also needed to improve milk composition. Ideally, breast milk composition would be regularly measured for individualized fortification to achieve the appropriate growth of preterm infants. However, when it is only possible to apply standard fortification, at least human milk composition in relation to lactation duration from birth should be taken into account. Targeting investigation should be performed and confirm the observed data.

Nutrients **2019**, *11*, 366

Author Contributions: All of the authors made significant contributions to the study: "conceptualization, J.-M.H., S.S. and L.-D.V.E.; methodology, mothers' interviews M.C. and C.P.; methodology, milk analysis with quality assessment and validation P.F.; validation, J.-M.H., S.S. and P.F.; formal analysis, J.-M.H.; investigation, M.C., C.P. and S.S.; writing—original draft preparation, J.-M.H.; writing—review and editing P.F., S.S., C.R. and L.-D.V.E.; supervision, J.-M.H.; funding acquisition, C.R. and L.-D.V.E.; finally, all authors read and approved the final version of the manuscript".

Funding: This study received an unrestricted grant and the loan of the MIRIS milk analyzer from Gallia laboratory.

Acknowledgments: We thank Inga C Teller, PhD, Nutricia Research, Utrecht, The Netherlands, for her significant help in reviewing and editing the manuscript.

Conflicts of Interest: The funders had no role in the design of the study; in the collection, analyses, or interpretation of data; in the writing of the manuscript, or in the decision to publish the results.

References

1. Agostoni, C.; Buonocore, G.; Carnielli, V.; De Curtis, M.; Darmaun, D.; Decsi, T.; Domellöf, M.; Embleton, N.; Fusch, C.; Genzel-Boroviczeny, O.; et al. Enteral Nutrient Supply for Preterm Infants: Commentary from the European Society of Paediatric Gastroenterology, Hepatology and Nutrition Committee on Nutrition. *J. Pediatr. Gastroenterol. Nutr.* **2010**, *50*, 85–91. [CrossRef] [PubMed]
2. Martin, C.R.; Brown, Y.F.; Ehrenkranz, R.A.; O'Shea, T.M.; Allred, E.N.; Belfort, M.B.; McCormick, M.C.; Leviton, A. Nutritional Practices and Growth Velocity in the First Month of Life in Extremely Premature Infants. *Pediatrics* **2009**, *124*, 649–657. [CrossRef] [PubMed]
3. McLeod, G.; Sherriff, J.; Nathan, E.; E Hartmann, P.; Simmer, K. Four-week nutritional audit of preterm infants born < 33 weeks gestation. *J. Paediatr. Child Health* **2013**, *49*, E332–E339. [PubMed]
4. Mahajan, S.; Chawla, D.; Kaur, J.; Jain, S. Macronutrients in breastmilk of mothers of preterm infants. *Indian Pediatr.* **2017**, *54*, 635–637. [CrossRef] [PubMed]
5. Embleton, N.E.; Pang, N.; Cooke, R.J. Postnatal Malnutrition and Growth Retardation: An Inevitable Consequence of Current Recommendations in Preterm Infants? *Pediatrics* **2001**, *107*, 270–273. [CrossRef] [PubMed]
6. Cooke, R.J.; Ainsworth, S.B.; Fenton, A.C. Postnatal growth retardation: A universal problem in preterm infants. *Arch. Dis. Child.-Fetal Neonatal Ed.* **2004**, *89*, 428–430. [CrossRef] [PubMed]
7. Bzikowska-Jura, A.; Czerwonogrodzka-Senczyna, A.; Olędzka, G.; Szostak-Węgierek, D.; Weker, H.; Wesołowska, A. Maternal Nutrition and Body Composition During Breastfeeding: Association with Human Milk Composition. *Nutrients* **2018**, *10*, 1379. [CrossRef] [PubMed]
8. Malý, J.; Burianova, I.; Vitkova, V.; Ticha, E.; Navratilova, M.; Cermáková, E. Preterm human milk macronutrient concentration is independent of gestational age at birth. *Arch. Dis. Child. Fetal Neonatal Ed.* **2018**, *104*, F50–F56. [CrossRef] [PubMed]
9. Buffin, R.; Decullier, E.; Loys, C.-M.; Hays, S.; Studzinsky, F.; Jourdes, E.; Picaud, J.-C.; De Halleux, V.; Rigo, J. Assessment of human milk composition using mid-infrared analyzers requires calibration adjustment. *Am. J. Perinatol.* **2017**, *37*, 552–557. [CrossRef] [PubMed]
10. Zhao, X.; Xu, Z.; Wang, Y.; Sun, Y. Studies of the relation between the nutritional status of lactating mothers and milk intake and growth of their infants in Beijing. Pt. 4. The protein and amino acid content of breast milk. *Ying Yang Xue Bao* **1989**, *11*, 227–332. [PubMed]
11. Mazurier, E.; Rigourd, V.; Perez, P.; Buffin, R.; Couedelo, L.; Vaysse, C.; Belcadi, W.; Sitta, R.; Nacka, F.; Lamireau, D.; et al. Effects of Maternal Supplementation with Omega-3 Precursors on Human Milk Composition. *J. Hum. Lact.* **2017**, *33*, 319–328. [CrossRef] [PubMed]
12. Hebel, P.; Francou, A.; Van Egroo, L.D.; Rougé, C.; Mares, P. Consommation alimentaire et apports nutritionnels chez les femmes allaitantes, en France. *Cah. Nutr. Diét.* **2018**, *25*, 303.
13. Underwood, M.A. Human Milk for the Premature Infant. *Pediatr. Clin. N. Am.* **2013**, *60*, 189–207. [CrossRef] [PubMed]
14. Charpak, N.; Ruiz, J.; M Med Sci on behalf of the KMC Team. M Med Sci on behalf of the KMC Team. Breast milk composition in a cohort of pre-term infants' mothers followed in an ambulatory programme in Colombia. *Acta Paediatr.* **2007**, *96*, 1755–1759.
15. Bauer, J.; Gerss, J. Longitudinal analysis of macronutrients and minerals in human milk produced by mothers of preterm infants. *Clin. Nutr.* **2011**, *30*, 215–220. [CrossRef] [PubMed]

16. Yoneyama, K.; Goto, I.; Nagata, H. Relationship between volume and concentrations of various components of breast milk in early and 2-6 months lactation. *Nihon Koshu Eisei Zasshi* **1994**, *41*, 157–164. [PubMed]
17. Rochow, N.; Fusch, G.; Choi, A.; Chessell, L.; Elliott, L.; McDonald, K.; Kuiper, E.; Purcha, M.; Turner, S.; Chan, E.; et al. Target Fortification of Breast Milk with Fat, Protein, and Carbohydrates for Preterm Infants. *J. Pediatr.* **2013**, *163*, 1001–1007. [CrossRef] [PubMed]
18. Bachour, P.; Yafawi, R.; Jaber, F.; Choueiri, E.; Abdel-Razzak, Z. Effects of Smoking, Mother's Age, Body Mass Index, and Parity Number on Lipid, Protein, and Secretory Immunoglobulin A Concentrations of Human Milk. *Breastfeed. Med.* **2012**, *7*, 179–188. [CrossRef] [PubMed]
19. Gidrewicz, D.A.; Fenton, T.R. A systematic review and meta-analysis of the nutrient content of preterm and term breast milk. *BMC Pediatr.* **2014**, *14*, 216. [CrossRef] [PubMed]

MDPI

Article

Higher Maternal Diet Quality during Pregnancy and Lactation Is Associated with Lower Infant Weight-For-Length, Body Fat Percent, and Fat Mass in Early Postnatal Life

Muna J. Tahir [1,*], Jacob L. Haapala [1], Laurie P. Foster [1], Katy M. Duncan [2], April M. Teague [2], Elyse O. Kharbanda [3], Patricia M. McGovern [4], Kara M. Whitaker [5,6], Kathleen M. Rasmussen [7], David A. Fields [2], David R. Jacobs, Jr. [1], Lisa J. Harnack [1] and Ellen W. Demerath [1]

[1] Division of Epidemiology and Community Health, University of Minnesota, Minneapolis, MN 55454, USA; haap0016@umn.edu (J.L.H.); fost0112@umn.edu (L.P.F.); jacob004@umn.edu (D.R.J.J.); harna001@umn.edu (L.J.H.); ewd@umn.edu (E.W.D.)
[2] Department of Pediatrics, University of Oklahoma Health Sciences Center, Oklahoma City, OK 73104, USA; Katy-Duncan@ouhsc.edu (K.M.D.); April-Teague@ouhsc.edu (A.M.T.); David-Fields@ouhsc.edu (D.A.F.)
[3] HealthPartners Institute, Minneapolis, MN 55425, USA; elyse.o.kharbanda@healthpartners.com
[4] Division of Environmental Health Sciences, University of Minnesota, Minneapolis, MN 55455, USA; pmcg@umn.edu
[5] Department of Health and Human Physiology, University of Iowa, Iowa City, IA 52242, USA; kara-whitaker@uiowa.edu
[6] Department of Epidemiology, University of Iowa, Iowa City, IA 52246, USA
[7] Division of Nutritional Sciences, Cornell University, Ithaca, NY 14850, USA; kathleen.rasmussen@cornell.edu
* Correspondence: mtahir@umn.edu; Tel.: +1-512-960-5035

Received: 12 February 2019; Accepted: 12 March 2019; Published: 15 March 2019

Abstract: Maternal pregnancy nutrition influences fetal growth. Evidence is limited, however, on the relationship of maternal diet during pregnancy and lactation on infant postnatal growth and adiposity. Our purpose was to examine associations between maternal diet quality during pregnancy and lactation with offspring growth and body composition from birth to six months. Maternal diet quality was serially assessed in pregnancy and at one and three months postpartum, using the Healthy Eating Index–2015 in a cohort of 354 fully breastfeeding mother–infant dyads. Infant length-for-age (LAZ), weight-for-age (WAZ), and weight-for-length (WLZ) Z-scores were assessed at birth, one, three, and six months. Infant body fat percent (BF%), fat mass (FM), and fat-free mass (FFM) were measured at six months using dual-energy X-ray absorptiometry. Higher maternal diet quality from pregnancy through three months postpartum was associated with lower infant WLZ from birth to six months ($p = 0.02$) and BF% at six months ($p \leq 0.05$). Higher maternal diet quality at one and three months postpartum was also associated with lower infant FM at six months ($p < 0.01$). In summary, maternal diet quality during pregnancy and lactation was inversely associated with infant relative weight and adiposity in early postnatal life. Additional research is needed to explore whether associations persist across the life course.

Keywords: maternal diet quality; pregnancy; lactation; infancy; growth; body composition

1. Introduction

The in utero environment is known to play an important role in fetal programming and subsequent offspring growth and development [1]. Intrauterine nutritional and environmental exposures may have adverse effects on lifelong organ structure and function, which ultimately influence offspring health and disease risk [1,2]. Within this context, research has shown that maternal pre-pregnancy body mass index (BMI) [3], gestational weight gain [3,4], and lifestyle during pregnancy [5] may influence offspring obesity. These effects may appear immediately in the form of macrosomia and birth weight large-for-gestational age, or may be latent, with adiposity appearing later in life [3]. The role of maternal diet during this window of vulnerability is of particular interest for obesity prevention [6]. Although maternal obesity is the strongest predictor of offspring obesity [3], the modifiability of maternal diet is potentially a crucial tool in combating obesity in offspring.

Accruing animal and human evidence also indicates that maternal nutrition beyond the pregnancy period may play a role in early life infant health and subsequent obesity and chronic disease risk [7–9]. For example, exclusive breastfeeding is recommended by the American Academy of Pediatrics and other as the optimal nutritional source for infants and is currently initiated by the majority of United States mothers [10,11]. It is possible that maternal diet influences breast milk composition, providing a pathway by which maternal diet may directly influence offspring growth [12]. It is known, for instance, that the composition of breast milk may alter epigenetic programming [13] and the infant gut microbiome and later health and obesity risk [14]. Maternal diet preferences also indirectly influence the general nutritional environment the infant is exposed to in the household through her purchasing and child feeding patterns [15].

Studies exploring the associations of maternal diet during pregnancy and lactation with offspring anthropometry or body composition have often focused on specific dietary components and the macronutrient composition of diet with inconsistent findings [8,16–18]. Dietary intake as measured by "single nutrients" is difficult to interpret given the interrelationships of most nutrients and biologically active components within and among foods and the associations of overall food-intake patterns with health [19]. Indices of diet quality, rather than specific nutrient intakes, may better reflect broader food-intake patterns, intakes of combinations of foods, nutrient-nutrient interactions, and may be more generalizable across populations [19]. A number of such indicators of diet quality have been developed and are increasingly used in health outcomes research [20].

There is a paucity of research assessing the relationship of maternal diet quality across the entire perinatal period (including both pregnancy and lactation) with offspring growth and body composition in early life [21–25]. Most studies have focused on anthropometry at birth [21,22] or in childhood [24], with few evaluating repeated measures of growth during infancy, a period of rapid change with potential effects extending across the lifespan [26]. Furthermore, limited research has examined the impact of maternal diet quality during pregnancy and lactation on measures of offspring body composition [23,25], which may provide different insights into growth trajectories and chronic disease risk [27].

The purpose of the present study is to evaluate associations of maternal diet quality during pregnancy and lactation with measures of infant growth (length-for-age (LAZ), weight-for-age (WAZ), and weight-for-length (WLZ) Z-scores from birth to six six months of age) and body composition (body fat percent (BF%)), fat mass (FM), and fat-free mass (FFM) at six months of age). We hypothesized that higher maternal diet quality (as measured by higher Healthy Eating Index–2015 (HEI–2015) total scores) would be associated with higher infant LAZ and lower infant WAZ and WLZ from birth to six months. We further hypothesized that higher maternal diet quality would be related to lower infant BF% and FM and higher infant FFM at six months of age.

2. Materials and Methods

2.1. Study Population

Mother–infant dyads included in these analyses were enrolled in the Mothers and Infants LinKed for Health (MILK) study, an ongoing prospective cohort study [28]. Mothers and their infants were recruited from Minneapolis, MN, and Oklahoma City, OK. A total of 367 pregnant women aged 21–45 years, with a pre-pregnancy BMI of 18.5–40.0 kg/m^2, an intention to exclusively breastfeed for at least 3 months, a healthy delivery (defined as <3 days in the hospital for vaginal deliveries and <5 days in the hospital for caesarean section deliveries), and who delivered singleton infants born at-term with a birth weight of ≥2500–≤4500 g were recruited into the study. Exclusion criteria for the study were as follows: (1) mothers consumed tobacco or >1 alcoholic drink per week during pregnancy/lactation; (2) mothers had a history of Type 1 or Type 2 diabetes or current gestational diabetes; (3) mothers had a congenital illness affecting infant feeding or growth; and/or (4) mothers were unable to speak or understand English. The final analytic sample included 354 mother–infant dyads who had maternal dietary data and infant growth measures on at least one of the time points when these items were measured.

The MILK study protocols were approved by the institutional review boards of the University of Minnesota, HealthPartners Institute and the University of Oklahoma Health Sciences Center. All women provided written informed consent at baseline and compensation was provided after each measurement visit.

2.2. Maternal Diet

Maternal dietary intake data was collected during the third trimester of pregnancy and at one and three months postpartum using the Diet History Questionnaire II (DHQ II), a food frequency questionnaire (FFQ) designed to assess frequency of intake and portion sizes for 134 food and beverage items and 8 dietary supplement items over the previous month [29]. A prior version of the DHQ II (DHQ I) with minimal modifications to food lists and nutrient databases has been evaluated for validity in prior research [29–31]. While one study showed that the DHQ I performed better than the Willett FFQ and Block FFQ at estimating energy and absolute nutrient intake [29], others showed that men and women under-reported energy and protein intakes on the DHQ I [30] which may lead to attenuation of estimated disease relative risks for these nutrients [31]. The DHQ II was analyzed using Diet Calc (National Cancer Institute, Bethesda, MD, USA), and food and nutrient values were generated using the United States Department of Agriculture (USDA) Food Patterns Equivalent Database and the Food and Nutrient Database for Dietary Studies.

Maternal diet quality during pregnancy and lactation was estimated using the HEI–2015, a scoring system designed to measure adherence to the 2015–2020 Dietary Guidelines for Americans (2015–2020 DGA) [32]. The psychometric properties of the HEI–2015 were evaluated using data from exemplary menus, a nationally representative sample and a prospective cohort study, with evidence supporting construct validity and reliability and criterion validity [33]. Previous studies have assessed the validity of prior HEI versions among pregnant women and showed that the index was useful in providing a composite dietary intake measure [34].

The HEI–2015 total score is the sum of 13 subcomponent scores that measure adequacy (Total Fruits, Whole Fruits, Total Vegetables, Greens and Beans, Whole Grains, Dairy Products, Total Protein Foods, Seafood and Plant Proteins, and Unsaturated:Saturated fats) and moderation (Refined Grains, Sodium, Added Sugars, and Saturated Fats). All subcomponents are scored from 0–5 or 0–10 based on intake between the minimum and maximum standards. Moderation components are reverse scored such that higher scores reflect lower intake. To account for variation in caloric intake between participants, all HEI–2015 components are standardized to 1000 kilocalories, except for the ratio of unsaturated to saturated fats. A higher HEI–2015 total score (out of 100) represents greater consistency of the diet with the 2015–2020 DGA (32). The code used to calculate the HEI–2015 scores was developed by the Division of Cancer Control and Population Sciences in the National Cancer Institute [35].

2.3. Infant Growth and Body Composition

At birth, one, three, and six months, infant length was measured with the infant undressed, but diapered using the Seca 416 infantometer (Seca, Birmingham, UK). Infant naked weight was measured using the high sensitivity scale embedded in the PEA POD (COSMED USA, Inc., Concord, CA, USA). Age and sex-specific LAZ, WAZ, and WLZ were subsequently calculated using the World

Health Organization Z-score classification system for term infants [36]. At six months, infant BF%, FM, and FFM were assessed utilizing dual X-ray energy absorptiometry (DXA), with the infant in a supine position wearing only a disposable diaper and swaddled in a light cotton blanket [37]. All measurements were obtained using standardized protocols and after cross-training of study staff [38].

2.4. Statistical Analyses

To provide a comprehensive description of the mother–infant dyads, demographic and clinical characteristics are presented as means ± standard deviations (SD) for continuous variables, and frequencies for categorical variables by tertiles of maternal HEI–2015 total scores during pregnancy. Differences in participant characteristics by maternal diet quality during pregnancy were evaluated using one-way ANOVAs and chi-square tests for continuous and categorical variables, respectively. Differences in participant characteristics among dyads who had complete dietary data during pregnancy and infant WLZ at birth ($N = 319$) versus those who had complete dietary data at three months postpartum and infant WLZ at six months ($N = 257$) were compared using *t*-tests and chi-square tests for continuous and categorical variables, respectively. Pearson correlations were calculated to test correlations between maternal HEI–2015 total scores during pregnancy and at one and three months postpartum.

Linear mixed effects models (PROC MIXED) were then used to evaluate the associations of repeated measures of maternal diet quality (HEI–2015 total scores) from pregnancy through three months postpartum with within-subject measures of infant LAZ, WAZ, and WLZ from birth to six months. The exposures were analyzed as continuous variables owing to the relatively limited sample size and after studying goodness of linear fit using scatter plots. The data were first adjusted for study site (Minneapolis, Oklahoma), maternal age (years), race (white, other), education (high school/GED/associates degree, Bachelor's degree, graduate degree), household income (<$60,000, $60,000–90,000, >$90,000), and total energy intake (time-dependent; kilocalories) (Model 1). The full models (Model 2) were additionally adjusted for the following potential confounders: pre-pregnancy BMI (kg/m^2), gestational weight gain (kilograms), parity (0, $\geq$1), mode of delivery (vaginal, cesarean section), duration of exclusive breastfeeding (one, three, six months), infant gestational age (weeks), and infant sex (male, female). The main exposures and all confounders were included as main effects, but interactions with time were only retained if the corresponding *p*-value estimates were significant at <0.05.

Next, linear regression models (PROC GLM) were run to examine separately the associations of maternal diet quality during pregnancy, one and three months postpartum with infant BF%, FM, and FFM at six months of age. The models were similarly adjusted for all aforementioned potential confounders (Models 1 and 2), with the addition of infant WLZ at birth and infant exact age at six months (Model 2). Given that infant fat mass index (FMI) and fat-free mass index (FFMI) scale to length (fat mass or fat-free mass/length^2) and have been proposed as more precise indicators of adiposity and nutrition status [39], we also tested the associations of maternal perinatal diet quality with these indices. We further explored associations of maternal diet during pregnancy and lactation with compartment specific fat mass (trunk, arm, and leg) to detect changes in the relative distribution of fat within the body in sub-analyses.

Lastly, in sensitivity analyses to examine if the relationship of maternal diet with infant growth and/or body composition is modified by exclusivity of breastfeeding, the full models for both infant growth and body composition were re-run to include only infants who were fully breastfed to six months. All statistical analyses were performed using SAS 9.4 (SAS Institute Inc., Cary, NC, USA).

3. Results

3.1. Participant Characteristics

The mothers included in the present study were mostly highly educated, white, multiparous, fully breastfeeding at six months with average gestational weight gain, and above average HEI–2015 scores during pregnancy and lactation (Table 1). Maternal education, parity, HEI–2015 scores at one and three months postpartum, and infant gestational age differed significantly by tertiles of maternal HEI–2015 scores during the third trimester of pregnancy, whereby mothers with higher pregnancy HEI–2015 scores were on average more educated, nulliparous, had higher postpartum HEI–2015 scores and delivered their infants later ($p < 0.05$). No significant differences were observed between characteristics of the mother–infant dyads who had complete dietary data during pregnancy/infant WLZ at birth and complete dietary data at three months postpartum/infant WLZ at six months postpartum ($p > 0.05$; data not shown). Maternal HEI–2015 scores during pregnancy were correlated with scores at one and three months postpartum ($r = 0.60$ and $r = 0.65$; $p < 0.001$, respectively). HEI–2015 scores at one and three months postpartum were also correlated with one another ($r = 0.63$, $p < 0.001$).

Table 1. Participant demographic and clinical characteristics by tertiles of maternal HEI–2015 total scores during the third trimester of pregnancy ($n = 329$).

Participant Characteristics	Total	HEI–2015 Tertiles (T) [a]			
		T1 ($n = 109$)	T2 ($n = 110$)	T3 ($n = 110$)	
		N (%) or Mean ± SD			*p*-Value
Study site					
Minnesota	233 (66)	69 (63)	72 (65)	78 (71)	0.47
Oklahoma	121 (34)	40 (37)	38 (35)	32 (29)	
Age, years	30.9 ± 4.1	30.2 ± 4.3	30.7 ± 4.4	31.4 ± 3.7	0.1
Race					
White	299 (86)	92 (85)	93 (87)	97 (89)	0.71
Other	49 (14)	16 (15)	14 (13)	12 (11)	
Education					
High school/GED/Associates degree	80 (24)	37 (36)	26 (24)	11 (11)	<0.001 *
Bachelor's degree	136 (40)	42 (41)	38 (36)	48 (46)	
Graduate degree	121 (36)	24 (23)	42 (40)	45 (43)	
Household income					
<$60,000	105 (31)	37 (36)	34 (32)	25 (24)	0.19
$60,000–90,000	85 (25)	30 (29)	24 (23)	27 (26)	
>90,000	147 (44)	36 (35)	48 (45)	52 (50)	
Parity					
0	146 (42)	33 (31)	54 (50)	52 (47)	0.01 *
≥1	204 (58)	74 (69)	55 (50)	58 (53)	
Pre-pregnancy BMI, kg/m²	26.4 ± 5.4	27.2 ± 6.2	26.4 ± 4.9	25.5 ± 5.0	0.06
Gestational weight gain					
Below or within IOM guidelines	200 (57)	54 (501)	67 (62)	63 (57)	0.23
Exceeding IOM guidelines	149 (43)	53 (50)	41 (38)	47 (43)	
Energy intake during pregnancy	1827 ± 530	1747 ± 550	1876 ± 535	1859 ± 498	0.15
Mode of delivery					
Vaginal	278 (80)	90 (83)	82 (76)	87 (81)	0.39
Caesarean section	70 (20)	18 (17)	26 (24)	21 (19)	
HEI–2015 score during third trimester of pregnancy	67.2 ± 8.7	57.5 ± 5.8	67.8 ± 2.1	76.1 ± 4.0	<0.001 *
HEI–2015 score at one month postpartum	65.9 ± 8.4	60.6 ± 7.5	65.2 ± 7.0	71.2 ± 7.6	<0.001 *
HEI–2015 score at three months postpartum	66.1 ± 8.7	60.5 ± 7.9	65.3 ± 7.2	71.5 ± 7.4	<0.001 *

Table 1. *Cont.*

Participant Characteristics	Total	T1 (*n* = 109)	T2 (*n* = 110)	T3 (*n* = 110)	
		HEI–2015 Tertiles (T) [a]			
		N (%) or Mean ± SD			*p*-Value
Duration of exclusive breastfeeding					
one month	21 (7)	7 (7)	10 (10)	2 (2)	0.16
three months	58 (18)	20 (22)	16 (16)	20 (19)	
six months	235 (75)	65 (71)	74 (74)	81 (79)	
Infant gestational age, weeks	39.8 ± 1.1	39.7 ± 1.0	39.6 ± 1.1	40.0 ± 1.1	0.04 *
Infant sex					
Male	178 (50)	53 (49)	57 (52)	53 (48)	0.84
Female	176 (50)	56 (51)	53 (48)	57 (52)	

Abbreviations: HEI = Healthy Eating Index; BMI = Body Mass Index; IOM = Institute of Medicine. [a] HEI–2015 T1: ≤63.9; T2: 64.0–70.9; T3: ≥70.9. * $p < 0.05$ for tests of differences in participant characteristics by tertiles of maternal HEI–2015 total scores during pregnancy using chi-square and one-way ANOVAs for categorical and continuous variables, respectively. Data are presented as mean ± SD or column percentages.

As expected, higher intake of all adequacy components (except Dairy Products) and lower intake of all moderation components were seen among mothers in the highest versus lowest tertile of HEI–2015 total scores (Supplementary file Table S1). Higher (better) scores for Whole and Total Fruits, Total Vegetables, Greens and Beans, Seafood and Plant Proteins, Total Protein Foods, Refined Grains, and Added Sugars appeared to drive higher average HEI–2015 total scores. HEI–2015 total scores were normally distributed at all time points and average scores were highest during the third trimester of pregnancy (Supplementary file Figure S1).

Among infants, on average, LAZ decreased from birth to six months, whereas WAZ and WLZ decreased from birth to three months, followed by a subsequent increase after three months (Table 2). The mean fat mass among infants at six months was approximately 2.8 kg (34% of total body weight), with most of the fat distributed in the legs (1.08 kg) and trunk (0.92 kg).

3.2. Maternal Diet Quality and Infant Growth

Maternal diet quality through pregnancy and lactation was inversely and significantly associated with WLZ, such that a 10-unit increase in the HEI–2015 total score from pregnancy through three months postpartum was associated with an approximately 0.12 lower infant WLZ from birth to six months in the fully-adjusted model ($\beta = -0.12$, $p = 0.02$) (Table 3). There was no significant interaction between maternal diet quality from pregnancy through three months postpartum and time, suggesting no difference between prenatal and postnatal exposure on these outcomes. No associations of maternal diet quality from pregnancy through three months postpartum with infant LAZ or WAZ were observed. In sensitivity analyses, findings for infant LAZ, WAZ, and WLZ were comparable when models were restricted to the 75% of study infants who were fully breastfed to six months.

3.3. Maternal Diet Quality and Infant Body Composition

Maternal diet quality during pregnancy and lactation was also inversely associated with infant adiposity (Table 4). Specifically, a 10-unit increase in HEI–2015 total scores during pregnancy was associated with an approximately 0.6% lower infant BF% at six months ($\beta = -0.58$, $p = 0.05$). A 10-unit increase in HEI–2015 total score at one month postpartum was associated with more than a 1% lower infant BF% ($\beta = -1.28$, $p < 0.001$) and 0.10 kg lower FM ($\beta = -0.13$, $p = 0.001$) at six months. Likewise, a 10-unit increase in HEI–2015 total score at three months postpartum was associated with an approximately 0.7% lower infant BF% ($\beta = -0.66$, $p = 0.01$) and 0.10 kg lower FM ($\beta = -0.10$, $p = 0.01$) at six months. Findings were similar for associations between maternal diet quality at one and three months postpartum and FMI/FFMI (data not shown). In sub-analyses, a 10-unit increase in HEI–2015 total scores at one and three months postpartum were associated with both lower trunk FM ($\beta = -0.06$,

p = 0.001 and β = −0.04, p = 0.03, respectively) and lower leg FM (β = −0.05, p = 0.01 and β = −0.04, p = 0.03, respectively). No other associations between maternal HEI–2015 scores during pregnancy or lactation and infant body composition were found. In sensitivity analyses that restricted the analysis to study infants that were fully breastfeeding at six months of age, associations between maternal HEI–2015 scores and infant body composition were similar to those incorporating all infants, except for associations between maternal HEI–2015 scores at three months postpartum and infant BF% and FM at six months, which were no longer significant (p < 0.10).

Table 2. Average infant growth measures from birth to six months and body composition measures at six months.

Infant Characteristics	N	Mean ± SD
Infant Growth from Birth to Six Months		
Weight-for-age, Z-scores		
Birth	350	0.46 ± 0.87
one month	343	0.13 ± 0.87
three months	332	−0.07 ± 0.88
six months	321	0.04 ± 0.96
Length-for-age, Z-scores		
Birth	342	1.17 ± 1.23
one month	346	0.06 ± 1.09
three months	332	−0.03 ± 1.06
six months	321	−0.20 ± 1.09
Weight-for-length, Z-scores		
Birth	341	−0.72 ± 1.40
one month	342	0.12 ± 1.08
three months	332	0.02 ± 0.98
six months	321	0.33 ± 1.07
Infant body composition at six months		
Total body fat, %	317	33.98 ± 3.76
Fat mass, kg	317	2.76 ± 0.48
Fat-free mass, kg	317	5.29 ± 0.67
Trunk fat mass, kg	317	0.92 ± 0.22
Arm fat mass, kg	317	0.38 ± 0.20
Leg fat mass, kg	317	1.08 ± 0.23

Abbreviations: SD = standard deviation.

Table 3. Associations of HEI–2015 total scores from the third trimester of pregnancy through three months postpartum with infant growth measures from birth to six months.

Infant Growth Measures from Birth to Six Months	Model 1				Model 2			
	N	β [a]	SE	p-Value	N	β [a]	SE	p-Value
LAZ	330	0.05	0.04	0.25	290	0.02	0.04	0.58
WAZ	330	−0.02	0.03	0.43	290	−0.04	0.03	0.15
WLZ	330	−0.13	0.05	**0.01**	290	−0.12	0.05	**0.02**

Abbreviations: SE = standard error; LAZ = length-for-age Z-score; WAZ = weight-for-age Z-score; WLZ = weight-for-length Z-score. [a] HEI–2015 total score was converted such that a 1-unit increase corresponds to a 10-point increase in HEI–2015 score. Model 1 = adjusted for study site, maternal age, race, education, household income and total energy intake during pregnancy. Model 2 = Model 1 + maternal pre-pregnancy BMI, gestational weight gain, mode of delivery, parity, breastfeeding exclusivity at six months, infant sex, and gestational age. Bolded values are statistically significant at p < 0.05.

Table 4. Associations of HEI–2015 total scores during the third trimester of pregnancy and at one and three months postpartum with infant body composition at six months.

Infant Body Composition Measures at Six Months	Model 1				Model 2			
	N	β [a]	SE	*p*-Value	N	β [a]	SE	*p*-Value
HEI–2015 total scores during pregnancy								
BF%	281	−0.72	0.28	**0.01**	262	−0.58	0.29	**0.05**
FM	281	−0.06	0.04	0.12	262	−0.03	0.04	0.36
FFM	281	0.002	0.05	0.97	262	−0.001	0.05	0.99
HEI–2015 total scores at one month postpartum								
BF%	254	−1.22	0.30	**<0.001**	235	−1.28	0.30	**<0.001**
FM	254	−0.10	0.04	**0.01**	235	−0.13	0.04	**0.001**
FFM	254	−0.003	0.05	0.94	235	−0.05	0.04	0.28
HEI–2015 total scores at three months postpartum								
BF%	248	−0.69	0.25	**0.01**	229	−0.66	0.26	**0.01**
FM	248	−0.09	0.04	**0.02**	229	−0.10	0.04	**0.01**
FFM	248	−0.04	0.05	0.52	229	−0.05	0.06	0.37

Abbreviations: BF% = body fat percent; FM = fat mass; FFM = fat free mass. [a] HEI–2015 total score was converted such that a 1-unit increase corresponds to a 10-point increase in HEI–2015 score. Model 1 = adjusted for study site, maternal age, race, education, household income, and total energy intake during pregnancy. Model 2 = Model 1 + maternal pre-pregnancy BMI, gestational weight gain, mode of delivery, parity, breastfeeding exclusivity at six months, infant sex, gestational age, weight-for-length Z-score, and exact age at body composition assessment. Bolded values are statistically significant at $p < 0.05$.

4. Discussion

In this prospective cohort study, we examined the associations of maternal diet quality during pregnancy and lactation with infant growth from birth to six months and body composition at six months. We found that higher maternal diet quality (as evidenced by higher HEI–2015 scores) from pregnancy through three months postpartum was inversely associated with infant WLZ from birth to six months and BF% at six months. Similarly, higher maternal diet quality during lactation (one and three months postpartum) was inversely related to infant FM (specifically trunk and leg fat mass) at six months. These associations occurred independently of important confounders, such as pre-pregnancy BMI and gestational weight gain. This research adds to accumulating animal and human evidence that suggests that not only the in utero but also the postnatal nutritional environment influences infant growth and body composition in infants who are breastfed and points to a potentially important role of dietary constituents and quality in shaping offspring future health [6].

Our findings are concordant with those of several other studies assessing mostly maternal prenatal diet quality and offspring growth and/or body composition at birth or later in life [21–25]. In line with our findings, Shapiro et al. [23] showed that maternal HEI–2010 scores of ≤57 (v > 57) measured via repeated 24-h recalls throughout pregnancy were associated with higher infant BF% ($\beta = 0.58$, $p < 0.05$) and FM ($\beta = 20.74$, $p < 0.05$) at birth among mother-offspring pairs in the Healthy Start Study in the United States. Likewise, in a pooled analysis of mother-child pairs from Project Viva (United States) and the Rhea study (Greece) [24], a three-point increase in the Mediterranean diet score during the first—second trimester of pregnancy (evaluated using an FFQ) was associated with lower offspring BMI Z-scores in mid-childhood at a magnitude similar to that found for WLZ in our study ($\beta = -0.14$, 95% CI = −0.15, −0.13).

Conversely, several studies have not found significant associations between maternal prenatal diet and infant growth [8,21,40]. For example, using data from the Infant Feeding Practices Study II, Poon et al. [21] showed that both the alternate Mediterranean diet score and the Alternative Healthy Eating Index for Pregnancy (AHEI-P) diet score (measured using an FFQ during the third trimester of pregnancy) were unrelated to infant birth weight, birth size (large or small for gestational age), birth

WLZ or change in WLZ at 4–6 months. Similarly, Rifas-Shiman et al. [40] reported a non-significant lower risk of giving birth to a small-for-gestational age infant with increasing AHEI-P scores calculated from responses to an FFQ during the first trimester of pregnancy among mother–infant dyads in Project Viva. The literature does not adequately assess the role of maternal postpartum diet on infant growth and/or body composition, which theoretically influences the infant via breast milk or the household nutritional environment. In one study assessing maternal macronutrient intake during the first trimester of pregnancy and at five years postpartum, Murrin et al. [8] demonstrated that higher prenatal sugar intake and both prenatal and postnatal saturated fat intake as measured by an FFQ were associated with a non-significant trend towards higher likelihood of having overweight/obese offspring. Although many diet quality scores were moderately to strongly correlated in various populations [41,42], use of the HEI–2015 as a measure of diet quality during pregnancy and postpartum may explain some discrepancies in findings and may limit our ability to compare findings to other studies using different diet quality scores or prior versions of the HEI.

The HEI–2015 aims to reflect the most recent DGA (2015–2020) and is thus generalizable to other studies using these guidelines [32]. The mean HEI–2015 scores across pregnancy and lactation ranged from 65.9 to 67.2 in our cohort. These scores are higher than the average scores for American adults (~58 out of 100) [43] but may reflect the potential improvement in dietary intake that women may pursue during the perinatal period [44]. Our study population is also more socioeconomically advantaged and healthy relative to pregnant and lactating women in the United States, as evidenced by a higher maternal education level, household income, lower pre-pregnancy BMI, and longer duration of exclusive breastfeeding. Given the dearth of studies that parallel the timing and methods of our exposure and outcome assessment and confounders considered, additional research is warranted to replicate our findings in larger, more diverse cohorts.

It is important to note that, although our findings are qualitatively similar to those of several other studies of maternal perinatal diet with infant growth and/or body composition [23,24], our effect sizes and those of related studies are small relative to the contribution of other maternal factors, such as obesity and gestational weight gain [45]. Even so, it is likely that infant adiposity is multifactorial, with several elements each contributing small effects to overall growth and fat accrual [46]. Recognizing and targeting these immediately modifiable factors, such as maternal diet in holistic interventions, could have a significant influence on offspring obesity throughout the lifespan [47]. For example, within the context of our study, a 10-unit increase in the HEI–2015 score could translate into increasing consumption of fruits from none to ≥0.8 cup equivalents to maximize points for this adequacy subcomponent (10 points). Alternatively, mothers could reduce consumption of refined grains from ≥4.3 oz equivalents to ≤1.8 oz equivalents to maximize points for this moderation subcomponent (10 points).

Maternal perinatal diet quality may impact infant growth and/or body composition through numerous mechanisms. Aside from direct influences on fetal growth, maternal prenatal and postnatal nutrition may lead to epigenetic modifications that affect offspring metabolic function, growth hormone secretion and appetite programming [13,48]. Higher HEI scores were associated with greater intake of fruits, fibers, folate, and vitamin C, higher plasma concentrations of carotenoids and vitamin C, and greater dietary variety in previous research [49]. In our study, we found that mothers in the highest tertile of HEI–2015 had especially high scores (scores that were greater than or equal to 80% of the maximum possible points for index components) on the Whole and Total Fruits, Total Vegetables, Greens and Beans, Seafood and Plant Proteins, Total Protein Foods, Refined Grains, and Added Sugars components. These beneficial characteristics of maternal diet are associated with reduced oxidative stress and inflammation, which may be reflected in the nutritional environment in the household during this critical period of offspring growth and development and throughout the life course [50]. Maternal diet may also alter the fatty acid profile [51] and the hormonal [52] and growth factor [53] content of breast milk the infants in this study were ingesting, which could theoretically influence growth and body composition [54].

Our study has several notable strengths, including the prospective cohort design with repeated measures of maternal diet from pregnancy through postpartum and infant growth from birth to six months. These factors provide a valuable opportunity to assess concurrent and prospective associations of maternal nutrition with offspring growth at a critical period of sensitivity in the life course. Our use of overall diet quality as an exposure (v specific macronutrients) may better mirror whole diet, accounting for the synergistic, interactive, and cumulative effects of multiple foods and nutrients, and may be a more practical tool for nutrition communication throughout pregnancy and lactation [19]. We also measured infant body composition at six months of age using DXA, which accurately measures fat mass and soft-tissue body components and serves as a marker of potential developmental programming [55]. We were able to account for the influence of a range of demographic, clinical, and lifestyle factors, which may confound the association of maternal diet and offspring growth and/or body composition.

We acknowledge potential limitations in our study. Maternal dietary intake was self-reported by mothers using the DHQ-II, which is subject to recall bias, measurement error, and exposure misclassification [56,57]. Nonetheless, this FFQ has been evaluated for validity in adult populations [29] and our models adjusted for total energy intake to reduce potential systematic measurement error. Our relatively small sample size and loss of data at later time points may have decreased statistical power. However, mother–infant dyads with complete exposure and outcome data during pregnancy/at birth were similar to those with complete exposure and outcome data at three months postpartum/six months. Our study population comprises predominantly non-Hispanic white women, which may limit generalizability across race/ethnicities. We did not follow-up infants beyond six months of age, which may limit our understanding of the long-term consequences of maternal nutrition on offspring growth, body composition and cardiometabolic diseases across the life course [58,59]. Important confounding variables may have not been controlled for in analyses. Most notably, it is possible that maternal diet is a marker for infant feeding practices that may influence child growth outcomes (including responsive infant feeding and timing of introduction of solid foods). Lastly, given the observational nature of the study, it is not possible to draw causal inferences from observed associations.

5. Conclusions

In conclusion, we found that higher maternal diet quality from pregnancy through three months postpartum was associated with lower infant WLZ from birth to six months and lower infant BF% at six months. Likewise, higher maternal diet quality at one and three months postpartum was associated with lower total body, trunk, and leg FM at six months. These findings point to the importance of focusing on maternal diet quality at critical periods of development, and the appreciable, albeit small, independent influence that maternal nutrition may play in optimal infant growth and fat development. Additional research is needed to determine the interplay between maternal perinatal diet quality, long-term offspring growth/body composition, and later chronic disease risk susceptibility in hopes of aiding current efforts to develop more specific dietary guidelines for pregnant and postpartum women.

Supplementary Materials: The following are available online at http://www.mdpi.com/2072-6643/11/3/632/s1, Table S1: Maternal HEI–2015 subcomponent scores and energy intake during pregnancy by tertiles of maternal HEI–2015 total scores during pregnancy (*n* = 329); Figure S1: Distribution of Maternal Healthy Eating Index-2015 (HEI–2015) scores: (a) during the third trimester of pregnancy (T1: ≤63.9; T2: 64.0–70.9; T3: ≥70.9); (b) at one month postpartum (T1: ≤61.8; T2: 61.9–69.7; T3: ≥69.8); and (c) at three months postpartum (T1: ≤62.8; T2: 62.9–70.2; T3: ≥70.3).

Author Contributions: Conceptualization, D.A.F. and E.W.D.; Data curation, J.L.H.; Formal analysis, M.J.T., D.R.J.J., and E.W.D.; Funding acquisition, D.A.F. and E.W.D.; Investigation, L.P.F., K.M.D., and A.M.T.; Supervision, E.W.D.; Writing—original draft, M.J.T.; Writing—review & editing, L.P.F., K.M.D., A.M.T., E.O.K., P.M.M., K.M.W., K.M.R., D.A.F., D.R.J.J., L.J.H., and E.W.D.

Funding: The research reported in this publication was supported by the National Institute of Child Health and Human Development (NICHD) of the National Institutes of Health under Award Number R01HD080444. MJT was supported by the National Heart, Lung, and Blood Institute of the National Institutes of Health under Award

Number T32HL007779. The content is solely the responsibility of the authors and does not necessarily represent the official views of the National Institutes of Health.

Acknowledgments: We would like to gratefully acknowledge and thank the mothers who gave us their time and agreed to participate in the MILK study.

Conflicts of Interest: The authors declare no conflict of interest.

References

1. Calkins, K.; Devaskar, S.U. Fetal origins of adult disease. *Curr. Probl. Pediatr. Adolesc. Health Care* **2011**, *41*, 158–176. [CrossRef] [PubMed]
2. Kwon, E.J.; Kim, Y.J. What is fetal programming?: A lifetime health is under the control of in utero health. *Obstet. Gynecol. Sci.* **2017**, *60*, 506–519. [CrossRef] [PubMed]
3. Williams, C.B.; Mackenzie, K.C.; Gahagan, S. The effect of maternal obesity on the offspring. *Clin. Obstet. Gynecol.* **2014**, *57*, 508–515. [CrossRef]
4. Kaar, J.L.; Crume, T.; Brinton, J.T.; Bischoff, K.J.; McDuffie, R.; Dabelea, D. Maternal obesity, gestational weight gain, and offspring adiposity: The exploring perinatal outcomes among children study. *J. Pediatr.* **2014**, *165*, 509–515. [CrossRef] [PubMed]
5. Mourtakos, S.P.; Tambalis, K.D.; Panagiotakos, D.B.; Antonogeorgos, G.; Arnaoutis, G.; Karteroliotis, K.; Sidossiset, L.S. Maternal lifestyle characteristics during pregnancy, and the risk of obesity in the offspring: A study of 5125 children. *BMC Pregnancy Childbirth* **2015**, *15*, 1–8. [CrossRef] [PubMed]
6. Parlee, S.D.; MacDougald, O.A. Maternal nutrition and risk of obesity in offspring: The Trojan horse of developmental plasticity. *Biochim. Biophys. Acta* **2014**, *1842*, 495–506. [CrossRef]
7. Bayol, S.A.; Farrington, S.J.; Stickland, N.C. A maternal 'junk food' diet in pregnancy and lactation promotes an exacerbated taste for 'junk food' and a greater propensity for obesity in rat offspring. *Br. J. Nutr.* **2007**, *98*, 843–851. [CrossRef] [PubMed]
8. Murrin, C.; Shrivastava, A.; Kelleher, C.C.; Lifeways Cross-Generation Cohort Study Steering Group. Maternal macronutrient intake during pregnancy and 5 years postpartum and associations with child weight status aged five. *Eur. J. Clin. Nutr.* **2013**, *67*, 670–679. [CrossRef] [PubMed]
9. Jackson, C.M.; Alexander, B.T.; Roach, L.; Haggerty, D.; Marbury, D.C.; Hutchens, Z.M.; Flynn, E.R.; Maric-Bilkan, C. Exposure to maternal overnutrition and a high-fat diet during early postnatal development increases susceptibility to renal and metabolic injury later in life. *Am. J. Physiol. Renal Physiol.* **2012**, *302*, F774–F783. [CrossRef] [PubMed]
10. Section on Breastfeeding. Breastfeeding and the use of human milk. *Pediatrics* **2012**, *129*, 827–841. [CrossRef] [PubMed]
11. Centers for Disease Control and Prevention. Breastfeeding Report Card—United States. 2018. Available online: https://www.cdc.gov/breastfeeding/pdf/2018breastfeedingreportcard.pdf (accessed on 8 December 2018).
12. Ballard, O.; Morrow, A.L. Human milk composition: Nutrients and bioactive factors. *Pediatr. Clin. N. Am.* **2013**, *60*, 49–74. [CrossRef] [PubMed]
13. Indrio, F.; Martini, S.; Francavilla, R.; Corvaglia, L.; Cristofori, F.; Mastrolia, S.A.; Neu, J.; Rautava, S.; Spena, G.R.; Raimondi, F.; et al. Epigenetic Matters: The Link between Early Nutrition, Microbiome, and Long-term Health Development. *Front. Pediatr.* **2017**, *5*, 1–14. [CrossRef] [PubMed]
14. Chu, D.M.; Meyer, K.M.; Prince, A.L.; Aagaard, K.M. Impact of maternal nutrition in pregnancy and lactation on offspring gut microbial composition and function. *Gut Microbes* **2016**, *7*, 459–470. [CrossRef] [PubMed]
15. Dhana, K.; Haines, J.; Liu, G.; Zhang, C.; Wang, X.; Field, A.E.; Chavarro, J.E.; Sun, Q. Association between maternal adherence to healthy lifestyle practices and risk of obesity in offspring: Results from two prospective cohort studies of mother-child pairs in the United States. *BMJ* **2018**, *362*, k2486. [CrossRef]
16. Tielemans, M.J.; Steegers, E.A.P.; Voortman, T.; Jaddoe, V.W.V.; Rivadeneira, F.; Franco, O.H.; Kiefte-de Jong, J.C. Protein intake during pregnancy and offspring body composition at 6 years: The Generation R Study. *Eur. J. Nutr.* **2017**, *56*, 2151–2160. [CrossRef] [PubMed]
17. Moon, R.J.; Harvey, N.C.; Robinson, S.M.; Ntani, G.; Davies, J.H.; Inskip, H.M.; Godfrey, K.M.; Dennison, E.M.; Calder, P.C.; Cooper, C.; et al. Maternal plasma polyunsaturated fatty acid status in late pregnancy is associated with offspring body composition in childhood. *J. Clin. Endocrinol. Metab.* **2013**, *98*, 299–307. [CrossRef] [PubMed]

18. Mckenzie, K.M.; Dissanayake, H.U.; McMullan, R.; ICaterson, I.D.; Celermajer, D.S.; Gordon, A.; Hyett, J.; Meroni, A.; Phang, M.; Raynes-Greenow, C.; et al. Quantity and Quality of Carbohydrate Intake during Pregnancy, Newborn Body Fatness and Cardiac Autonomic Control: Conferred Cardiovascular Risk? *Nutrients* **2017**, *9*, 1375. [CrossRef]
19. Tapsell, L.C.; Neale, E.P.; Satija, A.; Hu, F.B. Foods, Nutrients, and Dietary Patterns: Interconnections and Implications for Dietary Guidelines. *Adv. Nutr.* **2016**, *7*, 445–454. [CrossRef]
20. Schwingshackl, L.; Hoffmann, G. Diet quality as assessed by the Healthy Eating Index, the Alternate Healthy Eating Index, the Dietary Approaches to Stop Hypertension score, and health outcomes: A systematic review and meta-analysis of cohort studies. *J. Acad. Nutr. Diet.* **2015**, *115*, 780–800. [CrossRef]
21. Poon, A.K.; Yeung, E.; Boghossian, N.; Albert, P.S.; Zhang, C. Maternal Dietary Patterns during Third Trimester in Association with Birthweight Characteristics and Early Infant Growth. *Scientifica* **2013**, *2013*, 786409. [CrossRef]
22. Rodríguez-Bernal, C.L.; Rebagliato, M.; Iñiguez, C.; Vioque, J.; Navarrete-Muñoz, E.M.; Murcia, M.; Bolumar, F.; Marco, A.; Ballester, F. Diet quality in early pregnancy and its effects on fetal growth outcomes: The Infancia y Medio Ambiente (Childhood and Environment) Mother and Child Cohort Study in Spain. *Am. J. Clin. Nutr.* **2010**, *91*, 1659–1666. [CrossRef]
23. Shapiro, A.L.; Kaar, J.L.; Crume, T.L.; Starling, A.P.; Siega-Riz, A.M.; Ringham, B.M.; Glueck, D.H.; Norris, J.M.; Barbour, L.A.; Friedman, J.E.; et al. Maternal diet quality in pregnancy and neonatal adiposity: The Healthy Start Study. *Int. J. Obes.* **2016**, *40*, 1056–1062. [CrossRef]
24. Chatzi, L.; Rifas-Shiman, S.L.; Georgiou, V.; Joung, K.E.; Koinaki, S.; Chalkiadaki, G.; Margioris, A.; Sarri, K.; Vassilaki, M.; Vafeiadi, M.; et al. Adherence to the Mediterranean diet during pregnancy and offspring adiposity and cardiometabolic traits in childhood. *Pediatr. Obes.* **2017**, *12*, 47–56. [CrossRef] [PubMed]
25. Van den Broek, M.; Leermakers, E.T.; Jaddoe, V.W.; Steegers, E.A.; Rivadeneira, F.; Raat, H.; Hofman, A.; Franco, O.H.; Kiefte-de Jong, J.C. Maternal dietary patterns during pregnancy and body composition of the child at age 6 y: The Generation R Study. *Am. J. Clin. Nutr.* **2015**, *102*, 873–880. [CrossRef] [PubMed]
26. Gillman, M.W. Early infancy as a critical period for development of obesity and related conditions. In *Importance of Growth for Health and Development*; Nestec Ltd.: Vevey, Switzerland; S. Karger AG: Basel, Switzerland, 2010; Volume 65, pp. 13–20.
27. Rolland-Cachera, M.F.; Péneau, S. Growth trajectories associated with adult obesity. *World Rev. Nutr. Diet.* **2013**, *106*, 127–134. [PubMed]
28. Whitaker, K.M.; Marino, R.C.; Haapala, J.L.; Foster, L.; Smith, K.D.; Teague, A.M.; Jacobs, D.R., Jr.; Fontaine, P.L.; McGovern, P.M.; Schoenfuss, T.C.; et al. Associations of Maternal Weight Status Before, During, and After Pregnancy with Inflammatory Markers in Breast Milk. *Obesity* **2017**, *25*, 2092–2099. [CrossRef] [PubMed]
29. Subar, A.F.; Thompson, F.E.; Kipnis, V.; Midthune, D.; Hurwitz, P.; McNutt, S.; McIntosh, A.; Rosenfeld, S. Comparative validation of the Block, Willett, and National Cancer Institute food frequency questionnaires: The Eating at America's Table Study. *Am. J. Epidemiol.* **2001**, *154*, 1089–1099. [CrossRef] [PubMed]
30. Subar, A.F.; Kipnis, V.; Troiano, R.P.; Midthune, D.; Schoeller, D.A.; Bingham, S.; Sharbaugh, C.O.; Trabulsi, J.; Runswick, S.; Ballard-Barbash, R.; et al. Using intake biomarkers to evaluate the extent of dietary misreporting in a large sample of adults: The OPEN study. *Am. J. Epidemiol.* **2003**, *158*, 1–13. [CrossRef] [PubMed]
31. Kipnis, V.; Subar, A.F.; Midthune, D.; Midthune, D.; Schoeller, D.A.; Bingham, S.; Sharbaugh, C.O.; Trabulsi, J.; Runswick, S.; Ballard-Barbash, R.; et al. Structure of dietary measurement error: Results of the OPEN biomarker study. *Am. J. Epidemiol.* **2003**, *158*, 14–21. [CrossRef] [PubMed]
32. Krebs-Smith, S.M.; Pannucci, T.E.; Subar, A.F.; Kirkpatrick, S.I.; Lerman, J.L.; Tooze, J.A.; Wilson, M.M.; Reedy, J. Update of the Healthy Eating Index: HEI-2015. *J. Acad. Nutr. Diet.* **2018**, *118*, 1591–1602. [CrossRef]
33. Reedy, J.; Lerman, J.L.; Krebs-Smith, S.M.; Kirkpatrick, S.I.; Pannucci, T.E.; Wilson, M.M.; Subar, A.F.; Kahle, L.L.; Tooze, J.A. Evaluation of the Healthy Eating Index-2015. *J. Acad. Nutr. Diet.* **2018**, *118*, 1622–1633. [CrossRef]
34. Pick, M.E.; Edwards, M.; Moreau, D.; Ryan, E.A. Assessment of diet quality in pregnant women using the Healthy Eating Index. *J. Am. Diet. Assoc.* **2005**, *105*, 240–246. [CrossRef] [PubMed]
35. National Cancer Institute. SAS Code. US Department of Health and Human Services. 2018. Available online: https://epi.grants.cancer.gov/hei/sas-code.html (accessed on 22 January 2019).

36. WHO Multicentre Growth Reference Study Group. WHO Child Growth Standards based on length/height, weight and age. *Acta Paediatr. Suppl.* **2006**, *450*, 76–85.

37. Mazess, R.B.; Barden, H.S.; Bisek, J.P.; Hanson, J. Dual-energy X-ray absorptiometry for total-body and regional bone-mineral and soft-tissue composition. *Am. J. Clin. Nutr.* **1990**, *51*, 1106–1112. [CrossRef] [PubMed]

38. Lohman, T.M.R.; Roche, A.F. *Anthropometric Standardization Reference Manual*; Human Kinetics Books: Champaign, IL, USA, 2011.

39. Weber, D.R.; Leonard, M.B.; Zemel, B.S. Body composition analysis in the pediatric population. *Pediatr. Endocrinol. Rev.* **2012**, *10*, 130–139. [PubMed]

40. Rifas-Shiman, S.L.; Rich-Edwards, J.W.; Kleinman, K.P.; Oken, E.; Gillman, M.W. Dietary quality during pregnancy varies by maternal characteristics in Project Viva: A US cohort. *J. Am. Diet. Assoc.* **2009**, *109*, 1004–1011. [CrossRef]

41. George, S.M.; Ballard-Barbash, R.; Manson, J.E.; Reedy, J.; Shikany, J.M.; Subar, A.F.; Tinker, L.F.; Vitolins, M.; Neuhouser, M.L. Comparing indices of diet quality with chronic disease mortality risk in postmenopausal women in the Women's Health Initiative Observational Study: Evidence to inform national dietary guidance. *Am. J. Epidemiol.* **2014**, *180*, 616–625. [CrossRef]

42. Fallaize, R.; Livingstone, K.M.; Celis-Morales, C.; Macready, A.L.; San-Cristobal, R.; Navas-Carretero, S.; Marsaux, C.F.; O'Donovan, C.B.; Kolossa, S.; Moschonis, G.; et al. Association between Diet-Quality Scores, Adiposity, Total Cholesterol and Markers of Nutritional Status in European Adults: Findings from the Food4Me Study. *Nutrients* **2018**, *10*, 49. [CrossRef]

43. United States Department of Agriculture. The Center for Nutrition Policy and Promotion. HEI Scores for Americans. Available online: https://www.cnpp.usda.gov/hei-scores-americans (accessed on 22 January 2019).

44. Forbes, L.E.; Graham, J.E.; Berglund, C.; Bell, R.C. Dietary Change during Pregnancy and Women's Reasons for Change. *Nutrients* **2018**, *10*, 1032. [CrossRef]

45. Black, M.H.; Sacks, D.A.; Xiang, A.H.; Lawrence, J.M. The relative contribution of prepregnancy overweight and obesity, gestational weight gain, and IADPSG-defined gestational diabetes mellitus to fetal overgrowth. *Diabetes Care* **2013**, *36*, 56–62. [CrossRef] [PubMed]

46. Hruby, A.; Hu, F.B. The Epidemiology of Obesity: A Big Picture. *Pharmacoeconomics* **2015**, *33*, 673–689. [CrossRef] [PubMed]

47. Haire-Joshu, D.; Tabak, R. Preventing Obesity Across Generations: Evidence for Early Life Intervention. *Annu. Rev. Public Health* **2016**, *37*, 253–271. [CrossRef]

48. Mühlhäusler, B.S.; Adam, C.L.; McMillen, I.C. Maternal nutrition and the programming of obesity: The brain. *Organogenesis* **2008**, *4*, 144–152. [CrossRef]

49. Hann, C.S.; Rock, C.L.; King, I.; Drewnowski, A. Validation of the Healthy Eating Index with use of plasma biomarkers in a clinical sample of women. *Am. J. Clin. Nutr.* **2001**, *74*, 479–486. [CrossRef] [PubMed]

50. Shrivastava, A.; Murrin, C.; Sweeney, M.R.; Heavey, P.; Kelleher, C.C.; Lifeways Cross-generation Cohort Study Steering Group. Familial intergenerational and maternal aggregation patterns in nutrient intakes in the Lifeways Cross-Generation Cohort Study. *Public Health Nutr.* **2013**, *16*, 1476–1486. [CrossRef] [PubMed]

51. Bravi, F.; Wiens, F.; Decarli, A.; Dal Pont, A.; Agostoni, C.; Ferraroni, M. Impact of maternal nutrition on breast-milk composition: A systematic review. *Am. J. Clin. Nutr.* **2016**, *104*, 646–662. [CrossRef] [PubMed]

52. Kocaadam, B.; Köksal, E.; Türkyılmaz, C. Are breast milk adipokines affected by maternal dietary factors? *J. Pediatr. Endocrinol. Metab.* **2018**, *31*, 1099–1104. [CrossRef] [PubMed]

53. Lu, M.; Jiang, J.; Wu, K.; Li, D. Epidermal growth factor and transforming growth factor-α in human milk of different lactation stages and different regions and their relationship with maternal diet. *Food Funct.* **2018**, *9*, 1199–1204. [CrossRef]

54. Lind, M.V.; Larnkjær, A.; Mølgaard, C.; Michaelsen, K.F. Breastfeeding, Breast Milk Composition, and Growth Outcomes. In *Recent Research in Nutrition and Growth*; Nestlé Nutrition Institute: Vevey, Switzerland; S. Karger AG: Basel, Switzerland, 2018; Volume 89, pp. 63–77.

55. Demerath, E.W.; Fields, D.A. Body composition assessment in the infant. *Am. J. Hum. Biol.* **2014**, *26*, 291–304. [CrossRef]

56. Naska, A.; Lagiou, A.; Lagiou, P. Dietary assessment methods in epidemiological research: Current state of the art and future prospects. *F1000Research* **2017**, *6*, 1–8. [CrossRef]

57. Shim, J.S.; Oh, K.; Kim, H.C. Dietary assessment methods in epidemiologic studies. *Epidemiol. Health* **2014**, *36*, 1–8. [CrossRef] [PubMed]
58. Maslova, E.; Rytter, D.; Bech, B.H.; Henriksen, T.B.; Rasmussen, M.A.; Olsen, S.F.; Halldorsson, T.I. Maternal protein intake during pregnancy and offspring overweight 20 y later. *Am. J. Clin. Nutr.* **2014**, *100*, 1139–1148. [CrossRef] [PubMed]
59. Maslova, E.; Rytter, D.; Bech, B.H.; Henriksen, T.B.; Olsen, S.F.; Halldorsson, T.I. Maternal intake of fat in pregnancy and offspring metabolic health—A prospective study with 20 years of follow-up. *Clin. Nutr.* **2016**, *35*, 475–483. [CrossRef] [PubMed]

nutrients

MDPI

Article

Growth Patterns of Neonates Treated with Thermal Control in Neutral Environment and Nutrition Regulation to Meet Basal Metabolism

Shiro Kubota [1,2,†], Masayoshi Zaitsu [3,4,†] and Tatsuya Yoshihara [2,5,*]

1   Kubota Life Science Laboratory Co., Ltd., Saga 840-0535, Japan; kubotahp@gmail.com
2   Kubota Maternity Clinic, Fukuoka 810-0014, Japan
3   Department of Social and Behavioral Sciences, Harvard T.H. Chan School of Public Health,
    Boston, MA 02115, USA; m-zaitsu@m.u-tokyo.ac.jp
4   Department of Public Health, Graduate School of Medicine, The University of Tokyo, Tokyo 113-0033, Japan
5   Clinical Research Center, Fukuoka Mirai Hospital, Fukuoka 813-0017, Japan
*   Correspondence: tatsuya@clipharm.med.kyushu-u.ac.jp; Tel.: +81-92-662-3608
†   These authors contributed equally to this work.

Received: 8 February 2019; Accepted: 5 March 2019; Published: 11 March 2019

Abstract: Little is known about the growth patterns of low birth weight neonates (<2500 g) during standardized thermal control and nutrition regulation to meet basal metabolism requirements compared to those of non-low birth weight neonates (2500 g and above). We retrospectively identified 10,544 non-low birth weight and 681 low birth weight neonates placed in thermo-controlled incubators for up to 24 h after birth. All neonates were fed a 5% glucose solution 1 h after birth and breastfed every 3 h (with supplementary formula milk if applicable) to meet basal metabolism requirements. Maximum body-weight loss (%), percentage body-weight loss from birth to peak weight loss (%/day), and percentage body-weight gain from peak weight loss to day 4 (%/day) were assessed by multivariable linear regression. Overall, the growth curves showed a uniform J-shape across all birth weight categories, with a low mean maximum body-weight loss (1.9%) and incidence of neonatal jaundice (0.3%). The body-weight loss patterns did not differ between the two groups. However, low birth weight neonates showed significantly faster growth patterns for percentage body-weight gain: $\beta = 0.52$ (95% confidence interval, 0.46 to 0.58). Under thermal control and nutrition regulation, low birth weight neonates might not have disadvantages in clinical outcomes or growth patterns.

Keywords: growth chart; breastfeeding; physiological body-weight loss; thermal control; basal maintenance expenditure

1. Introduction

During the first days of life, infants show a physiological phenomenon of temporary body-weight loss. However, excess body-weight loss in the neonatal period tends to increase risks of hypoglycemia, hypernatremia, and hyperbilirubinemia, resulting in permanent neurological damage [1]. While the American Academy of Pediatrics criteria currently recommend a cut-off point of 7% for physiological body-weight loss [2], excess loss resulting from inadequate intake is an emerging concern among high-risk populations, such as (but not limited to) East-Asian neonates, even after controlling for genetic polymorphisms [3–6].

In a cohort of non-low birth weight (NLBW) neonates (birth weight 2500 g and above) in Japan not admitted to the neonatal intensive care unit (NICU), we previously reported a potential standardized neonatal regimen, a local neonatal protocol involving a combination of thermal control and nutrition regulation to meet basal metabolism requirements (~50 kcal/kg), in order to reduce the incidence of neonatal jaundice by preventing excess body-weight loss [7]. This combination method included a neutral thermal environment during the first hours after birth with an ambient room temperature of ~34 °C (93.2 °F) [8–10]. In contrast to ambient room temperature (~24 °C/75.2 °F), a neutral thermal environment maintained an optimal body temperature through circulatory stability and improved digestive function [8–10]. The combination method also included nutrition regulation to meet the basal metabolism requirement [11], in which neonates were fed a 5% glucose solution 1 h after birth, a convenient method to prevent hypoglycemia and neonatal jaundice [12], followed by breastfeeding every 3 h with supplementary formula milk if applicable. In previous findings, compared to either thermal control or nutrition regulation alone, the combination method had better safety profiles for neonatal growth (e.g., serum glucose levels, circulatory stability, digestive functions, and metabolism) [7,8].

In addition to gestational age and maternal factors (such as maternal smoking, preeclampsia, and gestational diabetes), birth weight plays a crucial role in neonatal body weight growth. Compared to NLBW neonates, low birth weight (LBW) neonates (birth weight <2500 g) are more likely to be at risk for disadvantages in clinical outcomes or growth patterns, such as excess body-weight loss, neonatal jaundice, and delayed body-weight growth [13–17]. However, few studies have focused on LBW neonates not admitted to the NICU, and little is known about the contribution of our combination method among LBW neonates.

Accordingly, the goal of the present study was to characterize the body-weight growth patterns and clinical outcomes in this population of neonates who had received the combination method. Using data from a cohort with NLBW and LBW neonates treated by the combination method, we examined the detailed patterns of body-weight growth among NLBW and LBW neonates during the first days of life. We also examined whether the body-weight growth pattern in LBW neonates differed from that in NLBW neonates under the combination method.

2. Materials and Methods

2.1. Study Settings

This retrospective, longitudinal study of neonates during the initial days of life used clinical data from a cohort of neonates born in Kubota Maternity Clinic in Fukuoka, Japan (1989 to 2017). The details of the database are described elsewhere [7]. Briefly, Kubota Maternity Clinic was a general obstetrics and gynecological hospital that provided general obstetrics care and closed in 2017. The characteristics of the population were similar to those reported in previous studies in Japan [3,18,19]. All neonates were treated with the combination method (standardized thermal control and nutrition regulation) and discharged by a physician on day 4 or later, when the serum bilirubin level was <15 mg/dL, body weight was recovered from its lowest level, and phototherapy was no longer required. Since the bilirubin level usually peaks at approximately day 4 among Japanese neonates [18,19], discharge on day 4 or later is the standard practice in Japan. Clinical information on all neonates and their mothers had been collected since 1989, except for those who had neonatal asphyxia, congenital heart failure or malformation, or were transferred to a tertiary hospital (such as a university hospital) for neonatal intensive care. We obtained a de-identified dataset through a research agreement with the hospital; the study was approved by the research ethics committees of Hakata Clinic and registered at UMIN-CTR (UMIN000030011).

We included all 11,445 eligible term and preterm neonates treated with the combination method who (1) had a birth weight 1700 g and greater, (2) had an Apgar score of 7 or higher, and (3) were not admitted to the NICU. We excluded 221 neonates with incomplete data (1.9%), yielding a study population of 11,224 neonates.

2.2. Thermal Control and Nutrition Regulation to Meet Basal Metabolism

The details of standardized thermal control and nutrition regulation to meet basal metabolism, a local neonatal regimen, have been described elsewhere [7,8]. Briefly, following delivery in a delivery room maintained at ~25 °C (77 °F), neonates were immediately wiped with cotton towels, administered intraoral suction on a warm bed at ~40 °C (104 °F), hugged by their mothers (skin-to-skin contact), and placed in a transparent thermo-controlled incubator (N-ideal H-2000, Nakamura Medical Industry, Tokyo, Japan) within 2 min of birth. Transparent thermo-controlled incubators were placed next to delivery beds in the delivery room (not placed in the NICU). Neonates remained visible and were continuously observed, not only by physicians and nurses (e.g., Apgar score), but also by their mothers. All neonates remained in incubators for at least 2 h. For the first hour, the incubators were set at 34 °C (93.2 °F); for the second hour, the temperature was turned down to 30 °C (86 °F) to help neonates adapt to normal room temperature. After the initial 2 h, NLBW neonates were transferred to a bed in a standard monitoring room set at ~24 to 26 °C (75.2 to 78.8 °F). LBW neonates with birth weights of 2000 to 2499 g stayed in incubators for an extra 2 to 12 h, while those with birth weights <2000 g stayed in incubators for an extra 12 to 24 h at 28 °C (82.4 °F) before transferring to a standard monitoring room. The optimal profiles of the neutral environment for thermogenesis, circulatory stability, and actual central/peripheral body temperatures have been validated in previous studies with human neonates [8–10].

For nutritional regulation, neonates were orally fed a 5% glucose solution (10 mL/kg) with 1 mL of vitamin K syrup containing 2 mg menatetrenone 1 h after birth and then breastfed every 3 h [7,12]. If breast milk production was insufficient, neonates were additionally bottle-fed formula milk until they were sated to maintain a 50 kcal/kg basal metabolism [11]. The oral 5% glucose solution has been shown to be a convenient method to prevent hypoglycemia, while allowing continuous breastfeeding [7,12]. In addition, supplemental bottle-feeding with formula milk was based on previous findings suggesting that the estimated calories provided by breast milk were below the basal metabolism a few days after birth among the study population and that nutrition regulation did not affect breastmilk production [7]. For NLBW neonates, high-calorie formula milk (16 kcal/20 mL) was administered for the first 48 h, followed by normal-calorie formula milk (13 kcal/20 mL) after 48 h. For LBW neonates, high-calorie formula milk was administered throughout the hospital stay.

2.3. Birth Weight Categories

Neonates, whose birth weights were normally distributed (Figure 1), were grouped into NLBW (2500 g and above; n = 10,544) and LBW (<2500 g; n = 680) neonates. Both groups were further categorized into a narrower range of birth weights: among LBW neonates, <2000 g (n = 20) and 2000 to 2499 g (n = 660); among NLBW neonates, 2500 to 2999 g (n = 5001), 3000 to 3499 g (n = 4752), 3,500 to 3999 g (n = 757), and 4000 g and above (n = 34).

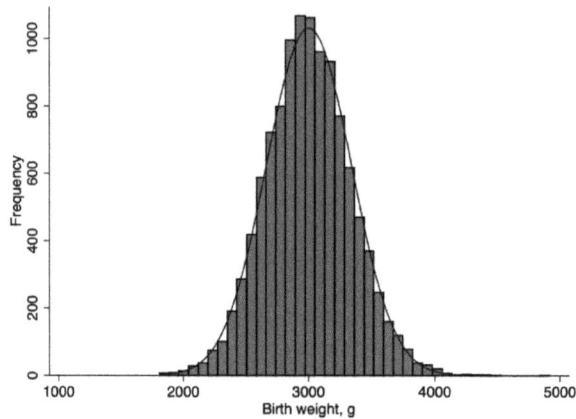

Figure 1. Distribution of birth weight among 11,224 neonates who received standardized thermal control and nutrition regulation to meet basal metabolism.

2.4. Assessment of Clinical Outcomes and Changes in Body Weight

Using body-weight measured at birth and every day during the first four days (SW-5200, Toitu, Tokyo, Japan), we determined the percentage body-weight growth every day [((weight each day − birth weight]/birth weight) × 100%) and maximum body-weight loss (([birth weight − minimum weight]/birth weight) × 100%). Excess body-weight loss was defined as a maximum body-weight loss of at least 7% [2]. The incidence of neonatal jaundice was defined as use of phototherapy, which was based on peak bilirubin levels ≥18 mg/dL along with other clinical findings [7].

Due to physiological body-weight loss, we divided the observation period into "decreasing" (before the day of peak weight loss) and "increasing" (after the day of peak weight loss) periods. For each period, we assessed the percentage body-weight change per day. During the first decreasing period from birth to peak weight loss, we determined the body-weight loss per day ((birth weight − minimum weight)/day of peak weight loss) and percentage body-weight loss per day (([birth weight − minimum weight]/birth weight) × (100%/day of peak weight loss)). During the latter increasing period, from peak weight loss to day 4, we determined the body-weight gain per day ((weight at day 4 − minimum weight)/(4 − day of peak weight loss)) and percentage body-weight gain per day ((weight at day 4 − minimum weight]/birth weight) × (100%/(4 − day of peak weight loss)).

2.5. Statistical Analysis

To produce growth curves in a growth chart among the 11,674 neonates, we plotted the mean body weights measured at birth and every day during the first four days separately by birth weight categories. We directly connected the mean values with lines, because fitted lines such as splines eliminated the initial drop of body weight in prior analyses (data not shown). We also drew a chart indicating the percentages of body-weight growth stratified by birth weight. In these charts, we additionally plotted the values measured at day 5 for neonates discharged at day 5 and later (10,336 neonates, 92%) and the values measured at one-month neonatal check-ups (approximately 30 days after birth; 8071 neonates, 72%).

The background demographics and clinical outcomes were then compared between the LBW and NLBW neonates by t- or chi-squared tests. In the decreasing period, the regression coefficient (β) and 95% confidence interval (CI) for the percentage body-weight loss per day against birth weight (referent group, NLBW neonates) were estimated by linear regression. Covariates included sex, gestational age, Apgar score, Cesarean delivery, maternal age, maternal body mass index, parity, hypertensive disorders of pregnancy, and birth year. In the increasing period, the same regression analysis was applied for the percentage body-weight gain per day. In stratified analyses using a continuous variable of birth weight, a linear regression model was applied for analysis among NLBW and NLBW neonates, respectively, to check the association between birth weight and percentage body-weight-growth.

The alpha value was set at 0.05 and all P-values were two-sided. Data were analyzed using STATA/MP 13.1 (Stata-Corp, College Station, TX, USA).

3. Results

Overall, the growth curves showed a uniform, J-shaped pattern across all birth weight categories (Figure 2). On average, the neonates started to recover their weights within two days of birth and the mean maximum body-weight loss (mean (SD)) was low (1.9% (1.5%)). The incidence of neonatal jaundice was 0.3% and the incidence of excess body-weight loss was 0.4% (Table 1). We did not observe any neonatal morbidity or mortality events, as well as readmission after discharge, during the study period. Most of the background distributions differed between the NLBW and LBW neonates, except for maternal age (Table 1).

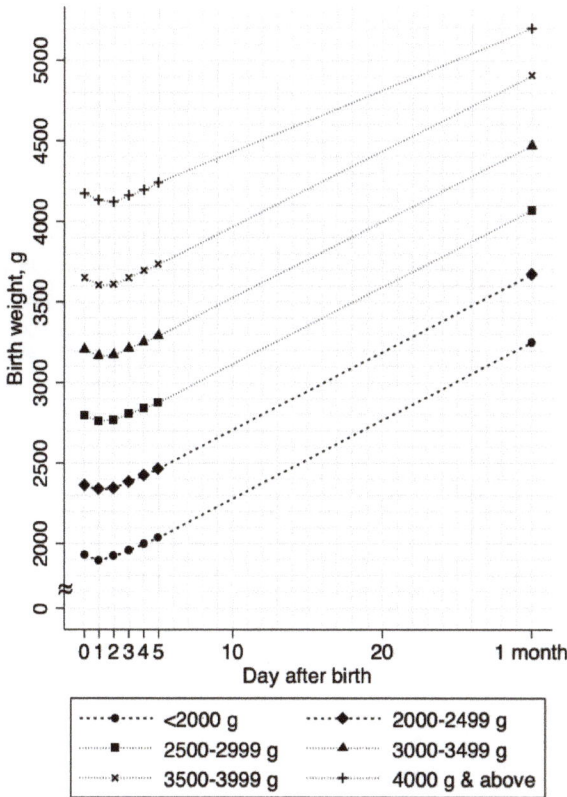

Figure 2. Body weight growth curves stratified by birth weights. The mean body weights are shown as connected lines stratified by birth weight categories. Values from day 0 through day 4 were estimated in 11,224 neonates; values at day 5 and 1 month (approximately 30 days of birth) were estimated in 10,336 neonates and 8071 neonates, respectively.

Table 1. Background and clinical characteristics of 11,224 neonates who received optimal thermal control with sufficient nutrition.

Characteristics	Mean (SD) or number (%)			*p*
	Total *n* = 11,224	NLBW Neonates (≥2500 g) *n* = 10,544	LBW Neonates (<2500 g) *n* = 680	
Background characteristics				
Female	5468 (49%)	5066 (48%)	402 (59%)	<0.001
Gestational age (weeks)	39.0 (1.2)	39.0 (1.1)	37.9 (1.4)	<0.001
Birth weight (g)	3000 (337)	3,043 (300)	2346 (136)	<0.001
Apgar score at 1 min	9.6 (0.6)	9.6 (0.6)	9.5 (0.6)	<0.001
Caesarean delivery	899 (8.0%)	810 (7.7%)	89 (13%)	<0.001
Maternal age (years)	31 (4)	31 (4)	31 (4)	0.58
Maternal BMI (kg/m^2)	20.0 (2.0)	20.1 (2.0)	19.7 (2.0)	<0.001
Multipara	5,256 (47%)	4,982 (47%)	274 (40%)	<0.001
Hypertensive disorders of pregnancy	43 (0.4%)	36 (0.3%)	7 (1.0%)	0.005
Birth year	2001 (8)	2001 (8)	2002 (8)	0.001
Clinical outcomes				
Maximum weight loss (%)	1.9 (1.5)	1.9 (1.5)	1.8 (1.5)	0.056
Excess weight loss 7% and above	44 (0.4%)	39 (0.4%)	5 (0.7%)	0.14
Day of peak weight loss	1.4 (0.9)	1.4 (0.9)	1.3 (0.9)	0.004
Body-weight loss per day (g)	40 (36)	41 (36)	31 (29)	<0.001
Percentage body-weight loss per day (%) [1]	1.3 (1.2)	1.3 (1.2)	1.3 (1.2)	0.52
Body-weight gain per day (g)	41 (24)	41 (25)	39 (21)	0.06
Percentage body-weight gain per day (%) [2]	1.4 (0.8)	1.4 (0.8)	1.7 (0.9)	<0.001
Peak bilirubin level at day 4 (mg/dL)	8.5 (2.7)	8.5 (2.7)	8.4 (2.7)	0.42
Phototherapy	30 (0.3%)	30 (0.3%)	0 (0.0%)	0.16

BMI, body mass index; NLBW, non-low birth weight; LBW, low birth weight. [1] Defined as ((birth weight − minimum weight)/birth weight) × (100%/day of peak weight loss). [2] Defined as ((weight at day 4 − minimum weight)/birth weight) × (100%/(4 − day of peak weight loss])).

Clinical outcomes, including maximum body-weight loss, bilirubin levels, and phototherapy, did not differ between NLBW and LBW neonates. Although the mean body-weight loss per day (grams) differed, the mean body-weight gain per day (grams) did not differ between groups (Table 1). Compared to those in NLBW neonates, the duration to peak weight loss was shorter (1.4 versus 1.3 days, *p* < 0.003) and the percentage of body-weight gain per day was higher (1.4 versus 1.7%, *p* < 0.001) in LBW neonates (Table 1). As birth weight decreased, the percentage body-weight gain increased (Figure 3).

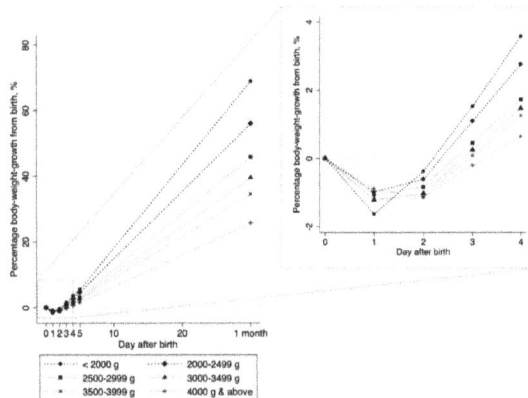

Figure 3. Daily percentages of bodyweight growth against birth weight. Each connected line shows the change in the percentage bodyweight growth from birth weight (([weight at every day − birth weight]/birth weight) × 100%), stratified by birth weight categories. Values from day 0 through day 4 were estimated in 11,224 neonates; values at day 5 and 1 month (approximately 30 days of birth) were estimated in 10,336 neonates and 8071 neonates, respectively.

Regression analyses in the decreasing period revealed that, compared to that in NLBW neonates, the percentage body-weight loss per day did not differ in LBW neonates (Table 2). In the increasing period, compared to that in NLBW neonates, the percentage body-weight gain per day was significantly higher in LBW neonates (Table 2): β = 0.52 (95% CI, 0.46 to 0.58). In stratified analyses in the decreasing period, a higher percentage of body-weight loss per day was only associated with higher birth weights in NLBW neonates (Table 3). In the increasing period, a higher percentage of body-weight gain per day was significantly associated with lower birth weights across all birth weight categories (Table 3).

Table 2. Body weight growth patterns against birth weight estimated by multivariable linear regression analysis.

Characteristics	Model 1 [3]		Model 2 [4]	
	β (95% CI)	*p*	β (95% CI)	*p*
Percentage body weight loss per day [1]				
LBW (<2500 g)	−0.03 (−0.12 to 0.06)	0.52	0.03 (−0.06 to 0.13)	0.48
Female			0.05 (0.004 to 0.09)	0.03
Gestational week			0.03 (0.01 to 0.05)	0.001
Apgar score			0.13 (0.09 to 0.17)	<0.001
Caesarean delivery			0.12 (0.03 to 0.20)	0.005
Maternal age			0.003 (−0.003 to 0.01)	0.34
Maternal body mass index			−0.03 (−0.04 to −0.01)	<0.001
Multipara			0.21 (0.16 to 0.26)	<0.001
Hypertensive disorders of pregnancy			0.24 (−0.11 to 0.59)	0.18
Birth year			−0.02 (−0.02 to −0.02)	<0.001
Percentage body weight gain per day [2]				
LBW (<2500 g)	0.33 (0.26 to 0.39)	<0.001	0.52 (0.46 to 0.58)	<0.001
Female			−0.12 (−0.15 to −0.10)	<0.001
Gestational week			0.15 (0.14 to 0.16)	<0.001
Apgar score			−0.02 (−0.04 to 0.01)	0.17
Caesarean delivery			0.001 (−0.05 to 0.05)	>0.99
Maternal age			−0.005 (−0.005 to 0.004)	0.82
Maternal body mass index			0.001 (−0.01 to 0.01)	0.97
Multipara			−0.08 (−0.11 to −0.05)	<0.001
Hypertensive disorders of pregnancy			0.05 (−0.18 to 0.29)	0.66
Birth year			−0.02 (−0.02 to −0.01)	<0.001

β, regression coefficient; CI, confidence interval; LBW, low birth weight. [1] Defined as ((birth weight − minimum weight)/birth weight) × (100%/day of peak weight loss). [2] Defined as ((weight at day 4 − minimum weight)/birth weight) × (100%/(4 − day of peak weight loss)). [3] Simple linear regression model for body-weight growth patterns against birth weight (referent group, NLBW neonates). Data were estimated in 10,544 non-low birth weight neonates and 680 low birth weight neonates. [4] Additionally adjusted for potential confounding variables of sex, gestational age, Apgar score, Cesarean delivery, maternal age, maternal body mass index, parity, hypertensive disorders of pregnancy, and birth year.

Table 3. Regression coefficients and 95% confidence intervals estimated with a continuous variable of birth weight in regression analyses, stratified by non-low birth weight and low birth weight neonates.

Characteristics	Percentage Body-Weight-Loss per Day [1]		Percentage Body-Weight-Gain per Day [2]	
	β (95% CI) [3]	*p*	β (95% CI) [3]	*p*
Overall (*n* = 11,224)				
Birth weight (per 100 g)	0.01 (−0.003 to 0.01)	0.06	−0.05 (−0.06 to −0.05)	<0.001
Non-LBW (2500 g and above, *n* = 10,544)				
Birth weight (per 100 g)	0.01 (0.001 to 0.02)	0.02	−0.04 (−0.05 to −0.03)	<0.001
LBW (<2500 g, *n* = 680)				
Birth weight (per 100 g)	0.003 (−0.07 to 0.08)	0.93	−0.16 (−0.21 to −0.11)	<0.001

Abbreviations: β, regression coefficient; CI, confidence interval; LBW, low birth weight. [1] Defined as ((birth weight − minimum weight)/birth weight) × (100%/day of peak weight loss). [2] Defined as ((weight at day 4 − minimum weight)/birth weight) × (100%/(4 − day of peak weight loss)). [3] Multivariable linear regression analysis for body-weight-growth patterns against birth weight, adjusted for sex, gestational age, Apgar score, cesarean delivery, maternal age, maternal body mass index, parity, hypertensive disorders of pregnancy, and birth year.

4. Discussion

Under a local neonatal regimen involving a standardized thermal control and nutrition regulation to meet basal metabolism, neonatal growth curves showed a uniform, J-shaped pattern with an initial physiological body-weight loss (~2% loss from birth weight), regardless of birth weight. The incidence of excess body-weight loss was low (<1%), with a short duration to peak weight loss (<2 days) and a low incidence of neonatal jaundice (0.3%). Compared to NLBW neonates, LBW neonates did not appear to have disadvantages in clinical outcomes, including excess body-weight loss and neonatal jaundice. LBW neonates also potentially showed faster body-weight gain patterns, with a trend suggesting that a higher percentage body-weight gain per day was associated with lower birth weight.

In the contemporary understanding of neonatal body-weight growth, the gradient of birth weight for the speed of body-weight gain is positive. That is, heavier birth weight neonates enjoy the advantages of faster body-weight gain and neonatal body-weight growth curves have continuously shown a divergent pattern across birth weight categories [13–17]. During the second half of the twentieth century, when perinatal and neonatal medicine and NICU were on the rise, this positive gradient in birth weight for neonatal body-weight gain was considered a rule of thumb and early illustrations of this positive gradient date as far back as Dancis et al. in 1948 [13]. Over the course of the twentieth century, this positive gradient persisted, reflecting larger insensible water loss in lower birth weight neonates (more premature neonates) [15,17,20]. Part of the reason for our observed pattern, which may potentially buck the contemporary trend, may be due to the neutral thermal environment that improves digestive function with circulatory stability, resulting in the neonatal capability to be fed sufficiently to meet basal metabolism and prolonged thermal regulation in LBW neonates [8–12]. Additionally, in a recent study with exclusively breastfed neonates, 50% of neonates recovered their weight around 10 days after birth [21]. By contrast, most of the neonates recovered their weight by 4 days after birth in our study, suggesting that our nutrition regulation with supplemental formula milk may successfully meet the basal metabolism in our high-risk population.

Body weight is a robust and straightforward anthropometric indicator to grow neonates. The criteria for body-weight-growth have been varied, which may partly due to differences in growth patterns of breastfeeding and formula-feeding based on calorific intakes. For example, in Western countries, studies suggest the potential cut-off point of ~5% to 12% for exclusively breastfed neonates and ~2% to 3% for formula-fed neonates [22–26], while the American Academy of Pediatrics recommends explicitly the cut-off point of 7% for excess body-weight loss among exclusively breastfed neonates [2]. By contrast, in Japan, although some studies suggest the potential cut-off point of ~4% to 10% [3,7], and other expert opinions even support the cut-off point of up to ~15%, any evidence-based clinical guidelines have yet to be established. Given the emerging concern for excess body-weight loss from inadequate intake in East-Asian countries including Japan [3–5], physiological body-weight loss would be a body-weight loss for neonates adequately fed to meet basal metabolism, at least in high-risk populations. Indeed, we observed an initial ~2% body-weight loss as the physiological loss with a low incidence of neonatal jaundice in our study population. Therefore, our growth chart may help not only healthcare providers but also mothers of neonates in that high-risk population.

Gestational age and maternal factors during pregnancy may affect neonatal growth pattern. In previous studies, gestational age ≤38 weeks was associated with an increased risk of neonatal jaundice [27]. Maternal smoking during pregnancy and preeclampsia were associated with decreased birth weight, whereas gestational diabetes was associated with increased birth weight [28–30]. In Finland, comprehensive screening of gestational diabetes was associated with decreased mean birth weight and macrosomia rates but was associated with an increased prevalence of neonatal hypoglycemia [30]. In our study, gestational age was consistently associated with neonatal growth pattern. However, significant effects of hypertensive disorders of pregnancy and maternal body mass index were inconsistent under the combination method (although we could not assess the impact of gestational diabetes due to the limitation of our data). Additionally, no hypoglycemia was observed in our high-risk population. Therefore, in addition to LBW neonates, neonates with gestational age ≤38 weeks, maternal smoking, preeclampsia, and gestational diabetes, might benefit the most from the combination method.

Some limitations should be noted. First, we did not have a control group with other nutrition treatments (exclusively breastfeeding and exclusively formula-feeding). In addition, we did not have hourly data of body weight and did not include sick infants who required intravenous infusion [13,25,31]. The thermal regulation also differed between birth weight categories, thereby introducing potential bias and limiting external generalizability. However, internal validity might be maintained for the comparison of NLBW and LBW neonates from the same source population. Second, other anthropometric parameters, such as body length and head circumference, and potential maternal confounding variables. such as diabetes, were not available [15–17,32]. However, body weight is a reliable indicator of neonatal growth [7,15]. Third, use of supplemental feeding is controversial due to the risk of breastfeeding failure, and excessive weight gain and the combination method might raise concerns for introducing unintended adverse events [1,33]. However, in previous studies, supplemental feeding did not affect breast milk production [7] and breastfeeding was not consistently associated with the reduction of obesity or body fat mass in children or adolescents [34]. In addition, a recent randomized clinical trial concluded that supplementing breastfeeding with early limited formula milk (10 mL) did not interfere with breastfeeding [35]. Fourth, we were not able to assess the impact of standardized early skin-to-skin contact on the incidence of hypothermia or hypoglycemia compared to our early skin-to-skin contact within 2 min. However, in previous studies, profiles of serum glucose levels and core/peripheral distribution of body temperature were likely better in neonates with thermal regulation compared with those without thermal regulation [7–9]. Therefore, future studies, including randomized controlled trials, are warranted to further understand how thermal control and nutrition regulation influence neonatal growth.

Lastly, in the era of early neonatal hospital discharge, standardized thermal regulation and nutrition regulation to meet basal metabolism (without NICU care) may reduce medical expenditures of readmission for phototherapy with costly NICU care among LBW neonates [1,36]. Even in our high-risk study population, we observed a very low incidence of neonatal jaundice (0.4%), which corresponds to the incidence in Western settings [27,37]. Given the very low incidence of neonatal jaundice, a potential reduction of medical expenditure would be expected by the combination method, in addition to improved education (clinicians, nurses, and parents) and systemic institutional and community-wide approaches [38,39]. Due to extra thermal regulation that might further stabilize cardiopulmonary circulation and improve digestive function, no phototherapy was required in NLBW neonates [8–10]. In addition, the combination method did not require intravenous infusion therapy and did not result in neonatal morbidity, mortality, or readmission after discharge, thereby potentially increasing the efficient use of NICU resources and potentially reducing clinical complications associated with intravenous infusion among LBW neonates. Furthermore, the initial 2 h spent in the incubator might help to avoid the substantially longer incubation required for phototherapy, thereby potentially increasing opportunities for physical contact between mothers and their babies.

In conclusion, neonatal body-weight growth patterns were characterized under standardized thermal control and nutrition regulation, with a low incidence of unfavorable clinical outcomes, regardless of birth weight category. Under this combination method, LBW neonates might not have disadvantages in clinical outcomes or growth patterns.

Author Contributions: Conceptualization, S.K.; data curation, S.K., M.Z., and T.Y.; formal analysis, M.Z. and T.Y.; investigation, S.K. and T.Y.; methodology, S.K., M.Z., and T.Y.; project administration, T.Y.; resources, S.K.; supervision, T.Y.; writing—original draft, M.Z.; writing—review and editing, S.K., M.Z., and T.Y.

Funding: This research received no external funding.

Acknowledgments: We would like to thank Editage (www.editage.jp) for English language editing.

Conflicts of Interest: The authors declare no conflict of interest.

References

1. Maisels, M.J.; McDonagh, A.F. Phototherapy for neonatal jaundice. *N. Engl. J. Med.* **2008**, *358*, 920–928. [CrossRef] [PubMed]

2. Gartner, L.M.; Morton, J.; Lawrence, R.A.; Naylor, A.J.; O'Hare, D.; Schanler, R.J.; Eidelman, A.I. American Academy of Pediatrics Section on Breastfeeding. *Pediatrics* **2005**, *115*, 496–506. [PubMed]

3. Sato, H.; Uchida, T.; Toyota, K.; Kanno, M.; Hashimoto, T.; Watanabe, M.; Nakamura, T.; Tamiya, G.; Aoki, K.; Hayasaka, K. Association of breast-fed neonatal hyperbilirubinemia with UGT1A1 polymorphisms: 211G>A (G71R) mutation becomes a risk factor under inadequate feeding. *J. Hum. Genet.* **2013**, *58*, 7–10. [CrossRef] [PubMed]

4. Chen, Y.J.; Chen, W.C.; Chen, C.M. Risk factors for hyperbilirubinemia in breastfed term neonates. *Eur. J. Pediatr.* **2012**, *171*, 167–171. [CrossRef] [PubMed]

5. Chen, C.F.; Hsu, M.C.; Shen, C.H.; Wang, C.L.; Chang, S.C.; Wu, K.G.; Wu, S.C.; Chen, S.J. Influence of breast-feeding on weight loss, jaundice, and waste elimination in neonates. *Pediatr. Neonatol.* **2011**, *52*, 85–92. [CrossRef]

6. Seske, L.M.; Merhar, S.L.; Haberman, B.E. Late-onset hypoglycemia in term newborns with poor breastfeeding. *Hosp. Pediatr.* **2015**, *5*, 501–504. [CrossRef] [PubMed]

7. Zaitsu, M.; Yoshihara, T.; Nakai, H.; Kubota, S. Optimal thermal control with sufficient nutrition may reduce the incidence of neonatal jaundice by preventing body-weight loss among NLBW infants not admitted to neonatal intensive care unit. *Neonatology* **2018**, *114*, 348–354. [CrossRef] [PubMed]

8. Kubota, S.; Koyanagi, T.; Hori, E.; Hara, K.; Shimokawa, H.; Nakano, H. Homeothermal adjustment in the immediate post delivered infant monitored by continuous and simultaneous measurement of core and peripheral body temperatures. *Biol. Neonate* **1988**, *54*, 79–85. [CrossRef] [PubMed]

9. Yoshimura, T.; Tsukimori, K.; Wake, N.; Nakano, H. The influence of thermal environment on pulmonary hemodynamic acclimation to extrauterine life in normal full-term neonates. *J. Perinat. Med.* **2007**, *35*, 236–240. [CrossRef] [PubMed]

10. Hey, E.N.; Katz, G. The optimum thermal environment for naked babies. *Arch. Dis. Child* **1970**, *45*, 328–334. [CrossRef]

11. Gross, S.J.; David, R.J.; Bauman, L.; Tomarelli, R.M. Nutritional composition of milk produced by mothers delivering preterm. *J. Pediatr.* **1980**, *96*, 641–644. [CrossRef]

12. Aoshima, N.; Fujie, Y.; Itoh, T.; Tukey, R.H.; Fujiwara, R. Glucose induces intestinal human UDP-glucuronosyltransferase (UGT) 1A1 to prevent neonatal hyperbilirubinemia. *Sci. Rep.* **2014**, *4*, 6343. [CrossRef] [PubMed]

13. Dancis, J.; O'Connell, J.R.; Holt, L.E., Jr. A grid for recording the weight of premature infants. *J. Pediatr.* **1948**, *33*, 570–572. [CrossRef]

14. Brosius, K.K.; Ritter, D.A.; Kenny, J.D. Postnatal growth curve of the infant with extremely LBW who was fed enterally. *Pediatrics* **1984**, *74*, 778–782. [PubMed]

15. Itabashi, K.; Takeuchi, T.; Okuyama, K.; Kuriya, N.; Ohtani, Y. Postnatal growth curves of very LBW Japanese infants. *Acta Paediatr. Jpn.* **1992**, *34*, 648–655. [CrossRef] [PubMed]

16. Itabashi, K.; Takeuchi, T.; Hayashi, T.; Okuyama, K.; Kuriya, N.; Otani, Y. Postnatal reference growth curves for very LBW infants. *Early Hum. Dev.* **1994**, *37*, 151–160. [CrossRef]

17. Diekmann, M.; Genzel-Boroviczeny, O.; Zoppelli, L.; von Poblotzki, M. Postnatal growth curves for extremely LBW infants with early enteral nutrition. *Eur. J. Pediatr.* **2005**, *164*, 714–723. [CrossRef] [PubMed]

18. Kuboi, T.; Kusaka, T.; Kawada, K.; Koyano, K.; Nakamura, S.; Okubo, K.; Yasuda, S.; Isobe, K.; Itoh, S. Hour-specific nomogram for transcutaneous bilirubin in Japanese neonates. *Pediatr. Int.* **2013**, *55*, 608–611. [CrossRef] [PubMed]

19. Itoh, S.; Kondo, M.; Kusaka, T.; Isobe, K.; Onishi, S. Differences in transcutaneous bilirubin readings in Japanese term infants according to feeding method. *Pediatr. Int.* **2001**, *43*, 12–15. [CrossRef]

20. Hammarlund, K.; Sedin, G.; Stromberg, B. Transepidermal water loss in newborn infants. VIII. Relation to gestational age and post-natal age in appropriate and small for gestational age infants. *Acta Paediatr. Scand.* **1983**, *72*, 721–728. [CrossRef]

21. Paul, I.M.; Schaefer, E.W.; Miller, J.R.; Kuzniewicz, M.W.; Li, S.X.; Walsh, E.M.; Flaherman, V.J. Weight Change Nomograms for the First Month After Birth. *Pediatrics* **2016**, *138*. [CrossRef] [PubMed]

22. American Academy of Pediatrics Subcommittee on Hyperbilirubinemia. Management of hyperbilirubinemia in the newborn infant 35 or more weeks of gestation. *Pediatrics* **2004**, *114*, 297–316. [CrossRef]

23. Laing, I.A.; Wong, C.M. Hypernatraemia in the first few days: Is the incidence rising? *Arch. Dis. Child. Fetal Neonatal Ed.* **2002**, *87*, F158–F162. [CrossRef]

24. Academy of Breastfeeding Medicine Protocol Committee. ABM clinical protocol #3: Hospital guidelines for the use of supplementary feedings in the healthy term breastfed neonate, revised 2009. *Breastfeed Med.* **2009**, *4*, 175–182.

25. Miller, J.R.; Flaherman, V.J.; Schaefer, E.W.; Kuzniewicz, M.W.; Li, S.X.; Walsh, E.M.; Paul, I.M. Early weight loss nomograms for formula fed newborns. *Hosp. Pediatr.* **2015**, *5*, 263–268. [CrossRef] [PubMed]

26. Martens, P.J.; Romphf, L. Factors associated with newborn in-hospital weight loss: Comparisons by feeding method, demographics, and birthing procedures. *J. Hum. Lact.* **2007**, *23*, 233–241. [CrossRef] [PubMed]

27. Maisels, M.J.; Kring, E. Length of Stay, Jaundice, and Hospital Readmission. *Pediatrics* **1998**, *101*, 995–998. [CrossRef]

28. Inoue, S.; Naruse, H.; Yorifuji, T.; Kato, T.; Murakoshi, T.; Doi, H.; Subramanian, S.V. Impact of Maternal and Paternal Smoking on Birth Outcomes. *J. Public Health (Oxf.)* **2017**, *39*, 1–10. [CrossRef]

29. Hung, T.H.; Hsieh, T.T.; Chen, S.F. Risk of Abnormal Fetal Growth in Women with Early- and Late-Onset Preeclampsia. *Pregnancy Hypertens.* **2018**, *12*, 201–206. [CrossRef]

30. Koivunen, S.; Torkki, A.; Bloigu, A.; Gissler, M.; Pouta, A.; Kajantie, E.; Vaarasmaki, M. Towards National Comprehensive Gestational Diabetes Screening—Consequences for Neonatal Outcome and Care. *Acta Obstet. Gynecol. Scand.* **2017**, *96*, 106–113. [CrossRef]

31. Fonseca, M.J.; Severo, M.; Santos, A.C. A new approach to estimating weight change and its reference intervals during the first 96 hours of life. *Acta Paediatr.* **2015**, *104*, 1028–1034. [CrossRef]

32. Villar, J.; Giuliani, F.; Bhutta, Z.A.; Bertino, E.; Ohuma, E.O.; Ismail, L.C.; Barros, F.C.; Altman, D.G.; Victora, C.; Noble, J.A.; et al. Postnatal growth standards for preterm infants: The Preterm Postnatal Follow-up Study of the INTERGROWTH-21(st) Project. *Lancet Glob. Health* **2015**, *3*, e681–e691. [CrossRef]

33. Victora, C.G.; Bahl, R.; Barros, A.J.; Franca, G.V.; Horton, S.; Krasevec, J.; Murch, S.; Sankar, M.J.; Walker, N.; Rollins, N.C. Lancet Breastfeeding Series Group: Breastfeeding in the 21st century: Epidemiology, mechanisms, and lifelong effect. *Lancet* **2016**, *387*, 475–490. [CrossRef]

34. Moschonis, G.; de Lauzon-Guillain, B.; Jones, L.; Oliveira, A.; Lambrinou, C.P.; Damianidi, L.; Lioret, S.; Moreira, P.; Lopes, C.; Emmett, P.; et al. The effect of early feeding practices on growth indices and obesity at preschool children from four European countries and UK schoolchildren and adolescents. *Eur. J. Pediatr.* **2017**, *176*, 1181–1192. [CrossRef]

35. Flaherman, V.J.; Narayan, N.R.; Hartigan-O'Connor, D.; Cabana, M.D.; McCulloch, C.E.; Paul, I.M. The Effect of Early Limited Formula on Breastfeeding, Readmission, and Intestinal Microbiota: A Randomized Clinical Trial. *J. Pediatr.* **2018**, *196*, 84–90. [CrossRef]

36. Suresh, G.K.; Clark, R.E. Cost-effectiveness of strategies that are intended to prevent kernicterus in newborn infants. *Pediatrics* **2004**, *114*, 917–924. [CrossRef]

37. Eggert, L.D.; Wiedmeier, S.E.; Wilson, J.; Christensen, R.D. The Effect of Instituting a Prehospital-Discharge Newborn Bilirubin Screening Program in an 18-Hospital Health System. *Pediatrics* **2006**, *117*, e855–e862. [CrossRef]

38. Flaherman, V.; Schaefer, E.W.; Kuzniewicz, M.W.; Li, S.X.; Walsh, E.M.; Paul, I.M. Health Care Utilization in the First Month After Birth and its Relationship to Newborn Weight Loss and Method of Feeding. *Acad. Pediatr.* **2018**, *18*, 677–684. [CrossRef]

39. Bhutani, V.K.; Johnson, L.H.; Schwoebel, A.; Gennaro, S. A Systems Approach for Neonatal Hyperbilirubinemia in Term and Near-Term Newborns. *J. Obstet. Gynecol. Neonatal Nurs.* **2006**, *35*, 444–455. [CrossRef]

Article

Vitamin B_{12} Status in Pregnant Adolescents and Their Infants

Julia L. Finkelstein [1,*], Ronnie Guillet [2], Eva K. Pressman [2], Amy Fothergill [1], Heather M. Guetterman [1], Tera R. Kent [1] and Kimberly O. O'Brien [1]

[1] Division of Nutritional Sciences, Cornell University, Ithaca, NY 14853, USA; af544@cornell.edu (A.F.); hg384@cornell.edu (H.M.G.); trk8@cornell.edu (T.R.K.); koo4@cornell.edu (K.O.O.)

[2] University of Rochester Medical Center, Rochester, NY 14642, USA; ronnie_guillet@urmc.rochester.edu (R.G.); eva_Pressman@urmc.rochester.edu (E.K.P.)

* Correspondence: jfinkelstein@cornell.edu; Tel.: +1-607-255-9180; Fax: +1-607-255-1033

Received: 20 December 2018; Accepted: 8 February 2019; Published: 13 February 2019

Abstract: Vitamin B_{12} deficiency has been associated with increased risk of adverse pregnancy outcomes. Few prospective studies have investigated the burden or determinants of vitamin B_{12} deficiency early in life, particularly among pregnant adolescents and their children. The objectives of this study were to determine the prevalence of vitamin B_{12} deficiency and to examine associations between maternal and neonatal vitamin B_{12} status in a cohort study of healthy pregnant adolescents. Serum vitamin B_{12} and folate concentrations were measured in adolescents at mid-gestation ($n = 124$; 26.4 ± 3.5 weeks) and delivery ($n = 131$; 40.0 ± 1.3 weeks), and in neonates at birth using cord blood. Linear regression was used to examine associations between maternal and neonatal vitamin B_{12} status. Although the prevalence of vitamin B_{12} deficiency (<148.0 pmol/L; 1.6%) in adolescents was low during pregnancy, 22.6% of adolescents were vitamin B_{12} insufficient (<221.0 pmol/L; 22.6%) at mid-gestation. Maternal vitamin B_{12} concentrations significantly decreased from mid-gestation to delivery ($p < 0.0001$), and 53.4% had insufficient vitamin B_{12} status at delivery. Maternal vitamin B_{12} concentrations ($p < 0.001$) and vitamin B_{12} deficiency ($p = 0.002$) at delivery were significantly associated with infant vitamin B_{12} concentrations in multivariate analyses, adjusting for gestational age, maternal age, parity, smoking status, relationship status, prenatal supplement use, pre-pregnancy body mass index, race, and intake of vitamin B_{12} and folate. Maternal vitamin B_{12} concentrations significantly decreased during pregnancy and predicted neonatal vitamin B_{12} status in a cohort of healthy pregnant adolescents.

Keywords: vitamin B_{12}; micronutrients; pregnancy; adolescents; folate

1. Introduction

Vitamin B_{12} deficiency (serum vitamin B_{12} <148.0 pmol/L) is a major public health problem globally [1,2]. Although the overall prevalence of vitamin B_{12} deficiency in the United States is estimated to be relatively low (6%), the burden of vitamin B_{12} deficiency is higher in the elderly, pregnant women, and young children (6–25%) [3]. Pregnant adolescents are at increased risk for a variety of micronutrient deficiencies and pregnancy complications, though there is limited data from this high-risk obstetric population.

Vitamin B_{12} deficiency in pregnancy has been associated with increased risk of pregnancy outcomes, including spontaneous abortion, pregnancy loss, intrauterine growth restriction, low birthweight (<2500 g), and neural tube defects (NTDs) [4–15]. Inadequate supply of vitamin B_{12} in pregnancy and early childhood can lead to long-term deficits in growth development in children [16,17].

Maternal vitamin B_{12} concentrations during pregnancy are thought to predict fetal [18–26] and early infant [25,27–29] vitamin B_{12} status. Previous cross-sectional studies in Norway, Turkey, Germany, United Kingdom, Serbia, and Brazil have noted a significant correlation between maternal and infant vitamin B_{12} status at delivery [18,20–22,30–33]; however, in one study in Belgium, maternal and infant vitamin B_{12} concentrations were not significantly correlated [34]. In one study in Germany, maternal serum vitamin B_{12} and holotranscobalamin (holoTC) concentrations at delivery were significantly correlated with cord blood holoTC concentrations ($p < 0.05$) [18]. In contrast, findings from cross-sectional studies examining the associations between maternal and infant vitamin B_{12} concentrations later in the postpartum period have been heterogeneous [35–40]. Maternal vitamin B_{12} and holoTC concentrations were significantly correlated with infant vitamin B_{12} concentrations in the first month (i.e., 2–30 days) postpartum in a study in Turkey [38]. In analyses in mother–infant dyads in the first 6 months postpartum, maternal and infant vitamin B_{12} concentrations were significantly associated in Canada and Cambodia (i.e., 3–27 weeks) [37], but not in India (i.e., 1–6 months) [35].

Prospective studies to date in The Netherlands, Norway, Turkey, India, and Spain have reported significant associations [19,23,24,26,29] between maternal vitamin B_{12} status during pregnancy and infant vitamin B_{12} status in cord blood or serum. In a prospective study in India, maternal vitamin B_{12} status during pregnancy was associated with infant vitamin B_{12} concentrations at 6 weeks of age [28]. In contrast, a study in Norway was conducted to examine the associations between maternal vitamin B_{12} biomarkers during pregnancy and vitamin B_{12} status in infants at birth and 6 months of age; maternal vitamin B_{12} concentrations did not significantly predict cord blood or infant vitamin B_{12} status, although there were significant associations noted for other biomarkers (i.e., maternal holoTC, holohaptocorrin (holoHC), and methylmalonic acid (MMA)) [25]. Although some studies to date have been conducted to examine vitamin B_{12} status in pregnant adolescents [41–43], most studies investigating the associations between maternal and infant vitamin B_{12} status have been conducted among adult pregnant women (i.e., 18 to 40 years). Of these studies, three cross-sectional studies reported participants which included adolescents, with age ranges of 15 to 38 years [32], 16 to 40 years [38], and 17 to 43 years [27]. However, adolescents comprised a small proportion (<15%) of the sample, and data presented were not stratified by age group, which constrained analysis and interpretation of findings for adolescents. There are limited prospective studies on the associations between maternal and infant vitamin B_{12} status conducted in high-risk obstetric groups such as adolescents.

Pregnant adolescents are at increased risk for a variety of micronutrient deficiencies and pregnancy complications [41,44]. The inadequate dietary intake of key nutrients among adolescents in industrialized countries [45], coupled with increased nutritional requirements for growth and development, warrants concern for health outcomes among pregnant adolescents. However, few data exist on the extent of vitamin B_{12} deficiency or its implications for fetal and child health in this high-risk obstetric population, which comprises over 5% of the US population and 11% globally [46,47]. Well-designed prospective studies are needed to elucidate the burden of vitamin B_{12} insufficiency in this key high-risk population and its implications for maternal and child health.

We, therefore, conducted a prospective observational analysis to: (1) determine the prevalence of vitamin B_{12} deficiency and insufficiency in pregnant adolescents and their infants; and (2) examine the associations of maternal and neonatal vitamin B_{12} status in healthy pregnant adolescents.

2. Materials and Methods

2.1. Study Population

Participants included in this study were enrolled in one of two prospective cohort studies funded by the United States Department of Agriculture (USDA). One study examined maternal and fetal bone health among pregnant adolescents ("bone health study") and collected maternal blood samples at mid-gestation and delivery, and cord blood samples at delivery. The other study evaluated iron status and anemia through gestation in pregnant adolescents aged 13 to 18 years and their infants ("anemia study"), and collected maternal and cord blood samples only at delivery. Both studies were observational cohort studies and not clinical trials (and, thus, do not need to be registered, as per protocol for clinical trials). Pregnant adolescents were recruited between 2006 and 2012, from the Rochester Adolescent Maternity Program (RAMP) in Rochester, New York.

Adolescents were eligible to participate if their pregnancies were 12 to 30 weeks in gestation at the time of the adolescents' enrollment in prenatal care at RAMP, and if the adolescents were healthy and carrying a single fetus. Adolescents were excluded if they had any known medical complications, including diabetes, preeclampsia, gestational hypertension, eating disorders, gastrointestinal diseases, HIV infection, or any other diagnosed medical conditions. Data on maternal and neonatal iron status [48,49] and on vitamin B_{12} transporters in placental tissue from this population [50] have been previously reported.

2.2. Ethics

Written informed consent was obtained from all study participants. The research protocol and study procedures were approved by the Institutional Review Boards (IRB) at Cornell University and the University of Rochester. The IRB approval included laboratory analyses of micronutrients in maternal and infant cord blood samples, including vitamin B_{12} and folate concentrations.

2.3. Follow-Up Procedures

Structured interviews were conducted to collect demographic information, including maternal age, educational level, socioeconomic status, and obstetric history at the baseline clinic visit. Detailed clinical, dietary (i.e., 24-h dietary recall), anthropometric, and biochemical data were collected at each visit. The participant recruitment and flow chart are presented in Figure 1. Of 251 participants who delivered at RAMP, a total of 194 participants ($n = 138$ participants in the bone study, recruited at mid-gestation; $n = 56$ participants in the anemia study, recruited at delivery) had archived blood samples available for analysis (Figure 1). All adolescents attending the Rochester Adolescent Maternity Program were prescribed a prenatal supplement as standard of care, which contained 27 mg iron, 12 µg vitamin B_{12}, 1000 µg folic acid, and other micronutrients (i.e., 1200 µg vitamin A, 120 mg vitamin C, 10 µg vitamin D_3, 22 mg vitamin E, 1.84 mg thiamin, 3 mg riboflavin, 20 mg niacin, 10 mg vitamin B_6, 200 mg calcium, 25 mg zinc, and 2000 µg copper).

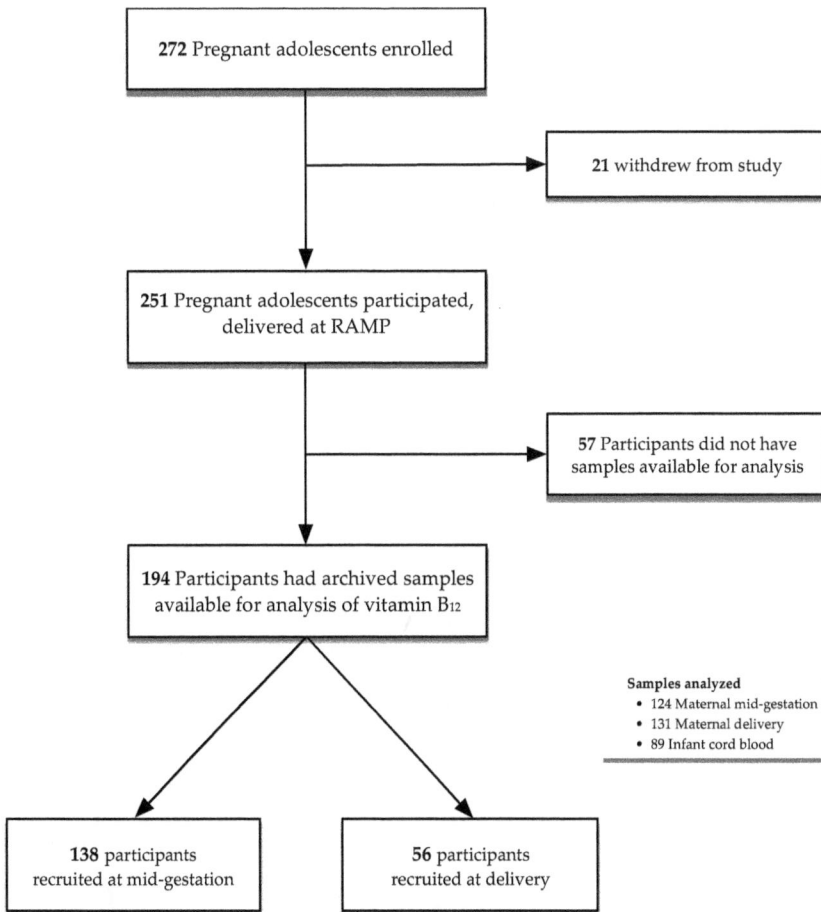

Figure 1. Participant flow diagram.

2.4. Laboratory Analyses

Non-fasting maternal venous blood samples (mid-gestation, delivery) and infant cord blood samples were allowed to clot at room temperature, separated by centrifugation, processed, and stored below −80 °C until analysis. A total of 124 maternal mid-gestation (26.4 ± 3.5 weeks), 131 maternal delivery (40.0 ± 1.3 weeks), and 89 infant cord blood samples were available for analysis.

Vitamin B_{12} concentrations were measured by electrochemiluminescence using the IMMULITE 2000 immunoassay system (Siemens Medical Solutions Diagnostics, Los Angeles, CA, USA). Three levels of controls (Bio-Rad) were used for serum vitamin B_{12}, with inter-assay coefficients of variation (CV) of 4.2% for Level 1 and 4.8% for Level 3. Serum folate concentrations were measured using the IMMULITE 2000 immunoassay system. The Bio-Rad Liquichek Immunoassay Plus Control (High & Low) were used as controls, with intra-assay precision of 6.7% and inter-assay precision of 6.6%.

2.5. Definitions of Outcomes

Conventional cutoffs were used to categorize variables where available; otherwise, medians of variables were defined based on their distributions in the population. Vitamin B_{12} deficiency and insufficiency were defined, following standard Centers for Disease Control and Prevention (CDC) definitions, as less than 148 pmol/L and less than 221.0 pmol/L, respectively [51]. Anemia was defined as hemoglobin <11.0 g/dL during the first and third trimesters, <10.5 g/dL during the second trimester, and <11.0 g/dL at delivery; and anemia status was adjusted for race [52]. Folate deficiency was defined as <6.8 nmol/L [51]. Maternal BMI was defined as the ratio of weight in kg to height in m^2 (kg/m^2), and categorized as <18.5, 18.5 to <25.0, 25.0 to <30.0, and $\geq$30.0 kg/m^2, in accordance with the CDC and World Health Organization (WHO) classifications [53]. Infant low birthweight was defined as <2500 g. Infant ponderal index was calculated as the ratio of weight in g to length in cm^3 ($g/cm^3 \times 100$).

2.6. Statistical Analyses

Binomial and linear regression models were used to examine the associations of maternal vitamin B_{12} status at mid-gestation and delivery with infant vitamin B_{12} status at birth. Binomial regression models were used to obtain risk ratio (RR) estimates for dichotomous variables [54–56]. Non-normally distributed variables were natural logarithmically transformed to ensure normality before further analysis. We also examined the associations between maternal and infant folate status. The values in Table 1 are presented as non-transformed values for interpretation purposes.

We explored potential nonlinearity of the relationships between covariates and outcomes nonparametrically, using stepwise restricted cubic splines [57,58]. If nonlinear associations were not reported, they were not significant. The Rothman and Greenland approach was used to evaluate and adjust for confounding, in which all known or suspected risk factors for the outcome which lead to a >10% change-in-estimate were included in the models [59]. Observations with missing data for covariates were retained in analyses using the missing indicator method [60]. Statistical analyses were conducted using SAS software, version 9.4 (SAS Institute, Inc., Cary, NC, USA).

Table 1. Characteristics of the study population.

Variables [a]	Original Cohort (n = 251)	Current Study (n = 194)	Recruited at Mid-Gestation (n = 138)	Recruited at Delivery (n = 56)
Maternal				
Age at enrollment, years	17.3 (16.5, 18.1)	17.3 (16.5, 18.1)	17.3 (16.4, 18.1)	17.3 (16.6, 18.1)
Age at delivery, years	17.5 (16.7, 18.3)	17.6 (16.8, 18.4)	17.6 (16.7, 18.4)	17.4 (16.9, 18.2)
<16 years, % (n)	12.0 (30)	9.8 (19)	9.4 (13)	10.7 (6)
Gestational age at delivery, weeks	39.9 (38.7, 40.7)	40 (39.0, 40.9)	40.0 (38.9, 40.9)	40.0 (39.2, 41.0)
Pre-term (<37 weeks), % (n)	8.0 (20)	7.8 (15)	8.8 (12)	5.4 (3)
Parity ≥1, % (n)	17.3 (43)	15.1 (29)	8.7 (12)	30.9 (17)
Smoking at enrollment, % (n)				
Never a smoker	77.8 (189)	78.5 (150)	77.5 (107)	81.1 (43)
Past smoker	15.2 (37)	14.4 (27)	12.3 (17)	18.9 (10)
Current smoker	7.0 (17)	7.3 (14)	10.1 (14)	0.0 (0)
Relationship status [b], % (n)	13.5 (33)	10.5 (20)	1.5 (2)	34.0 (18)
WIC [c] program participant	60.9 (148)	63.2 (120)	80.0 (100)	37.7 (20)
Self-reported prenatal supplement use, % (n)				
≥2 pills per week	54.1 (131)	55.5 (106)	56.6 (77)	52.7 (29)
Dietary folate, µg/day	617.2 (397.0, 948.9)	617.2 (400.8, 950.45)	692.7 (464.2, 1020.6)	415.3 (283.9, 624.7)
Dietary vitamin B_{12}, µg/day	4.6 (2.7, 6.5)	4.6 (2.7, 6.6)	5.0 (3.7, 6.9)	2.8 (1.4, 5.2)
Pre-pregnancy BMI, kg/m^2	23.5 (20.8, 28.0)	23.7 (20.8, 28.0)	23.3 (20.8, 28.1)	24.7 (20.8, 27.9)
<18.5 kg/m^2, % (n)	6.9 (17)	7.3 (14)	6.62 (9)	9.1 (5)
≥18.5 to <25 kg/m^2, % (n)	54.3 (133)	52.4 (100)	55.2 (75)	45.5 (25)
≥25.0 to <30 kg/m^2, % (n)	20.8 (51)	21.5 (41)	19.9 (27)	25.5 (14)
≥30 kg/m^2, % (n)	18.0 (44)	18.9 (36)	18.4 (25)	20.0 (11)
Gestational weight gain (GWG), kg	15.9 (11.8, 20.5)	16.4 (11.8, 20.5)	15.5 (11.8, 20.5)	17.3 (12.3, 21.4)
Inadequate [d] GWG, % (n)	15.0 (36)	13.9 (26)	14.3 (19)	13.0 (7)
Within IOM range, % (n)	22.9 (55)	24.0 (45)	26.3 (35)	18.5 (10)
Excessive GWG, % (n)	62.1 (149)	62.0 (116)	59.0 (79)	68.5 (37)
Race, % (n)				
Caucasian	27.9 (70)	29.4 (57)	33.3 (36)	19.6 (11)
African American	71.3 (179)	69.6 (135)	65.2 (90)	80.4 (45)
Native American	0.8 (2)	1.0 (2)	1.5 (2)	0.0 (0)
Ethnicity, % (n)				
Hispanic	24.3 (61)	26.3 (51)	24.6 (34)	30.4 (17)
Infant				
Birthweight, g	3206.0 (2904.0, 3550.0)	3266.0 (2928.0, 3581.0)	3258.0 (2892.0, 3581.0)	3318.5 (3055.5, 3584.0)
Birth length, cm	51.0 (49.0, 52.7)	51.3 (49.5, 52.9)	51.0 (49.5, 52.5)	52.0 (50.0, 53.5)
Weight-for-length z-score < −2, % (n)	27.0 (60)	27.0 (47)	27.3 (35)	26.1 (12)
Ponderal index, $g/cm^3 \times 100$	2.4 (2.2, 2.7)	2.4 (2.3, 2.7)	2.4 (2.2, 2.7)	2.4 (2.3, 2.6)
Male sex, % (n)	52.8 (132)	51.0 (099)	50.7 (70)	51.8 (29)

[a] Values are median interquartile range (IQR) and % (n); [b] Data presented are adolescents that report being in a relationship during pregnancy vs. single; [c] The Special Supplemental Nutrition Program for Women, Infants, and Children (WIC); [d] Gestational Weight Gain: categorized as inadequate or excessive, using Institute of Medicine (IOM) recommendations that vary based on pre-partum body mass index (BMI).

3. Results

3.1. Baseline Characteristics

The characteristics of participants in this study are presented in Table 1. Participants and their infants enrolled in the overall cohort studies and in the current study (i.e., with available serum vitamin B_{12} data) were similar in terms of baseline characteristics, including maternal age, socioeconomic characteristics, and nutritional status. A total of 194 participants had archived samples available for analysis; 138 of these participants were recruited at mid-gestation (bone health study), and 56 participants were recruited at delivery (anemia study) (Figure 1). We also examined potential differences in demographic, socioeconomic, and nutritional factors between participants in the two cohort studies. These variables were identified *a priori* as potential confounders and were considered and adjusted for in all of the multivariate analyses. Vitamin B_{12} and folate concentrations were analyzed in maternal samples that were collected at mid-gestation ($n = 124$) and delivery ($n = 131$); and in infant cord blood samples ($n = 89$).

3.2. Maternal and Neonatal Vitamin B_{12} Status

Maternal and neonatal vitamin B_{12} status are presented in Table 2. At the mid-gestation visit ($n = 124$; 26.4 ± 3.5 weeks gestation), 1.6% of women were vitamin B_{12} deficient ($n = 2/124$; <148.0 pmol/L), and 22.6% were vitamin B_{12} insufficient ($n = 28/124$; <221.0 pmol/L). Maternal serum vitamin B_{12} concentrations significantly decreased from mid-gestation to delivery ($n = 61$; 39.9 ± 1.0 weeks; mid-gestation: median = 358.9, interquartile range (IQR) = 233.9, 400.7 vs. delivery: median = 226.2, IQR = 185.2, 311.8; $p < 0.0001$).

The prevalence of maternal vitamin B_{12} insufficiency at delivery ($n = 70/131$; 53.4%) was significantly higher than at mid-gestation ($n = 28/124$; 22.6%, $p < 0.05$). The prevalence of vitamin B_{12} insufficiency was low in infants at birth: 0.0% were vitamin B_{12} deficient (<148.0 pmol/L), and 2.3% were vitamin B_{12} insufficient (<221.0 pmol/L). No mothers or infants were folate deficient (<6.8 nmol/L) or insufficient (<10.0 nmol/L) during this study.

The associations between maternal and infant serum vitamin B_{12} concentrations are presented in Table 3. Maternal vitamin B_{12} status at mid-gestation was not significantly associated with infant serum vitamin B_{12} concentrations ($p > 0.05$).

At delivery, maternal serum vitamin B_{12} concentrations ($p < 0.001$) and vitamin B_{12} deficiency ($p < 0.0001$) were significantly associated with infant serum vitamin B_{12} concentrations in multivariate analyses, adjusting for gestational age at sample collection, maternal age, parity, smoking status, relationship status, reported prenatal supplement use, pre-pregnancy BMI, race, and intake of vitamin B_{12} and folate. Similarly, maternal vitamin B_{12} insufficiency at delivery was significantly associated with infant serum vitamin B_{12} concentrations ($p < 0.01$) in multivariate analyses, adjusting for gestational age at sample collection, maternal age, parity, smoking status, relationship status, reported prenatal supplement use, pre-pregnancy BMI, race, and intake of vitamin B_{12} and folate. Maternal serum folate concentrations were not significantly associated with infant serum vitamin B_{12} concentrations ($p > 0.05$).

The associations between maternal vitamin B_{12} and folate statuses and infant serum folate concentrations are presented in Table 4. Maternal serum folate concentrations at mid-gestation were not significantly associated with infant serum folate concentrations ($p > 0.05$).

Table 2. Maternal and infant vitamin B[12] and folate status.

	Maternal				Infant		
	Mid-Gestation	**Delivery**			**Cord Blood**		
Variables [a]	**Total**	**Total**	**Recruited at Mid-Gestation**	**Recruited at Delivery**	**Total**	**Mothers Recruited at Mid-Gestation**	**Mothers Recruited at Delivery**
n	124	131	75	56	89	58	31
Serum vitamin B[12], pmol/L	343.7 (237.8, 400.7)	216.2 (161.6, 297.8)	216.2 (173.4, 311.8)	211.2 (158.7, 267.0)	597.0 (471.6, 796.3)	569.4 (478.6, 844.3)	602.9 (406.6, 722.1)
<148.0 pmol/L	1.6 (2)	15.3 (20)	14.7 (11)	16.1 (9)	0.0 (0)	0.0 (0)	0.0 (0)
≥148 to <221.0 pmol/L	21.0 (26)	38.2 (50)	37.3 (28)	39.3 (22)	2.3 (2)	0.0 (0)	6.5 (2)
≥221 pmol/L	77.4 (96)	46.6 (61)	48.0 (36)	44.6 (25)	99.8 (87)	100.0 (58)	93.5 (29)
n	122	130	74	56	86	55	31
Serum folate, nmol/L	39.3 (31.7, 50.5)	39.7 (31.8, 50.4)	42.8 (32.2, 51.4)	37.7 (28.8, 48.4)	66.7 (53.1, 85.5)	66.3 (52.1, 84.4)	67.7 (55.5, 98.4)
≤29.45 [b] nmol/L	19.7 (24)	20.0 (26)	13.5 (10)	28.6 (16)	2.3 (2)	3.6 (2)	0.0 (0)
>29.45, ≤35.79 nmol/L	20.5 (25)	16.9 (22)	18.9 (14)	14.3 (8)	2.3 (2)	3.6 (2)	0.0 (0)
>35.79, ≤43.94 nmol/L	19.7 (24)	20.0 (26)	17.6 (13)	23.2 (13)	4.7 (4)	5.5 (3)	3.2 (1)
>43.94, ≤52.66 nmol/L	19.7 (24)	22.3 (29)	29.7 (22)	12.5 (7)	14.0 (12)	12.7 (7)	16.1 (5)
>52.66 nmol/L	20.5 (25)	20.8 (27)	20.3 (15)	21.4 (12)	76.7 (66)	74.6 (41)	80.7 (25)

[a] Values are median and interquartile range (IQR) and (%) *n*. [b] Note: No values of serum folate were <6.8 nmol/L; the cut-offs presented for serum folate are quintiles based on the distribution of serum folate concentrations at mid-gestation.

Table 3. Associations between maternal vitamin B_{12} and folate status with infant serum vitamin B_{12} concentrations.

Maternal Variables	Time-Point	n	Univariate [b]		Multivariate [c]		Multivariate [d]	
			β (SE)	p-Value	β (SE)	p-Value	β (SE)	p-Value
Serum vitamin B_{12},[a] pmol/L	Mid-gestation	54	0.29 (0.17)	0.09	0.28 (0.16)	0.08	0.31 (0.16)	0.06
	Delivery (All)	64	0.85 (0.12)	<0.0001	0.74 (0.12)	<0.0001	0.77 (0.12)	<0.001
	Delivery (Recruited at mid-gestation)	33	0.57 (0.20)	0.004	0.53 (0.18)	0.003	0.53 (0.16)	0.001
	Delivery (Recruited at delivery)	31	1.09 (0.13)	<0.0001	0.97 (0.14)	<0.0001	1.02 (0.12)	<0.001
<148.0 pmol/L	Mid-gestation	54	n/a	n/a	n/a	n/a	n/a	n/a
	Delivery (All)	64	−0.65 (0.16)	<0.0001	−0.54 (0.14)	0.0002	−0.62 (0.15)	<0.0001
	Delivery (Recruited at mid-gestation)	33	−0.63 (0.22)	0.004	−0.60 (0.19)	0.002	−0.56 (0.18)	0.002
	Delivery (Recruited at delivery)	31	−0.72 (0.22)	0.001	−0.59 (0.22)	0.008	−0.67 (0.21)	0.002
<221.0 pmol/L	Mid-gestation	54	−0.16 (0.15)	0.28	−0.18 (0.14)	0.20	−0.18 (0.14)	0.21
	Delivery (All)	64	−0.42 (0.12)	0.0007	−0.30 (0.12)	0.01	−0.33 (0.12)	0.008
	Delivery (Recruited at mid-gestation)	33	−0.23 (0.17)	0.17	−0.19 (0.16)	0.22	−0.26 (0.15)	0.07
	Delivery (Recruited at delivery)	31	−0.61 (0.17)	0.0004	−0.44 (0.17)	0.01	−0.41 (0.19)	0.03
Serum folate [a], nmol/L	Mid-gestation	53	−0.24 (0.14)	0.09	−0.24 (0.14)	0.09	−0.28 (0.15)	0.06
	Delivery (All)	64	0.12 (0.16)	0.47	0.07 (0.16)	0.68	0.09 (0.17)	0.61
	Delivery (Recruited at mid-gestation)	33	0.09 (0.22)	0.69	0.12 (0.23)	0.60	0.16 (0.26)	0.54
	Delivery (Recruited at delivery)	31	0.12 (0.25)	0.63	0.06 (0.23)	0.81	0.11 (0.22)	0.62
<40.0 nmol/L	Mid-gestation	53	0.10 (0.12)	0.40	0.11 (0.12)	0.37	0.14 (0.13)	0.28
	Delivery (All)	64	−0.06 (0.13)	0.65	0.02 (0.13)	0.87	0.02 (0.13)	0.88
	Delivery (Recruited at mid-gestation)	33	−0.05 (0.17)	0.76	−0.08 (0.17)	0.64	0.04 (0.20)	0.86
	Delivery (Recruited at delivery)	31	−0.08 (0.21)	0.69	0.11 (0.18)	0.56	0.15 (0.17)	0.39

[a] Statistical analyses: Linear regression models were used to examine associations between maternal vitamin B_{12} and folate status and infant serum vitamin B_{12} concentrations; vitamin B_{12} and folate concentrations were natural logarithmically transformed prior to analyses; [b] Adjusted for gestational age of sample collection; [c] Adjusted for gestational age of sample collection, maternal age at delivery, parity (≥1 vs. 0), ever smoked (yes vs. no), relationship status (single vs. married/in a relationship), self-reported prenatal supplement use (≥2 vs. <2 pills/week), pre-pregnancy BMI, and race (African American vs. other); [d] Adjusted for gestational age of sample collection, maternal age at delivery, parity (≥1 vs. 0), ever smoked (yes vs. no), relationship status (single vs. married/in a relationship), self-reported prenatal supplement use (≥2 vs. <2 pills/week), pre-pregnancy BMI, race (African American vs. other), intake of vitamin B_{12}, and intake of folate.

Table 4. Associations between maternal Vitamin B$_{12}$ and folate status with infant serum folate concentrations.

Maternal Variables	Time-Point	n	Univariate [b] β (SE)	p-Value	Multivariate [c] β (SE)	p-Value	Multivariate [d] β (SE)	p-Value
Serum vitamin B$_{12}$ [a] pmol/L	Mid-gestation	51	−0.04 (0.16)	0.79	−0.19 (0.14)	0.17	−0.16 (0.13)	0.22
	Delivery (All)	61	−0.02 (0.11)	0.88	−0.08 (0.11)	0.48	−0.08 (0.11)	0.45
	Delivery (Recruited at mid-gestation)	30	−0.20 (0.15)	0.18	−0.20 (0.13)	0.13	−0.22 (0.12)	0.07
	Delivery (Recruited at delivery)	31	0.14 (0.16)	0.37	0.04 (0.15)	0.78	0.06 (0.15)	0.67
<148.0 pmol/L	Mid-gestation	51	n/a	n/a	n/a	n/a	n/a	n/a
	Delivery (All)	61	0.07 (0.13)	0.60	0.16 (0.12)	0.16	0.18 (0.12)	0.14
	Delivery (Recruited at mid-gestation)	30	0.26 (0.17)	0.13	0.26 (0.14)	0.07	0.38 (0.13)	0.005
	Delivery (Recruited at delivery)	31	−0.10 (0.18)	0.57	0.07 (0.17)	0.65	0.03 (0.17)	0.88
<221.0 pmol/L	Mid-gestation	51	−0.05 (0.14)	0.71	0.01 (0.13)	0.91	0.07 (0.12)	0.54
	Delivery (All)	61	−0.01 (0.09)	0.90	0.05 (0.09)	0.58	0.07 (0.09)	0.47
	Delivery (Recruited at mid-gestation)	30	0.18 (0.12)	0.12	0.22 (0.11)	0.04	0.27 (0.10)	0.006
	Delivery (Recruited at delivery)	31	−0.19 (0.13)	0.15	−0.09 (0.12)	0.45	−0.12 (0.13)	0.37
Serum folate [a], nmol/L	Mid-gestation	50	0.27 (0.14)	0.06	0.05 (0.13)	0.69	0.003 (0.13)	0.98
	Delivery (All)	61	0.54 (0.09)	<0.0001	0.47 (0.10)	<0.0001	0.50 (0.10)	<0.0001
	Delivery (Recruited at mid-gestation)	30	0.54 (0.11)	<0.0001	0.55 (0.13)	<0.001	0.53 (0.15)	0.0003
	Delivery (Recruited at delivery)	31	0.57 (0.13)	<0.0001	0.45 (0.13)	0.0005	0.44 (0.12)	0.0003
<40.0 nmol/L	Mid-gestation	50	−0.25 (0.12)	0.03	−0.13 (0.10)	0.21	−0.09 (0.10)	0.39
	Delivery (All)	61	−0.42 (0.08)	<0.0001	−0.40 (0.08)	<0.0001	−0.42 (0.08)	<0.0001
	Delivery (Recruited at mid-gestation)	30	−0.44 (0.09)	<0.0001	−0.43 (0.10)	<0.0001	−0.51 (0.12)	<0.0001
	Delivery (Recruited at delivery)	31	−0.41 (0.12)	0.0006	−0.32 (0.12)	0.01	−0.33 (0.12)	0.006

[a] Statistical analyses: linear regression models were used to examine associations between maternal vitamin B$_{12}$ and folate status and infant serum vitamin B$_{12}$ concentrations; vitamin B$_{12}$ and folate concentrations were natural logarithmically transformed prior to analyses; [b] Adjusted for gestational age of sample collection; [c] Adjusted for gestational age of sample collection, maternal age at delivery, parity (≥1 vs. 0), ever smoked (yes vs. no), relationship status (single vs. married/in a relationship), self-reported prenatal supplement use (≥2 vs. <2 pills/week), pre-pregnancy BMI, and race (African American vs. other); [d] Adjusted for gestational age of sample collection, maternal age at delivery, parity (≥1 vs. 0), ever smoked (yes vs. no), relationship status (single vs. married/in a relationship), self-reported prenatal supplement use (≥2 vs. <2 pills/week), pre-pregnancy BMI, race (African American vs. other), intake of vitamin B$_{12}$, and intake of folate

Nutrients **2019**, *11*, 397

Maternal serum folate concentrations at delivery were significantly associated with infant serum folate concentrations ($p < 0.0001$) in multivariate analyses, adjusting for gestational age at sample collection, maternal age, parity, smoking status, relationship status, prenatal supplement use, pre-pregnancy BMI, race, and intake of vitamin B_{12} and folate. Similarly, lower maternal serum folate concentrations (<40.0 nmol/L) at delivery were associated with lower infant serum folate concentrations ($p < 0.0001$) in multivariate analyses, adjusting for gestational age at sample collection, maternal age, parity, smoking status, relationship status, prenatal supplement use, pre-pregnancy BMI, race, and intake of vitamin B_{12} and folate.

4. Discussion

In this prospective analysis among pregnant adolescents, maternal vitamin B_{12} concentrations significantly decreased during pregnancy and predicted neonatal vitamin B_{12} status. Although the prevalence of vitamin B_{12} deficiency (<148.0 pmol/L; 1.6%) was low in adolescents during pregnancy, 22.6% of adolescents were vitamin B_{12} insufficient (<221.0 pmol/L; 22.6%) at mid-gestation. Maternal serum vitamin B_{12} concentrations decreased significantly during pregnancy, and at delivery, 15.3% of mothers were vitamin B_{12} deficient and 53.4% were vitamin B_{12} insufficient (Table 2).

This is among the first studies conducted to date to examine the burden of vitamin B_{12} deficiency in pregnant adolescents and its association with neonatal vitamin B_{12} status in this high-risk obstetric population. The prevalence of vitamin B_{12} deficiency in this study was low (1.6% mid-gestation, 15.3% delivery) and similar to a previous study conducted in Spain among pregnant adolescents (vitamin B_{12} deficiency, T1: 8.3%) [42]. However, the prevalence of vitamin B_{12} deficiency noted in this study was lower than previous studies conducted in pregnant adolescents in Canada (median, T3: 158 pmol/L, IQR: 114, 207 pmol/L; vitamin B_{12} <148.0 pmol/L: 43%) [43] and in Venezuela (vitamin B_{12} <200.0 pg/mL (<148.0 pmol/L), T1: 50.0%, T2: 58.8%, T3: 72.5%) [61]. Maternal vitamin B_{12} concentrations in our study were also higher than in a previous study in pregnant adolescents in the United Kingdom (geometric mean, Trimester 3 (T3): 177 pmol/L, 95% CI: 169, 185 pmol/L) [41].

The prevalence of vitamin B_{12} insufficiency (<221.0 pmol/L), however, was high in this study at both mid-gestation (22.6%) and delivery (53.4%). Although all participants were prescribed prenatal vitamins containing vitamin B_{12} and folic acid, self-reported adherence to prenatal supplements was low. Additionally, while most participants reported dietary intake of vitamin B_{12} at or above the RDA for this group (i.e., median (IQR): 4.5 (2.6, 6.6) μg/day vs. RDA: 2.6 μg/day), approximately 25% of participants reported dietary intake below the RDA. In addition to low dietary intake of vitamin B_{12}, vitamin B_{12} absorption could also be impaired by inadequate bioavailability, losses from processing and cooking animal-source foods, high dose folic acid, metabolic changes during pregnancy (e.g., hemodilution, fetal transfer), gastrointestinal symptoms, infections, and medications [4,62]. For example, since vitamin B_{12} is bound to protein carriers in the food matrix, vitamin B_{12} bioavailability may vary by food source [62,63].

The decline in maternal vitamin B_{12} concentrations during gestation in this study is also consistent with previous studies in adult pregnant women in Canada [64], Spain [26], Norway [25,29], and India [28,65], and in 12 of 13 longitudinal studies included in a systematic review of vitamin B_{12} status and birthweight in adult pregnant women [66]. The observed decrease in vitamin B_{12} concentrations throughout pregnancy could be due to hemodilution, increased protein synthesis, increased requirements for methyl donors during gestation, or a low intake or adherence to prenatal supplements to meet increased requirements [67]. However, there are limited data from pregnant adolescents, who have higher nutritional requirements for their own growth.

The prevalence of vitamin B_{12} deficiency and insufficiency in infants was low in this study (0–3%). Infant vitamin B_{12} concentrations were 2.5-fold higher than maternal vitamin B_{12} concentrations at delivery. These findings are consistent with previous studies in adult pregnant women, which have reported neonatal vitamin B_{12} concentrations 27% to 100% higher than maternal concentrations at delivery [18,19,21,26,30,31,33] and mid-gestation [23], although this has not been reported in all studies [22,25,28,29]. Higher vitamin B_{12} concentrations in offspring indicate active transfer to the fetus, which may occur due to upregulation of placental B_{12} transporter proteins or other active transport mechanisms that have yet to be established.

In this study, maternal vitamin B_{12} status at delivery, but not at mid-gestation, was significantly associated with infant vitamin B_{12} status. Maternal vitamin B_{12} status at delivery has been associated with vitamin B_{12} status in offspring at birth in previous cross-sectional studies [18,20–22,30–33]. There are, however, limited prospective data on maternal vitamin B_{12} status during pregnancy and its association with infant vitamin B_{12} status—particularly among adolescents—to compare findings. Evidence from studies in adult pregnant women have reported significant correlations between maternal vitamin B_{12} status during pregnancy and their infants [19,23,24,26,29]. Few prospective analyses to date have considered potential confounders of these associations in multivariate analyses [25,28]. In a recent study in adult pregnant women (median age = 22, IQR = 20–24 years) in Southern India, maternal vitamin B_{12} status during each trimester was associated with infant vitamin B_{12} status at 6 weeks of age [28], even after adjusting for maternal vitamin B_{12} supplementation. Similarly, a study conducted among pregnant women (mean age = 29.9, SD = 4.4 years) in Norway found that maternal vitamin B_{12} levels did not significantly predict cord blood or infant vitamin B_{12} status, although other vitamin B_{12} biomarkers (i.e., maternal holoTC, holoHC, MMA) were associated [25].

This study has several limitations. Neonatal micronutrient status was assessed at a single time point from cord blood, precluding our ability to evaluate longer-term impacts on infant vitamin B_{12} status or functional outcomes. Longitudinal data on maternal vitamin B_{12} concentrations were available only from a subset of participants in the parent cohort studies, limiting our ability to examine changes in vitamin B_{12} concentrations during pregnancy. Although participants in both cohort studies had similar sociodemographic characteristics (e.g., maternal age, gestational age at initiation of prenatal care, adherence to prenatal vitamins, gestational age at delivery), participants enrolled at mid-gestation (bone study) were more likely to be participants in the Special Supplemental Nutrition Program for Women, Infants, and Children (WIC) program, current smokers, Caucasian, primiparous, and had higher self-reported dietary intake of vitamin B_{12} and folate, compared to participants who were recruited at delivery (anemia). All of these variables were identified *a priori* as potential confounders and were considered and adjusted for in multivariate analyses; however, there may be residual confounding due to additional factors that were not evaluated or adjusted for in these studies. Vitamin B_{12} concentrations assessed at mid-gestation may not reflect vitamin B_{12} status during the relevant etiologic period periconceptionally or for maternal–fetal transfer of cobalamin and subsequent infant status and perinatal outcomes [68]. Additionally, serum folate is a biomarker of short-term dietary intake and does not reflect longer-term or usual intake. Vitamin B_{12} and folate assessments were also based on a single biomarker (i.e., total serum vitamin B_{12} and serum folate concentrations). Inclusion of additional circulating (i.e., holo-transcobalamin) and functional (i.e., methylmalonic acid) biomarkers of vitamin B_{12} metabolism and erythrocyte folate concentrations would improve assessment and interpretation of findings in mother–infant dyads [4]. Additionally, while the low prevalence of vitamin B_{12} deficiency in this study is similar to previous research in pregnant adolescents in Canada and the United Kingdom, a study population of generally adequate vitamin B_{12} status limits the generalizability of results to other populations that may be at greater risk for vitamin B_{12} deficiency, particularly in resource-limited settings [41,43]. Findings should also be interpreted in the context of a folate-replete population (i.e., among participants prescribed high-dose prenatal folic acid (1000 µg) and in a population exposed to folic acid fortification); this also limits the generalizability of findings to other settings. Finally, although findings from this study demonstrated an association of maternal and infant vitamin B_{12} status at delivery, the interpretation of these findings is not causal. Future prospective studies are needed to examine mechanisms of vitamin B_{12} transfer to the fetus and to determine the impact of vitamin B_{12} status on maternal and child health outcomes.

5. Conclusions

In summary, in this cohort of healthy pregnant adolescents, maternal vitamin B_{12} concentrations significantly decreased during pregnancy and predicted infant vitamin B_{12} status. This is one of the first prospective studies to date to evaluate the burden of vitamin B_{12} insufficiency in pregnant adolescents and their infants, a population that is at high risk for both micronutrient deficiencies and pregnancy complications. Findings suggest that vitamin B_{12} deficiency is an important public health problem in this high-risk obstetric population. Future research is needed to increase vitamin B_{12} status and improve the health of adolescent mothers and their children.

Author Contributions: Conceptualization—parent cohort studies, K.O.O., R.G., and E.K.P.; conceptualization—vitamin B_{12} sub-study, J.L.F. and K.O.O.; analysis, J.L.F. and A.F.; writing—original draft manuscript, J.L.F.; writing—review and editing, J.L.F., K.O.O., H.M.G., and A.F.; supervision, R.G., E.K.P., and K.O.O.; project administration, E.K.P.; laboratory analyses, T.R.K.; funding acquisition, K.O.O. and J.L.F. All authors contributed to the development of this manuscript and read and approved the final version.

Funding: This research was funded by the United States Department of Agriculture (USDA) 2005-35200-15218, USDA 2010-34324-20769, and AFRI/USDA 2011-03424; and the Division of Nutritional Sciences, Cornell University.

Acknowledgments: The authors are grateful to the mothers and children, and the midwives of the Strong Midwifery Group, who made this research possible.

Conflicts of Interest: The authors have no conflicts of interest to disclose. The funders had no role in the design of the study, in the collection, analyses, interpretation of data, or in the writing of the manuscript.

References

1. Allen, L.H. How common is vitamin B-12 deficiency? *Am. J. Clin. Nutr.* **2009**, *89*, 693S–696S. [CrossRef] [PubMed]
2. McLean, E.; de Benoist, B.; Allen, L.H. Review of the magnitude of folate and vitamin B_{12} deficiencies worldwide. *Food Nutr. Bull.* **2008**, *29*, S38–S51. [CrossRef] [PubMed]
3. Allen, L.H.; Rosenberg, I.H.; Oakley, G.P.; Omenn, G.S. Considering the case for vitamin B_{12} fortification of flour. *Food Nutr. Bull.* **2010**, *31*, S36–S46. [CrossRef] [PubMed]
4. Finkelstein, J.L.; Layden, A.J.; Stover, P.J. Vitamin B-12 and Perinatal Health. *Adv. Nutr.* **2015**, *6*, 552–563. [CrossRef] [PubMed]
5. Ratan, S.K.; Rattan, K.N.; Pandey, R.M.; Singhal, S.; Kharab, S.; Bala, M.; Singh, V.; Jhanwar, A. Evaluation of the levels of folate, vitamin B_{12}, homocysteine and fluoride in the parents and the affected neonates with neural tube defect and their matched controls. *Pediatr. Surg. Int.* **2008**, *24*, 803–808. [CrossRef]
6. Adams, M.J., Jr.; Khoury, M.J.; Scanlon, K.S.; Stevenson, R.E.; Knight, G.J.; Haddow, J.E.; Sylvester, G.C.; Cheek, J.E.; Henry, J.P.; Stabler, S.P.; et al. Elevated midtrimester serum methylmalonic acid levels as a risk factor for neural tube defects. *Teratology* **1995**, *51*, 311–317. [CrossRef] [PubMed]
7. Wilson, A.; Platt, R.; Wu, Q.; Leclerc, D.; Christensen, B.; Yang, H.; Gravel, R.A.; Rozen, R. A common variant in methionine synthase reductase combined with low cobalamin (vitamin B_{12}) increases risk for spina bifida. *Mol. Genet. Metab.* **1999**, *67*, 317–323. [CrossRef]
8. Gu, Q.; Li, Y.; Cui, Z.L.; Luo, X.P. Homocysteine, folate, vitamin B_{12} and B6 in mothers of children with neural tube defects in Xinjiang, China. *Acta Paediatr.* **2012**, *101*, e486–e490. [CrossRef]
9. Ray, J.G.; Wyatt, P.R.; Thompson, M.D.; Vermeulen, M.J.; Meier, C.; Wong, P.Y.; Farrell, S.A.; Cole, D.E. vitamin B_{12} and the risk of neural tube defects in a folic-acid-fortified population. *Epidemiology* **2007**, *18*, 362–366. [CrossRef]
10. Molloy, A.M.; Kirke, P.N.; Troendle, J.F.; Burke, H.; Sutton, M.; Brody, L.C.; Scott, J.M.; Mills, J.L. Maternal vitamin B_{12} status and risk of neural tube defects in a population with high neural tube defect prevalence and no folic Acid fortification. *Pediatrics* **2009**, *123*, 917–923. [CrossRef]
11. Rowland, A.S.; Baird, D.D.; Shore, D.L.; Weinberg, C.R.; Savitz, D.A.; Wilcox, A.J. Nitrous oxide and spontaneous abortion in female dental assistants. *Am. J. Epidemiol.* **1995**, *141*, 531–538. [CrossRef] [PubMed]

12. Hubner, U.; Alwan, A.; Jouma, M.; Tabbaa, M.; Schorr, H.; Herrmann, W. Low serum vitamin B_{12} is associated with recurrent pregnancy loss in Syrian women. *Clin. Chem. Lab. Med.* **2008**, *46*, 1265–1269. [CrossRef] [PubMed]

13. Reznikoff-Etievant, M.F.; Zittoun, J.; Vaylet, C.; Pernet, P.; Milliez, J. Low Vitamin B(12) level as a risk factor for very early recurrent abortion. *Eur. J. Obs. Gynecol. Reprod. Biol.* **2002**, *104*, 156–159. [CrossRef]

14. Hogeveen, M.; Blom, H.J.; den Heijer, M. Maternal homocysteine and small-for-gestational-age offspring: Systematic review and meta-analysis. *Am. J. Clin. Nutr.* **2012**, *95*, 130–136. [CrossRef] [PubMed]

15. Muthayya, S.; Kurpad, A.V.; Duggan, C.P.; Bosch, R.J.; Dwarkanath, P.; Mhaskar, A.; Mhaskar, R.; Thomas, A.; Vaz, M.; Bhat, S.; et al. Low maternal vitamin B_{12} status is associated with intrauterine growth retardation in urban South Indians. *Eur. J. Clin. Nutr.* **2006**, *60*, 791–801. [CrossRef] [PubMed]

16. Black, M.M. Effects of vitamin B_{12} and folate deficiency on brain development in children. *Food Nutr. Bull.* **2008**, *29*, S126–S131. [CrossRef]

17. Pepper, M.R.; Black, M.M. B12 in fetal development. *Semin. Cell Dev. Biol.* **2011**, *22*, 619–623. [CrossRef]

18. Obeid, R.; Morkbak, A.L.; Munz, W.; Nexo, E.; Herrmann, W. The cobalamin-binding proteins transcobalamin and haptocorrin in maternal and cord blood sera at birth. *Clin. Chem.* **2006**, *52*, 263–269. [CrossRef]

19. Muthayya, S.; Dwarkanath, P.; Mhaskar, M.; Mhaskar, R.; Thomas, A.; Duggan, C.; Fawzi, W.W.; Bhat, S.; Vaz, M.; Kurpad, A. The relationship of neonatal serum vitamin B_{12} status with birth weight. *Asia Pac. J. Clin. Nutr.* **2006**, *15*, 538–543.

20. Balci, Y.I.; Ergin, A.; Karabulut, A.; Polat, A.; Dogan, M.; Kucuktasci, K. Serum vitamin B_{12} and folate concentrations and the effect of the Mediterranean diet on vulnerable populations. *Pediatr. Hematol. Oncol.* **2014**, *31*, 62–67. [CrossRef]

21. Adaikalakoteswari, A.; Vatish, M.; Lawson, A.; Wood, C.; Sivakumar, K.; McTernan, P.G.; Webster, C.; Anderson, N.; Yajnik, C.S.; Tripathi, G.; et al. Low maternal vitamin B_{12} status is associated with lower cord blood HDL cholesterol in white Caucasians living in the UK. *Nutrients* **2015**, *7*, 2401–2414. [CrossRef] [PubMed]

22. Radunovic, N.; Lockwood, C.J.; Stanojlovic, O.; Steric, M.; Kontic-Vucinic, O.; Sulovic, N.; Hrncic, D.; Ackerman Iv, W.E. Fetal and maternal plasma homocysteine levels during the second half of uncomplicated pregnancy. *J. Matern.-Fetal Neonatal Med.* **2014**. [CrossRef] [PubMed]

23. Bergen, N.E.; Schalekamp-Timmermans, S.; Jaddoe, V.W.; Hofman, A.; Lindemans, J.; Russcher, H.; Tiemeier, H.; Steegers-Theunissen, R.P.; Steegers, E.A. Maternal and neonatal markers of the homocysteine pathway and fetal growth: The Generation R Study. *Paediatr. Perinat. Epidemiol.* **2016**, *30*, 386–396. [CrossRef] [PubMed]

24. Kalay, Z.; Islek, A.; Parlak, M.; Kirecci, A.; Guney, O.; Koklu, E.; Kalay, S. Reliable and powerful laboratory markers of cobalamin deficiency in the newborn: Plasma and urinary methylmalonic acid. *J. Matern.-Fetal Neonatal Med.* **2016**, *29*, 60–63. [CrossRef]

25. Hay, G.; Clausen, T.; Whitelaw, A.; Trygg, K.; Johnston, C.; Henriksen, T.; Refsum, H. Maternal folate and cobalamin status predicts vitamin status in newborns and 6-month-old infants. *J. Nutr.* **2010**, *140*, 557–564. [CrossRef]

26. Murphy, M.M.; Molloy, A.M.; Ueland, P.M.; Fernandez-Ballart, J.D.; Schneede, J.; Arija, V.; Scott, J.M. Longitudinal study of the effect of pregnancy on maternal and fetal cobalamin status in healthy women and their offspring. *J. Nutr.* **2007**, *137*, 1863–1867. [CrossRef] [PubMed]

27. Deegan, K.L.; Jones, K.M.; Zuleta, C.; Ramirez-Zea, M.; Lildballe, D.L.; Nexo, E.; Allen, L.H. Breast milk vitamin B-12 concentrations in Guatemalan women are correlated with maternal but not infant vitamin B-12 status at 12 months postpartum. *J. Nutr.* **2012**, *142*, 112–116. [CrossRef] [PubMed]

28. Finkelstein, J.L.; Kurpad, A.V.; Thomas, T.; Srinivasan, K.; Duggan, C. vitamin B_{12} status in pregnant women and their infants in South India. *Eur. J. Clin. Nutr.* **2017**, *71*, 1046–1053. [CrossRef]

29. Varsi, K.; Ueland, P.M.; Torsvik, I.K.; Bjorke-Monsen, A.L. Maternal serum cobalamin at 18 weeks of pregnancy predicts infant cobalamin status at 6 months—A prospective, observational study. *J. Nutr.* **2018**, *148*, 738–745. [CrossRef]

30. Lindblad, B.; Zaman, S.; Malik, A.; Martin, H.; Ekstrom, A.; Amu, S.; Holmgren, A.; Norman, M. Folate, vitamin B_{12}, and homocysteine levels in South Asian women with growth-retarded fetuses. *Acta Obstet. Gynecol. Scand.* **2005**, *84*, 1055–1061. [CrossRef]

31. Bjorke Monsen, A.L.; Ueland, P.M.; Vollset, S.E.; Guttormsen, A.B.; Markestad, T.; Solheim, E.; Refsum, H. Determinants of cobalamin status in newborns. *Pediatrics* **2001**, *108*, 624–630. [CrossRef]

32. Guerra-Shinohara, E.M.; Paiva, A.A.; Rondo, P.H.; Yamasaki, K.; Terzi, C.A.; D'Almeida, V. Relationship between total homocysteine and folate levels in pregnant women and their newborn babies according to maternal serum levels of vitamin B_{12}. *BJOG* **2002**, *109*, 784–791. [CrossRef] [PubMed]

33. Koc, A.; Kocyigit, A.; Soran, M.; Demir, N.; Sevinc, E.; Erel, O.; Mil, Z. High frequency of maternal vitamin B_{12} deficiency as an important cause of infantile vitamin B_{12} deficiency in Sanliurfa province of Turkey. *Eur. J. Nutr.* **2006**, *45*, 291–297. [CrossRef]

34. Jacquemyn, Y.; Ajaji, M.; Karepouan, N.; Jacquemyn, N.; Van Sande, H. vitamin B_{12} and folic acid status of term pregnant women and newborns in the Antwerp region, Belgium. *Clin. Exp. Obstet. Gynecol.* **2014**, *41*, 141–143. [PubMed]

35. Mittal, M.; Bansal, V.; Jain, R.; Dabla, P.K. Perturbing Status of vitamin B_{12} in Indian Infants and Their Mothers. *Food Nutr. Bull.* **2017**, *38*, 209–215. [CrossRef]

36. Dilli, D.; Dogan, N.N.; Orun, U.A.; Koc, M.; Zenciroglu, A.; Karademir, S.; Akduman, H. Maternal and neonatal micronutrient levels in newborns with CHD. *Cardiol. Young* **2018**, *28*, 523–529. [CrossRef] [PubMed]

37. Chebaya, P.; Karakochuk, C.D.; March, K.M.; Chen, N.N.; Stamm, R.A.; Kroeun, H.; Sophonneary, P.; Borath, M.; Shahab-Ferdows, S.; Hampel, D.; et al. Correlations between maternal, breast milk, and infant vitamin B_{12} concentrations among mother-infant dyads in Vancouver, Canada and Prey Veng, Cambodia: An exploratory analysis. *Nutrients* **2017**, *9*, 270. [CrossRef] [PubMed]

38. Coban, S.; Yilmaz Keskin, E.; Igde, M. Association between maternal and infantile markers of cobalamin status during the first month post-delivery. *Indian J. Pediatr.* **2018**, *85*, 517–522. [CrossRef]

39. Williams, A.M.; Stewart, C.P.; Shahab-Ferdows, S.; Hampel, D.; Kiprotich, M.; Achando, B.; Lin, A.; Null, C.A.; Allen, L.H.; Chantry, C.J. Infant serum and maternal milk vitamin b-12 are positively correlated in Kenyan infant-mother dyads at 1–6 months postpartum, irrespective of infant feeding practice. *J. Nutr.* **2018**, *148*, 86–93. [CrossRef]

40. Bellows, A.L.; Smith, E.R.; Muhihi, A.; Briegleb, C.; Noor, R.A.; Mshamu, S.; Sudfeld, C.; Masanja, H.; Fawzi, W.W. Micronutrient deficiencies among breastfeeding infants in Tanzania. *Nutrients* **2017**, *9*, 1258. [CrossRef]

41. Baker, P.N.; Wheeler, S.J.; Sanders, T.A.; Thomas, J.E.; Hutchinson, C.J.; Clarke, K.; Berry, J.L.; Jones, R.L.; Seed, P.T.; Poston, L. A prospective study of micronutrient status in adolescent pregnancy. *Am. J. Clin. Nutr.* **2009**, *89*, 1114–1124. [CrossRef] [PubMed]

42. Baron, M.A.; Solano, L.; Pena, E.; Moron, A. (Nutritional status of folate, vitamin B_{12} and iron in pregnant adolescents). *Arch. Latinoam. Nutr.* **2003**, *53*, 150–156. [PubMed]

43. Gadowsky, S.; Gale, K.; Wolfe, S.A.; Jory, J.; Gibson, R.; O'Connor, D. Biochemical folate, B12, and iron status of a group of pregnant adolescents accessed through the public health system in southern Ontario. *J. Adolsc. Health* **1995**, *16*, 465–474. [CrossRef]

44. Davis, L.M.; Chang, S.C.; Mancini, J.; Nathanson, M.S.; Witter, F.R.; O'Brien, K.O. Vitamin D insufficiency is prevalent among pregnant African American adolescents. *J. Pediatr. Adolesc. Gynecol.* **2010**, *23*, 45–52. [CrossRef] [PubMed]

45. Moran, V. A systematic review of dietary assessments of pregnant adolescents in industrialised countries. *Br. J. Nutr.* **2007**, *97*, 411–425. [CrossRef] [PubMed]

46. Klein, J.D. American Academy of Pediatrics Committee on, A. Adolescent pregnancy: Current trends and issues. *Pediatrics* **2005**, *116*, 281–286. [CrossRef] [PubMed]

47. Sedgh, G.; Finer, L.B.; Bankole, A.; Eilers, M.A.; Singh, S. Adolescent Pregnancy, Birth, and abortion rates across countries: Levels and recent trends. *J. Adolesc. Health* **2015**, *56*, 223–230. [CrossRef]

48. Lee, S.; Guillet, R.; Cooper, E.M.; Westerman, M.; Orlando, M.; Kent, T.; Pressman, E.; O'Brien, K.O. Prevalence of anemia and associations between neonatal iron status, hepcidin, and maternal iron status among neonates born to pregnant adolescents. *Pediatr. Res.* **2016**, *79*, 42–48. [CrossRef]

49. Lee, S.; Guillet, R.; Cooper, E.M.; Westerman, M.; Orlando, M.; Pressman, E.; O'Brien, K.O. Maternal inflammation at delivery affects assessment of maternal iron status. *J. Nutr.* **2014**, *144*, 1524–1532. [CrossRef]

50. Layden, A.J.; O'Brien, K.O.; Pressman, E.K.; Cooper, E.M.; Kent, T.R.; Finkelstein, J.L. vitamin B_{12} and placental expression of transcobalamin in pregnant adolescents. *Placenta* **2016**, *45*, 1–7. [CrossRef]

51. Yetley, E.A.; Pfeiffer, C.M.; Phinney, K.W.; Bailey, R.L.; Blackmore, S.; Bock, J.L.; Brody, L.C.; Carmel, R.; Curtin, L.R.; Durazo-Arvizu, R.A.; et al. Biomarkers of vitamin B-12 status in NHANES: A roundtable summary. *Am. J. Clin. Nutr.* **2011**, *94*, 313S–321S. [CrossRef] [PubMed]

52. Sullivan, K.M.; Mei, Z.; Grummer-Strawn, L.; Parvanta, I. Haemoglobin adjustments to define anaemia. *Trop. Med. Int. Health* **2008**, *13*, 1267–1271. [CrossRef] [PubMed]

53. WHO. *Physical Status: The Use and Interpretation of Anthropometry*; Report of a WHO Expert Committee; World Health Organization: Geneva, Switzerland, 1995; p. 329.

54. Spiegelman, D.; Hertzmark, E. Easy SAS calculations for risk or prevalence ratios and differences. *Am. J. Epidemiol.* **2005**, *162*, 199–200. [CrossRef] [PubMed]

55. Wacholder, S. Binomial regression in GLIM: Estimating risk ratios and risk differences. *Am. J. Epidemiol.* **1986**, *123*, 174–184. [CrossRef] [PubMed]

56. Zou, G. A modified poisson regression approach to prospective studies with binary data. *Am. J. Epidemiol.* **2004**, *159*, 702–706. [CrossRef] [PubMed]

57. Durrleman, S.; Simon, R. Flexible regression models with cubic splines. *Stat. Med.* **1989**, *8*, 551–561. [CrossRef] [PubMed]

58. Govindarajulu, U.S.; Spiegelman, D.; Thurston, S.W.; Ganguli, B.; Eisen, E.A. Comparing smoothing techniques in Cox models for exposure-response relationships. *Stat. Med.* **2007**, *26*, 3735–3752. [CrossRef]

59. Greenland, S. Modeling and variable selection in epidemiologic analysis. *Am. J. Public Health* **1989**, *79*, 340–349. [CrossRef]

60. Miettinen, O. *Theoretical Epidemiology*; John Wiley & Sons: New York, NY, USA, 1985; Volume 107.

61. Garcia-Casal, M.N.; Osorio, C.; Landaeta, M.; Leets, I.; Matus, P.; Fazzino, F.; Marcos, E. High prevalence of folic acid and vitamin B_{12} deficiencies in infants, children, adolescents and pregnant women in Venezuela. *Eur. J. Clin. Nutr.* **2005**, *59*, 1064–1070. [CrossRef]

62. Green, R.; Allen, L.H.; Bjorke-Monsen, A.L.; Brito, A.; Gueant, J.L.; Miller, J.W.; Molloy, A.M.; Nexo, E.; Stabler, S.; Toh, B.H.; et al. vitamin B_{12} deficiency. *Nat. Rev. Dis. Prim.* **2017**, *3*, 17040. [CrossRef]

63. Tucker, K.L.; Rich, S.; Rosenberg, I.; Jacques, P.; Dallal, G.; Wilson, P.W.; Selhub, J. Plasma vitamin B-12 concentrations relate to intake source in the Framingham Offspring study. *Am. J. Clin. Nutr.* **2000**, *71*, 514–522. [CrossRef] [PubMed]

64. Visentin, C.E.; Masih, S.P.; Plumptre, L.; Schroder, T.H.; Sohn, K.J.; Ly, A.; Lausman, A.Y.; Berger, H.; Croxford, R.; Lamers, Y.; et al. Low serum vitamin b-12 concentrations are prevalent in a cohort of pregnant Canadian women. *J. Nutr.* **2016**, *146*, 1035–1042. [CrossRef]

65. Wadhwani, N.S.; Pisal, H.R.; Mehendale, S.S.; Joshi, S.R. A prospective study of maternal fatty acids, micronutrients and homocysteine and their association with birth outcome. *Matern. Child Nutr.* **2015**, *11*, 559–573. [CrossRef] [PubMed]

66. Sukumar, N.; Rafnsson, S.B.; Kandala, N.B.; Bhopal, R.; Yajnik, C.S.; Saravanan, P. Prevalence of vitamin B-12 insufficiency during pregnancy and its effect on offspring birth weight: A systematic review and meta-analysis. *Am. J. Clin. Nutr.* **2016**, *103*, 1232–1251. [CrossRef] [PubMed]

67. Rush, E.C.; Katre, P.; Yajnik, C.S. vitamin B_{12}: One carbon metabolism, fetal growth and programming for chronic disease. *Eur. J. Clin. Nutr.* **2013**, *68*, 2–7. [CrossRef] [PubMed]

68. De-Regil, L.M.; Harding, K.B.; Roche, M.L. Preconceptional nutrition interventions for adolescent girls and adult women: Global guidelines and gaps in evidence and policy with emphasis on micronutrients. *J. Nutr.* **2016**, *146*, 1461S–1470S. [CrossRef] [PubMed]

nutrients

MDPI

Review

The Function and Alteration of Immunological Properties in Human Milk of Obese Mothers

Ummu D. Erliana * and Alyce D. Fly

Indiana University Bloomington School of Public Health, Bloomington, IN 47405, USA; afly@indiana.edu
* Correspondence: uerliana@indiana.edu; Tel.: +1-812-778-0040

Received: 29 April 2019; Accepted: 1 June 2019; Published: 6 June 2019

Abstract: Maternal obesity is associated with metabolic changes in mothers and higher risk of obesity in the offspring. Obesity in breastfeeding mothers appears to influence human milk production as well as the quality of human milk. Maternal obesity is associated with alteration of immunological factors concentrations in the human milk, such as C-reactive protein (CRP), leptin, IL-6, insulin, TNF-Alpha, ghrelin, adiponectin, and obestatin. Human milk is considered a first choice for infant nutrition due to the complete profile of macro nutrients, micro nutrients, and immunological properties. It is essential to understand how maternal obesity influences immunological properties of human milk because alterations could impact the nutrition status and health of the infant. This review summarizes the literature regarding the impact of maternal obesity on the concentration of particular immunological properties in the human milk.

Keywords: maternal obesity; gestational weight gain; immunological properties; human milk; nutrition; health

1. Introduction

In 2014, the number of pregnant women with overweight and obesity were estimated at 38.9 million and 14.6 million, respectively, worldwide [1]. The prevalence of women of childbearing age with obesity in the US from 1976 to 2014 increased four-fold, from 7.4% to 27.5% [2,3]. In addition, the prevalence of pre-pregnancy obesity of mothers in the US was more than 20% [4,5]. According to 38 jurisdictions of the District of Columbia, New York City, and 48 states in the United States, the prevalence of mothers that had a normal body weight before pregnancy decreased from 47.3% to 45.1% from 2011 to 2015 [6]. A review by Fields et al. (2016) brought attention to the association between several bioactive components of human milk and adiposity in infancy [7]. There is great interest in understanding the contributions of maternal obesity to changes in the composition or functional properties of human milk. Obesity in lactating mothers was found to be related to changes in the concentration of several bioactive components of their milk [8]. For example, an increased leptin concentration found in human milk is noteworthy because leptin may contribute to maternal obesity [9–14]. Alterations of immunological constituents may influence the genetic, metabolic, and epigenetic processes in the child [15]. The excessive weight gain of an obese child was found to be related the increase in leptin and adiponectin concentrations in the milk from obese mothers [16,17]. Furthermore, increased insulin-like growth factor 1 (IGF-I) and ghrelin in human milk were also correlated with the increased growth rate of an obese infant [16–19].

Human milk consists of nutrients and active factors. In addition to nutritive functions, some milk constituents also have bioactive properties such as whey proteins (immunoglobulin, lactoferrin, and alpha-lactalbumin) and casein proteins (κ-casein and β-casein) [20]. Other bioactive proteins in human milk include immunological factors such as antibodies, live cells, cytokines or signaling molecules; enzymes such as lactoferrin, lysozyme, and bile salt stimulated lipase; glycoproteins or oligosaccharides (oligosaccharide-enriched fraction along with secretory IgA), glycolipids, and high molecular weight protein), alpha-lactalbumin, gut microflora like prebiotic, haptocorrin (vitamin B12-binding protein), and nutrients for the infants' immune system [20]. The activities of these bioactive proteins need further study.

Moreover, human milk composition, both over a single feeding and over the duration of lactation is unique and dynamic. Colostrum is the first milk produced (30 mL/24 h) from 30 to 40 h until a few days postpartum [21,22]. Transitional milk is the milk that is produced from 5 days to 2 weeks postpartum and mature milk is the milk that is produced after 2 weeks postpartum [22]. Mature milk has two types, foremilk and hindmilk. Foremilk is the initial milk of a feeding, while hindmilk is the last milk of feeding (which contains milk fat up to three-fold more than in foremilk [23]. Some characteristics of human milk depend on a multitude of factors of the mother. The DARLING (Davis Area Research on Lactation, Infant Nutrition and Growth) study showed important determinants of human milk composition to include maternal ideal body weight (%IBW), protein intake, nursing frequency, menstruation, and parity [24]. The composition of protein in human milk is also significantly influenced by weaning because weaning was shown to decrease the volume of milk production [25]. Furthermore, this study also showed that milk volume and protein level were inversely associated; if the milk production volume was decreased from >500 to <300 mL, the protein concentration would be increased from 1.23 to 2.01 g/100 mL [25]. The influence of obesity on the quality and quantity of human milk has been demonstrated in several studies. A significantly different and less diverse microbiome in human milk has been demonstrated over lactation period, maternal body mass index (BMI), and delivery mode [26]. The ratio of omega-6 to omega-3 of obese breastfeeding mother is increased, while the concentration of fatty acids (DHA, EPA, DPA,) and carotenoids (lutein) are decreased [15]. Alterations of bioactive properties in the human milk of obese mothers could increase the incidence of obesity, insulin resistance, type 2 diabetes, and other adverse metabolic outcomes [27].

The primary aim of this review is to describe the immunological functions of specific factors in human milk and the alterations that occur in conditions of overweight and obesity mothers. The secondary aim is to explore the other influential factors that stimulate alterations of specific immunological properties that also could influence their functions in human milk.

2. The Function and Alteration of Immunological Properties in Human Milk

Breast milk contains essential bioactive components that have been demonstrated by numerous studies. Many components studied have been found to provide a function, for example to prevent infections, heal diseases, and improve health status [28]. Many bioactive components still remain that are not well studied. Studies are needed to understand the implications of the alteration in the specific bioactive components to maternal and infant health. One study demonstrated that the alteration of specific bioactive components in human milk would influence infant health outcome in both the short term and long term [29]. The understanding of this dynamic variation of human milk is needed because it may suggest practices needed for breastfeeding management and distribution of milk donations [22]. Modification of dietary intake and immunization can optimize the concentration of bioactive components in human milk [30,31]. Table 1 lists the bioactive components of human milk, and whether studies have reported any alterations in the concentration of milk of obese mothers. The functions of these bioactives and details about the changes are described further in the text.

Table 1. Alteration of immunological properties in human milk of obese mothers.

No	Bioactive Components	Alterations	References
1.	Antimicrobial		
	a. Lactoferrin	Increase	Houghton et al., 1985 [32]
	b. Lactadherin	None reported	-
	c. Lactoperoxidase	None reported	-
	d. Lysozyme	None reported	-
	e. Mucins (MUC1 and MUC4)	None reported	-

Table 1. *Cont.*

No	Bioactive Components	Alterations	References
2.	Cells		
	a. Lymphocytes	None reported	-
	b. Macrophages	None reported	-
	c. Neutrophils	Increase	Islam et al., 2006 [33]
	d. Stem cells	Decrease	Twigger et al., 2015 [34]
3.	Chemokines		
	a. Granulocyte Colony Stimulating Factor	None reported	-
	b. Macrophage migration inhibitory factor (MIF)	None reported	-
	c. Chemokine receptors (CXCR1/CXCR2)	None reported	-
	d. CXCL-9 (MIIP)	None reported	-
4.	Cytokines		
	a. Interleukin-1 beta (IL-1β)	None reported	-
	b. Interleukin-2 (IL-2)	Increase	Collado et al., 2012 [35]
	c. Interleukin-4 (IL-4)	Increase (colostrum) Decrease (1-month milk)	Collado et al., 2012 [35]
	d. Interleukin-6 (IL-6)	No alteration Increase	Whitaker et al., 2017 [8] Collado et al., 2012 [35]
	e. Interleukin-7 (IL-7)	None reported	-
	f. Interleukin-8 (IL-8)	None reported	-
	g. Interleukin-10 (IL-10)	Increase	Collado et al., 2012 [35]
	h. Interferon gamma-induced protein 10 (IP-10)	None reported	-
	i. Monocyte chemoattractant protein-1 (MCP-1)	None reported	-
	j. Interferon Gamma (IFN-γ)	Increase (colostrum) Decrease (1-month milk)	Collado et al., 2012 [35]
	k. Transforming growth factor beta (TGF-β)	Decrease No alteration	Collado et al., 2012 [35] Fields et al., 2017 [18]
	l. Tumor necrosis factor-alpha (TNF-α)	Increase (colostrum) Decrease (1-month milk)	Collado et al., 2012 [35]
5.	Cytokines inhibitors		
	a. Tumor necrosis factor receptor-I (TNFR I)	None reported	-
	b. Tumor necrosis factor receptor-II (TNFR II)	None reported	-
6.	Growth Factors		
	a. Epidermal growth factor (EGF)	Decrease	Khodabakhshi et al., 2015 [19]
	b. Heparin-binding EGF-like growth factor (HB-EGF)	None reported	-
	c. Insulin-like growth factor 1 (IGF-1)	Increase	Khodabakhshi et al., 2015 [19]
	d. Nerve growth factor (NGF)	None reported	-
	e. Vascular endothelial growth factor (VEGF)	None reported	-
7.	Hormones		
	a. Adiponectin	Increase	Martin et al., 2006 [36]
	b. Calcitonin	None reported	-
	c. Erythropoietin (Epo)	None reported	-
	d. Insulin	Increase	Fields et al., 2017 [18]
	e. Ghrelin	Decrease	Khodabakhshi et al., 2015 [19]; Zhang N, et al., 2011 [37]
	f. Leptin	Increase	De Luca et al., 2016 [9]; Lemas et al., 2016 [10]; Uysal et al., 2002 [11]; Andreas et al., 2014 [12]; Quinn et al., 2014 [13]; Eilers et al., 2002 [14];
	g. Obestatin	Decrease	Aydin et al., 2008 [38]
	h. Resistin	No alteration	Andreas et al., 2016 [39]; Savino et al., 2012 [40]
	i. Somatostatin	None reported	-
8.	Immunoglobulins		
	a. IgA	Increase (colostrum and serum)	Miranda et al.,1983 [41]; Islam et al., 2006 [33]; Fujimori et al., 2015 [42]
	b. IgG	Increase Decrease Constant (colostrum) Decrease (serum)	Miranda et al., 1983 [41] Islam et al., 2006 [33] Fujimori et al., 2015 [42]
	c. IgM	Increase Increase (colostrum) Decrease (serum)	Islam et al., 2006 [33] Fujimori et al., 2015 [42]
	d. Insulin-like-growth factor-binding proteins (IGFBP)	Decrease	Khodabakhshi et al., 2015 [19]

Table 1. *Cont.*

No	Bioactive Components	Alterations	References
9	Lipids	Increase (triglyceride and cholesterol)	Fujimori et al., 2015 [42]
		Increase (saturated fatty acids)	
		Decrease (n-3 fatty acidss)	
		Decrease (unsaturated to saturated fatty acid ratio)	Makela et al., 2013 [43]
		Increase (ratio of n-6 to n-3)	
		Increase (ratio of n-6 to n-3)	
		Decrease (lutein, docosahexaenoic acid, eicosapentaenoic acid, and docasapentaenoic acid)	Panagos et al., 2016 [15]
10.	Microbiota	*Bifidobacterium* (decrease) *Staphylococcus* (increase)	Delzenne & Cani, 2011 [44]; Collado et al., 2012 [35]
11.	Nucleic Acids	None reported	-
12.	Oligosaccharide and Glycans		
	a. Human milk oligosaccharides (HMOs)	No alteration	Azad et al. 2018 [45]
	b. Gangliosides	None reported	-
	c. Glycosaminoglycans (GAGs)	Decrease	Cerdo et al., 2018 [46]
	d. Osteoprotegerin	None reported	-
	e. Soluble CD14s (SCD14s)	Decrease	Collado et al., 2012 [35]
13.	Other Proteins		
	a. Alpha-Lactalbumin (LALBA)	Increase (6–15 days postpartum)	Sanchez-Poso et al., 1987 [47]
		Decrease (16–30 days postpartum)	
		Decrease (31–60 days postpartum)	
		Increase (>60 days postpartum)	
	b. Alpha-1 Antitrypsin (AAT)	None reported	-
	c. Alpha-Amylase (α-amylase)	None reported	-
	d. Bile Salt Stimulated Lipase (BSSL)	None reported	-
	e. Casein	Decrease	Jevitt et al., 2007 [48]
	f. C-Reactive Protein (CRP)	Increase	Whitaker, 2017 [8]
	f. Folate-Binding Protein (FBP)	None reported	-
	g. Haptocorrin	None reported	-

2.1. Anti-Microbial

Lactoferrin is one of the major iron-binding glycoproteins in colostrum that is immunomodulatory [49]. Lactoferrin is bound to iron and helps in uptake of iron in the cell by specific receptors in the infant intestinal tract [50]. Lactoferrin also functions at the level of DNA, as a transcription factor that influences for example, cell signaling proteins and immune protein synthesis [20,51]. Lactoferrin acts to constrain bacterial growth by binding iron, limiting iron needed for growth of bacteria and other pathogens [52]. The formation of lactoferrin without iron has also been found in human milk. This form was demonstrated to eradicate *Candida albicans*, *Escherichia coli*, *Pseudomonas aeruginosa*, *Streptococcus mutans*, *Streptococcus pneumoniae*, and *Vibrio cholerae* [53]. Lactoferrin concentration was significantly greater in colostrum of mothers who were greater than the 90% Weight for Height, an older surrogate measure for obesity (that corresponds to a BMI > 30 kg/m^2) [32].

Lactadherin is a glycoprotein associated with milk fat globules of human milk together with mucins, xanthine oxidase, and butyrophylin [20]. The main role of lactadherin may be to protect the newborn infant from rotaviral infection, a common cause of diarrheal disease and gastroenteritis [54]. Lactadherin works against the infection by creating apoptosis in infected cells of the infant and decreases inflammation by inhibition of TLR4 and the NF-κB signaling cascade [55–57]. Lactadherin is not digested in the stomach and passes to the intestine to maintain gut health by ameliorating inflammation [57–59]. A study in Mexico with 200 infants demonstrated that infants who were breastfed by milk that contained a low concentration of lactadherin developed severe diarrhea [54]. Other infants who received high levels lactadherin from human milk were asymptomatic of diarrhea.

The catalytic reactions of lactoperoxidase have bactericidal effects that kill Gram-positive and Gram-negative bacteria [60,61]. This bactericidal effect results from catalytic oxidation of substrate such as thiocyanates with hydrogen peroxide resulting in hypothiocyanite ion (OSCN$^-$). The concentration of lactoperoxidase of human milk at the first 6 months was found to be constant between 1–1.5 units/mL [62]. Lactoperoxidase in human milk was shown to detoxify H_2O_2 both in the infant gut and mammary gland of mothers, with additional anti-microbial functions. Whether lactoperoxidase concentration in the human milk of obese mothers is altered has not yet been determined.

Another antimicrobial constituent of milk, lysozyme, is present at a concentration that is 3000-fold higher in human milk than cow's milk. It is an active enzyme that works together with lactoferrin to effectively kill Gram-negative bacteria [63]. A study of the concentration of lysozyme in human milk of obese mothers was not found.

Mucins, another antimicrobial factor, make up one of three major protein fractions of human milk, along with casein and whey [64]. Mucins are a type of glycoprotein that consist of up to 80% carbohydrate, including mannose, as well as sulfonic acid [65]. MUC1 and MUC4 were identified in human milk for the first time by Liu et al. (2012) [66]. Furthermore, their study demonstrated that MUC1 was better than MUC4 in protecting the human epithelium cells (FHs 74 Int cells, CaCo-2 cells) from invasion by Salmonella [66]. Mucins are part of a passive immunity in human milk that protect the infant small intestine and stomach by inhibiting the binding of pathogens [67,68]. There was no study found that compared mucins concentration in the human milk of obese mothers to mothers with normal BMI.

2.2. Cells

The existence of lymphocytes in human milk was first discovered in colostrum [69]. Another study also showed that GFP+ leukocytes were transferred from mothers to their infants through breast milk [70]. T lymphocyte cells (CD8+, CD4+, and CD19+) are produced by GFP+ leukocytes in the Peyer's patches (PPs) [71]. Moreover, Th-2 lymphocytes were shown to contribute to produce specific cytokines such as IL-4, IL-13, IL-15 [56]. This study also demonstrated that the composition of lymphocytes in human milk and blood were different [72]. There were no studies found that compared lymphocyte concentrations in milk of obese mothers with other mothers.

Macrophages and the mammary endothelium support the production of TNF-Alpha in human milk [73,74]. Macrophages protect the infant from the infection by pathogens by activation of T-cells [75,76]. The cells move into the maternal blood and then are shifted to human milk by the mammary epithelial cells [22]. There were no studies found that reported comparisons of macrophage concentration in milk between women with obesity and normal weight conditions.

Neutrophils are another type of leukocyte that are abundant in the colostrum [77]. The three activities of neutrophils in human milk are bactericidal, phagocytic, and enzymatic [78]. Microbiocidal activity occurs through the production of oxidants and granule enzyme activities. In addition, neutrophils demonstrated significant bactericidal activity against *Staphylococcus aureus* [78]. This study by Grazioso and Buescher (1996) also demonstrated the effective ratio for phagocytosis between neutrophils and staphylococci cells at 2:1. A study by Islam et al. (2006) demonstrated that the percentages of breast milk phagocytes (neutrophil-macrophages) of a cell count in mothers with underweight, normal BMI, and overweight were 46.6% ± 19.35, 60.24% ± 6.93, and 55.55% ± 16.16, respectively [33].

Stem cells have been found in human milk that differentiate into luminal, basal, and myoepithelial layers [79]. These stem cells can regulate an octamer-binding transcription factor 4 expression that repairs human cells. The stem cells in human milk protect the cells of the infant [80]. The stem cells in breast milk were hypothesized to function as a neuroprotective factor in a study that used intra-nasal application of breast milk for rescuing preterm neonates from brain injury and to improve long-term neurocognitive improvement of preterm infants [81]. A review by Kakulas (2015) concluded that the number of lactocytes (mature mammary epithelial cells) in obese mothers was lower than other reports of lactocytes in healthy mature milk likely due to lower milk supply of obese mothers. In addition, a study by Twigger et al. (2015) in a review by Kakulas (2015) explained that the number of lactocytes in mothers with obesity was decreased because of the lower expression CK18 expression. Moreover, the lower expression of CK18 expression (a marker of luminal epithelial cell activity such as lactocytes) would reduce epithelial tissue capability to synthesize milk in mammary glands in obese mothers [34,80].

2.3. Chemokines

Chemokines are a type of cytokine that may stimulate cell movement [22]. Chemokine production is influenced by lactoferrin and a derivative of lactoferrin [82]. These factors have not been studied with regard to maternal body weight and lactation.

G-CSF functions as a hematopoietic growth factor, to stimulate the proliferation of clonal and neutrophil progenitor differentiation [83]. A study by Wallace et al. (1997) showed that the concentration of G-CSF in 30 samples of mothers with normal BMI ranged from 14 to >2500 pg/mL [84]. Again, there was no information found as to how high maternal body weight affected this factor.

The production of macrophage migration inhibitory factor (MIF) is equivocal with some reports that MIF is not produced by specific human cells [85]. Another study demonstrated that MIF is produced by a large number of T-cells [86]. The specific interleukins found to induce MIF production are tumor necrosis factor-alpha (TNF-α) and interferon gamma (IFNγ) [85,87]. There was no information found regarding how or whether MIF is affected by maternal body weight.

2.4. Cytokines

Cytokines interact with the other cells and cross the intestinal barrier to enhance or defend, and reduce inflammation [22]. The cytokines in human milk have roles as immunomodulatory and anti-inflammatory to reduce infection in the infant [88]. Additionally, mammary glands need cytokines for growth and development. Cytokines regulate proliferation and inflammation of cells [89].

Both regulation and development of mammary gland function are mediated by IL-6 and TNF-Alpha [90]. In addition, fever and systemic inflammation also correlate with IL-6, for example IL-6 was higher when mastitis occurred in the lobes [91,92]. IL-6, IL-10, and TGF-β are maternal cytokines associated with maturation of the intestinal immune system. These cytokines also regulate IgA production, development, and differentiation of cells [93]. A study by Whitaker et al. (2017) found no significant association between the maternal weight and IL-6 level [8]. Another study by Fields et al. (2017) also demonstrated similar results of no association between BMI of maternal mothers and IL-6 concentration in their milk [18]. However, a study by Collado (2012) showed contrasting results. IL-6 concentration was higher in colostrum of overweight and obese mothers (81.85 pg/mL) than in colostrum of healthy BMI mothers (62.86 pg/mL), while IL-6 concentrations in 1-month milk of overweight and obese mothers (13.22 pg/mL) was lower than IL-6 concentration of normal BMI mothers (22.12 pg/mL) [35].

The main functions of IL-7 are to support lymphoid development, produce T-cells in the thymus, and assist in T-cell survival at secondary of peripheral lymphoid tissue [94]. IL-7 from human milk is absorbed by the infant through the intestine.

In a study of mothers with varying BMI, the concentrations of TNFα in colostrum and 1-month milk of mothers with normal BMI were reported at 9.87 and 10.60 pg/mL, respectively [35]. However, these values differ greatly from the study by Meki et al. (2003) [91]. This study concluded that TNFα is a proinflammatory cytokine that increased over the course of lactation; however, the values seem to be similar with no standard deviation indicated. These authors also reported that the TNFα concentration in colostrum and 1-month milk of mothers with BMI ≤25 kg/m^2 and BMI >25 kg/m^2 were 11.41 and 10.23 pg/mL, respectively. However, there was no statistically significant difference in the TNFα content of the milk in these mothers [35]. The values for this cytokine are logarithmically different than those reported by others in the literature [35]. Therefore, while no differences were detected by Meki et al. the inconsistency with the literature makes it hard to conclude with certainty that the concentration of TNFα is not influenced by maternal BMI.

2.5. Growth Factors

Epidermal growth factor (EGF) is a growth modulator that can be found in amniotic fluid [95]. The function of EGF in human milk, as a growth and weight regulator, is related to appetite of the infant [19]. Another role of EGF is to support the maturation of infant gut by development, from a high intestinal permeability to become more selectively permeable [95]. A study by Khodabakhshi (2015) demonstrated that the EGF concentration in milk of obese mothers (0.038 ng/mL) was significantly lower ($p = 0.013$) than that of mothers with normal BMI (0.040 ng/mL). Furthermore, low concentration of EGF in the mother's milk was significantly associated with higher infant body weight [19].

HB-EGF is a growth factor that can be found in both amniotic fluid and human milk [96,97]. HB-EGF appears to protect the infant intestine, specifically the intestinal epithelium from cytokine-induced apoptosis and hypoxic necrosis by decreasing the level of nitrogen and reactive oxygen species produced [98–101].

The Insulin-like Growth Factors (IGF)-I in human blood were found in 1968 by Jacobs et al. [102]. IGF-1 has an important role in infant growth where the high concentration of IGF-1 in the human milk promotes rapid growth to the infants that may be associated with overweight and obesity in later life [103]. IGF-1 in breast milk is thought to protect premature infants from retinopathy of prematurity (ROP) in the first 6 weeks of life; IGF-1 deficiency during the first weeks of neonatal life was associated with ROP [104]. In addition, infants who consumed only breast milk had a lower incidence of ROP than the infants who consume infant formula [104]. The concentration of IGF-1 in the human milk of obese mothers (89.63 ng/mL) and normal mothers (75.09) was not significantly different ($p = 0.787$).

Specific neurons of the peripheral nervous system need nerve growth factor (NGF) to survive [105]. NGF has been shown to stimulate the phosphorylation of tyrosine on the TrkA protein receptor in human mast cells-1 (HMC-1) [106]. There was no literature found that described the effect of overweight or obesity in mothers on milk content or functions of NGF.

Human milk contains vascular endothelial growth factor (VEGF) and epidermal growth factor (EGF) at concentrations that are 100 times higher than blood of lactating mothers [107]. The concentration of VEGF from 33 samples of lactating mothers ranged from 12.6 to 155.0 (median 50.0) ng/mL. The concentration of VEGF was significantly positively correlated with the EGF concentration, but negatively correlated with hepatic growth factor (HGF) concentration [107]. VEGF receptors, such as fetal liver kinase 1, are expressed primarily in endothelial cells [108]. There were no studies found that investigated the effect of maternal overweight or obesity on concentrations of VEGF and HGF.

2.6. Hormones

The role of adiponectin is to regulate metabolism of the body, specifically affecting satiety, tissue sensitivity to insulin, stimulate glucose uptake, and decrease energy expenditure [109–111]. Studies by Martin et al. (2006) showed a positive correlation between maternal BMI and the concentration of adiponectin [36]. Furthermore, higher adiponectin in milk increases the risk of being overweight in childhood [109]. In contrast, another study specifically demonstrated lower weight gain and leaner body of neonates [112].

Calcitonin has functions to inhibit gastric acid production, or regulate fluid balance, food intake, and gastrointestinal motility [113–115]. The calcitonin concentration level in the human milk was found to be 10-40-fold greater than the concentration in the serum [116]. Lactating mothers have elevated serum calcitonin which is released into their milk [117]. The effects of maternal obesity on milk calcitonin have not been reported.

Ghrelin was identified by Kojima et al. (1999), as a 28-amino-acid-peptide. It was found to be predominantly expressed in the stomach [118], but it is also expressed in the pituitary, hypothalamus, and pancreas and functions to regulate food intake, sleep, behavior, gastric acid secretion and motility, glucose metabolism, exocrine and endocrine pancreatic function, cardiac performances and vascular resistance, energy metabolism, cell proliferation and survival, stimulates of ghrelin (GH), Arginine vasopressin (AVP), prolactin (PRL), adrenocorticotropic hormone (ACTH) secretion, and inhibits the gonadotropin secretion [119]. Ghrelin in the human milk functions as a growth factor that stimulates feeding by the infant [120]. A study by Khodabakhsh (2015) showed a negative correlation between obese mothers and the concentration of ghrelin in milk [19]. A similar result also was also demonstrated in a study by Zhang et al. (2011), where the concentration of ghrelin in human milk of normal weight mothers was higher than that in mothers with obesity [37].

Insulin in human milk functions as a hormone, and also influences the microbiome community in the lumen of the gastrointestinal tract [10]. Mothers with obesity accumulate more insulin in their human milk than mothers with normal body weight [109,121]. A study by Ley et al. (2012) specifically showed that the insulin concentrations were higher in the milk from obese mothers than mothers with normal BMI at 3 months post-partum [121]. The impact of higher insulin and glucose in human milk is more rapid growth of the infant which may increase risk for childhood obesity because of the higher energy content of diabetic breast milk than breast milk of healthy mothers [122,123]. In addition, Plagemann et al. (2002) observed a positive correlation between infant overweight and volume of diabetic breast milk (DBM) consumed at 2 years of age compared to infants who consumed banked breast milk from non-diabetic mothers [124]. This study showed early neonatal consumption of breastmilk from diabetic mothers was positively correlated to risk of overweight, and impaired glucose tolerance (IGT) during childhood [124].

Leptin is an adipocyte-derived hormone that was first found in 1994 [125]. Opposing the adiponectin function, leptin functions to increase appetite and suppress energy expenditure. It is considered a regulator of long-term energy balance [126]. The mean concentration of leptin at 1 month postpartum in the human milk of 50 normal weight mothers was 2.5 ± 1.5 ng/mL [9]. The concentration of milk leptin significantly decreased, up to 33.7% after 6 months of lactation, this may be in response to the decrease in fat mass of the mothers [18]. A study by Fields et al. (2017) demonstrated other determinants of leptin concentration in human milk such as BMI category, and gender of infant [18]. This study also showed a significant positive correlation between maternal BMI and leptin concentration in human milk. Furthermore, another study demonstrated higher concentrations of leptin in the milk of obese mothers, up to three times the concentration in milk of normal weight mothers [10]. Uysal et al. (2002) also found a significant correlation between high leptin concentration and maternal obesity [11]. The positive association between maternal obesity and leptin concentration in milk also noted by Andreas et al. (2014) and Quinn et al. (2014) [12,13]. At the 3rd and 28th day after delivery, the leptin concentration in milk of mothers who had BMI over 25 showed a significantly higher concentration than the milk of normal weight mothers [14]. Other data supporting this finding, are that increasing concentrations of leptin were found in milk of obese mothers at 1 month postpartum (4.8 ± 2.7 ng/mL) [9].

Obestatin is a hormone that was found in 2005 in the gastrointestinal tract [127]. Obestatin in human milk may control the appetite and gastrointestinal function of infant as the infant adapts to receiving milk [38,128]. The concentration of obestatin is negatively correlated with the BMI of mothers [38]. Therefore, the concentration of obestatin in the milk of overweight and obese mothers will be lower than the concentration of obestatin in milk of normal weight mothers.

Resistin is a hormone in breast milk that regulates fetal growth, appetite, and metabolic development of the infant [129]. The median concentrations of resistin in 23 samples of breast milk were 0.18 ng/mL (IR = 0.44) [40]. Study by Savino et al. (2012) and Andreas at al. (2016) found that the concentration of resistin was not altered in milk of mothers with obese BMI [39,40]. However, the concentration of resistin was higher in the serum of mothers with obese BMI [39]. The higher concentration of resistin in the serum and milk would increase the concentration of hormones (cortisol, estradiol, leptin, progesterone, prolactin, triiodothyronine, and thyroxine) and inflammatory marker (C-reactive protein) [130].

Somatostatin functions to inhibit gastrin secretion [131]. The highest concentration of somatostatin was found on the first day postpartum, in colostrum [132]. The concentration of milk somatostatin is 7.2 times higher than the concentration in the plasma of mothers, due to an active transport process moving somatostatin from blood to mammary gland. Reports of maternal obesity on somatostatin concentration have not been found.

2.7. Immunoglobulins

The major immunoglobulin found in human milk is Secretory IgA (sIgA, 80%–90%). This antibody is transferred from the milk to the infant, at the rate of 0.3 g/kg/day. About 10% is absorbed from the intestines and enters the infant blood [133]. A deficiency of sIgA in the mucous membranes can be substituted by a high concentration of IgM antibodies, particularly for infants with selective IgA deficiency [134]. Carbonare (1997) showed that sIgA fully protects the intestinal tract from enteropathogenic *E. coli* antigens, and is not destroyed by digestion [135]. A study by Islam et al. (2006) also compared the sIgA concentrations among three BMI groups. The study showed that IgA concentration in the human milk of overweight (5.60 ± 1.47 g/L) and normal mothers (5.67 ± 165 g/L) were higher than IgA concentration in malnourished mothers (5.22 ± 1.68 g/L) [63]. The standard deviation of the IgA concentration in the milk of the mothers with normal BMI may be a typographical error as it seems very high.

Unlike IgA, IgG and IgM are completely digested in the small intestine [20]. Thus, intact IgG is not present in significant quantities at the intestinal mucosal surface of infants. IgG is present at a low concentration (0.1 mg/mL) in human milk though it has a role in activating the complement system and antibody-dependent cytotoxicity [136]. The concentrations of IgG in breast milk of mothers with underweight, normal BMI, and overweight were 0.095 ± 024 g/L, 0.096 ± 024 g/L, and 0.093 ± 020 g/L, respectively [33].

IgM concentration in the colostrum of human milk is the second highest of the immunoglobulins. The mean concentration of IgM is 0.47 ± 0.10 g/L [33]. IgM protects the infant intestinal mucosal surface from viruses and bacteria [136]. Islam et al. (2006) demonstrated the mean concentration of overweight mothers and normal mothers were 0.47 ± 0.01 g/L and 0.47 ± 0.09 g/L, respectively.

Insulin-like growth factor-binding proteins can be found in human milk and include insulin growth factor I (IGF-1) and insulin growth factor 2 (IGF-2). Colostrum contains the highest concentration of IGF compared to the later milks [137]. IGF-1 functions to protect the enterocytes in the intestine from damage by oxidative stress, stimulates erythropoiesis, and is related to increasing hematocrit [137]. IGF-binding protein 2 is higher in preterm milk than term milk [138]. IGF-1 concentration is influenced by BMI of mothers whereby IGF-1 concentration was found to decrease significantly in the obese breastfeeding mothers [19].

2.8. Lipids

Lipids contribute up to 44% of the total energy in human milk [139]. Lipids can have an antimicrobial function in the infant intestine [140]. In addition, the activity and development of antimicrobial function in infant's gut is supported by milk fats through provision of nutrients [141]. Cell membranes of bacteria were shown to be damaged by antimicrobial activity of free fatty acids and monoglycerides [142]. A study by Fujimori et al. (2015) demonstrated that both the triglyceride and cholesterol concentration in the colostrum of overweight and obese mothers were higher than concentrations in colostrum of normal weight mothers [42]. Another study by Makela et al. (2013) showed that the fatty acid composition in human milk of overweight mothers was significantly higher in total saturated fatty acids and lower in n-3 fatty acids than in milk of normal weight mothers [43]. Furthermore, unsaturated to saturated fatty acid ratios in milk and the ratio of n-6 to n-3 was higher in overweight mothers versus normal weight mothers. A complete analysis of the various in fatty acids concentration in obese and normal mothers has been conducted by Panagos et al. (2016) [15]. A meta-analysis and a randomized controlled trial (RCT) showed associations between fatty acid composition in human milk and child growth were not significant [143–146].

2.9. Microbiota

Human milk supports proliferation of the milk microbiota by regulating the balance of specific microbiota such as, *Lactobacillus, Bacteroides*, and *Bifidobacteria*. The proliferation of these microbiota stimulates the activation of T regulatory cells and transform intrauterine TH2 predominant to rebalance TH1/TH2 [147]. About 200 kinds of important bacteria grow to constitute the first gut microbiota in the newborn [148]. *Bifidobacterium* and *Lactobacillus* are the primary probiotic in the gut that provide nutrients to the newborn intestine and protect it by establishing an acidic environment rich in short chain fatty acids (SCFAs) [148,149]. Dendritic cells (DCs) and CD18+ cells function to capture the microorganism from the mother's intestinal bacteria and translocate them to lactating mammary glands, therefore the number of DCs and CD18+ cells will increase late in pregnancy and during lactation [150]. A decreasing number of *Bifidobacterium* and increasing number of *Staphylococcus* have been demonstrated in human milk of obese and overweight mothers [35,44]. Furthermore, the higher concentration of *Lactobacillus* and the lower of concentration of *Bifidobacterium* for 3 months may contribute to increased risk of overweight and obesity to the infant [151]. Another study also demonstrated that obese infant and obese children have significantly different microbial flora because of different colonization through breast milk bacteria [152]. The alteration of the microbiota is not the only influence on nutritional status of the infant, but may be related to increased risk of asthma, allergies, inflammatory bowel disease, and type 1 diabetes. Thus, by Turnbaugh et al. (2012), Dibaise et al. (2009), Karvonen et al. (2014), Kostic et al. (2015), Frank et al. (2007) and Cabrera-Rubio et al. (2012) concluded that the bacteria in human milk of obese mothers may influence the health status of their infants [153–158].

2.10. Nucleic Acids

The degradation of milk nucleic acid polymers (RNA) provides nucleotides, which have been found to improve immune function, enhance the bioavailability of iron, adjust the microflora in the intestine, alter plasma lipids, and support growth and maturation of the gut [159–163]. Production of nucleotides in human milk needs further study because there is no study yet that compares the concentration of nucleotides in the milk of mothers with obese and normal BMI.

2.11. Oligosaccharides and Glycans

Human milk oligosaccharides (HMOs) protect infants from microbial infections, protect gut microbiota, prevent microbial adhesion, and invasion in the intestinal mucosa [164]. HMOs are small carbohydrates that bind pathogen bacteria and also facilitate the establishment of a protective microbiome in the intestinal tract of the infant [165]. HMOs are the main glycan and influential in the human milk because they inhibit the growth of pathogenic bacteria by binding to the pathogen and preventing pathogen binding to the intestinal epithelium [165]. HMOs prevent the attachment of viruses and bacteria surface on intestinal epithelium [166]. HMO is involved in development of the submucosa lymphoid structure and intestinal epithelial cell (IEC) barrier [166]. Colostrum contains two kinds of HMOs, 2′-fucosyllactose and 3′-galacto-syllactose, that reduce cytokine production and the inflammatory reactions [167,168]. A study by Azad et al. (2018) showed no significant correlation between the maternal BMI and the concentration of HMO in human milk [45]. This is of interest because a variation of HMO composition in human milk could influence the gut microbiome of the infant and variation in HMO has been shown to be significantly related to growth and body composition of the infant [169].

Prebiotics such as glycosaminoglycans, oligosaccharides, glycoprotein, and glycolipids are needed to support growth of the microbiota in the intestinal tract [166]. Glycosaminoglycans may be protective to infants, particularly the newborn, because the number of these polysaccharides are highest at the first month of lactation [170]. Furthermore, a study by Coppa et al. (2012) also demonstrated that the number of glycosaminoglycans in preterm human milk was higher than in term milk. A study of the association between BMI mothers and metabolic capacity of glycosaminoglycans was shown by Cardo et al. (2018). They found significant degradation of glycosaminoglycan activity in human milk of obese mothers compared to normal BMI mothers [46].

The concentration of soluble CD14 (sCD14) in the colostrum is 20-fold greater than the concentration in maternal serum [171]. A function of sCD14 in the infant intestine is to modulate the innate and adaptive immune response of bacterial colonization [171]. A study by Collado (2012) found that sCD14 in colostrum (28.22 µg/mL) and 1-month milk (5.54 µg/mL) of normal BMI mothers was higher than that of overweight and obese mothers, which contained 23.21 and 4.35 µg/mL, respectively [35]. These differences were not significant.

Osteoprotegerin is part of the TNF super family components that inhibit the activation of TNF-induction to prevent proliferation process of T cells. The concentration of osteoprotegerin in epithelial cells of the human mammary gland and human milk is 1000 times higher than the concentration in human serum [172]. Osteoprotegerin function in the Th1 cells may re-balance the concentration of Th1 and Th2 of newborn infants [173]. However, there was no study found that investigated maternal obesity and overweight on osteoprotegerin concentration in human milk.

2.12. Other Proteins

Alpha-lactalbumin of human milk is a bioactive protein that can be found in the whey fraction of milk [174]. Alpha-lactalbumin is present in both cow milk and human milk, and contains 123 amino acids [175]. Alpha-lactalbumin makes up 22% of total protein and 36% of whey protein in human milk [176,177]. The alpha-lactalbumin functions in synthesis of lactose and regulation of water circulation to human milk through osmotic systems [176]. The level of alpha-lactalbumin in milk is influenced by genetic, environment, and dietary factors [178]. For these reasons, alpha-lactalbumin plays a vital role as a bioactive protein. A previous study showed that the concentration of alpha-lactalbumin in human milk was influenced by nutritional status of mothers. Standard weight (SW) of mothers significantly influenced the concentration of alpha-lactalbumin in human milk [47]. Specifically, the concentration of alpha-lactalbumin in human milk of mothers who had SW, <10% SW, 10%–25% SW, and >25% SW were significantly different at 16–30 days and 31–60 days after delivery.

α1-Antitrypsin is a protease inhibitor in human milk. It has been shown to be an influential protein together with antichymotrypsin [179]. α1-Antitrypsin in human milk can be detected in the stool of human milk-fed infants [180,181]. The function of α1-antitrypsin and antichymotrypsin may be to permit absorption of certain bioactive proteins by limiting the digestibility of certain proteins in the infant gut [182]. Furthermore, α1-antitrypsin and antichymotrypsin affected the total nitrogen balance of infants who are fed by human milk. The inhibitory activity of α1-antitrypsin was not affected by pH (down to pH 2) or temperature [183]. It may be relatedly a protector of bioactive proteins (such as lactoferrin) at the small intestine, particularly in the upper part [183,184]. There was no study found that explored maternal obesity or overweight and α1-Antitrypsin.

Alpha-amylase is another significant bioactive protein in human milk [185]. The newborn infant has low concentrations of salivary amylase and pancreatic amylase activity. Milk amylase therefore may function to compensate for the low infant alpha-amylase from human milk [186]. Future studies should be conducted to better understand the significance of milk alpha-amylase for utilization of carbohydrate in mixed diet-fed infants [182]. There was no report that explored alpha-amylase concentration in milk of overweight and obese mothers.

Bile Salt Stimulated Lipase (BSSL) has been identified in human milk but not cow milk, where it contributes to 1%–2% of total milk protein. BSSL is influential as an active enzyme to digest lipid in the lumen of the infant gut. It hydrolyzes the esters of vitamin A and cholesterol, lyso-phospholipids, and other milk fats such as monoglycerides, diglycerides, and triglycerides [20]. A study by Andersson et al. (2007) demonstrated that BSSL can be destroyed or deactivated by the pasteurization process. Therefore, preterm infants who consume pasteurized milk from donor mothers show reductions in lipid absorption. Anthropometric measurements of length, weight, heel-to-knee of infants who consumed non-pasteurized milk increased more in these growth markers than infants who consumed pasteurized milk [187].

Human milk proteins are often classified according the nature of how the colloids are distributed in milk, as either two components, whey or casein. Casein proteins are distinguished by their arrangement in micelles. Human milk contains 20%–40% caseins [20]. The concentration of β-casein is the highest among other human caseins. β-casein is a highly phosphorylated protein [188]. MUC2 functions to protect the human small intestine layer by multiple mechanisms [189]. The adhesion of *Helicobacter pylori* is prevented by κ-casein activity at the gastric mucosa. More specifically, κ-casein at the epithelial cell surface has role as a soluble receptor analogue which prevents the binding of mucosal epithelium to bacteria [182]. The ratio of the concentration of casein and whey decreases over time [190]. At the first 2 weeks after delivery, the ratio concentration of casein and whey was found to be 20:80. After 2 weeks postpartum the ratio concentration of casein and whey changed to 35:65 and stayed constant. The beta-casein expression was reduced in milk of overfed rats, which suggests a similar effect might occur on beta-casein concentration in the human milk of overweight and obese mothers [48].

C-reactive protein is an inflammatory protein marker that is primarily produced by the liver [191]. Maternal obesity was found to be associated with an alteration of CRP. A study by Whitaker et al. (2017) showed that the concentrations of CRP in the human milk of 126 obese and or mothers with excessive weight gain were significantly higher than mothers with the normal weight gain or lower weight. High concentrations of CRP level in infant serum can increase the risk of cardiovascular disease in the long term due to CRP level functions to control serum cholesterol level [192].

Folate-binding protein regulates the metabolism of the vitamin folic acid, through distribution, absorption, and retention [193]. The soluble folate-binding protein (FBP) function is a glycosylated compound that binds the vitamin in a way that prevents proteolytic degradation in the low gastric pH. The glycosylation appears to provide protection from digestion [194]. Furthermore, folate bound to FBP is gradually released in the small intestine mucosa [195]. There was no study found that reported effects of maternal obesity or overweight on FBP concentration in milk.

Haptocorrin is an anti-microbial substance that was found to reduce *E. coli* and is stable against proteolytic digestive enzymes [196]. In addition, haptocorrin remains undigested at pH 3.5 by pepsin and pancreatic enzymes [196]. High concentrations of haptocorrin are not needed for pathogenic activity. Another function of haptocorrin in human milk is as a substitute for intrinsic factor, that helps in the absorption of Vitamin B_{12} in the newborn (acting as a vitamin B_{12}-binding protein) [197]. There was no study found that reported effects of maternal obesity or overweight on haptocorrin concentration in milk.

3. The Influence of Other Factors on the Levels of Immunological Properties in Human Milk

The composition of human milk is also influenced by other maternal factors, such as diet, age, ethnicity, metabolic health, type of delivery, smoking, amount of sleep, stress, and physical activity. Additionally, it is influenced by physiological factors such as diurnal variation, stage of lactation and stage of nursing, as well as factors related to the infant, including gender, birthweight, body composition, and behavioral factors, such as ad libitum feeding and time between feedings [7]. The main influential factors on human milk composition are preterm delivery and stage of lactation [198]. Studies that demonstrate an influence of other factors that affect immunological properties of milk are reported in Table 2.

Table 2. Alteration of immunological properties by other influential factors.

Influence Factors	Immunological Properties	Alteration
Malnourishment	Lysozyme	A study by Hennart et al. (1991) demonstrated that malnourished mothers had up to four times higher concentration of lysozyme than well-nourished mothers [199].
	Lactoperoxidase	A study by Chang (1990) showed that the concentration of lactoperoxidase in human milk of malnourished Chinese women decreased up to 50% [200].
	sIgA	The concentration of sIgA in milk of malnourished mothers was lower than the concentration of sIgA in milk of normal weight mothers [41].
	IgG	A study by Miranda et al. (1983), with malnourished Colombian mothers, demonstrated lower IgG concentrations compared to milk of well-nourished mothers [41].
	Alpha-lactalbumin	A study by Lonnerdal et al. (1976) in Ethiopian and Swedish mothers demonstrated that alpha-lactalbumin concentration in well-nourished mothers was higher than in milk of malnourished mothers [178].
Lactation Stages	Lymphocyte	The mean quantity of lymphocyte plasma cells in colostrum of mothers with low BMI mothers was higher than that of mothers with normal BMI [33].
	Macrophages	The composition of colostrum is high in macrophages (30%–50% in leukocytes) [201]. However, macrophage concentration was found to decrease after 1 month of lactation and maturation of human milk [22,202].
	G-CSF	A study by Calhoun et al. (2000) demonstrated that the highest concentration of G-CSF in human milk was 1–2 days postpartum. MIF concentration is highest at the first month of lactation and continues to decrease with time of lactation [203].
	IL-6	The IL-6 levels were at similar levels in milk samples taken at 1 month and 3 months postpartum [8].
	IL-7	A showed that mean concentration of IL-7 was influenced by the age of infants. IL-7 concentration in human milk of 6 months breastfed infants (103.5 ± 37.8 pg/dL) was higher than IL-7 concentration of 2 months of breastfed infants (69.8 ± 40 pg/dL) [204]. In a study of healthy mothers' milk over time, the concentration of TNFα in colostrum was higher than the concentration in transitional and mature milk [205]. Another study showed that the TNFα concentration in colostrum (402.80 ± 29.65 pg/mL) and mature milk (178.30 ± 14.41 pg/mL) were higher than the TNFα concentration in transitional milk (135.50 ± 8.26 pg/mL) [91].
	EGF	In milk of 33 mothers at day 1–7 postpartum, the concentration of EGF was reported to range from 33.3 to 184.3 (median 71.2) ng/mL [107].
	IGF-1	A study by Erikkson et al. (1993) showed that the concentration of IGF-1 in the colostrum to transitional milk (decreased five-fold day 1 to day 3, and then remained relatively constant through day 8 of lactation [206].
	NGF	A study by Ai et al. (2012) measured NGF levels of seven healthy breastfeeding mothers on the first 3 days postpartum. Measurement of NGF levels showed very high variation and NGF levels postpartum milk were, 236 ± 332 ng/L, 173 ± 113 ng/L, and 178 ± 248 ng/L, for 1, 2 and 3 days, respectively, but were not significantly different over time [207].
	VEGF	A study by Kobata et el. (2008) demonstrated decreasing concentrations of VEGF from 1 to 7 days postpartum [107].
	Ghrelin	While the ghrelin concentration was high in breast milk, there were significant increases in infants' weight gain at 4 months of age [208]. The negative correlation between ghrelin concentration in serum and infants BMI occurred at the first month of life [146]. Another study also showed that low concentration of ghrelin in infants would slow the weight gain during the first 3 months [209]. Transitional milk and colostrum contain small amounts of Immunoglobulin G (IgG) while the amount of IgG is higher in mature milk [20].
	Lipid	Colostrum, transition milk, and mature milk contains lipid concentrations of 3–4, 7.2, and 56.2 g/L respectively [210].
	Gangliosides	Around 6%–10% of total lipid mass in human infant brain consists of gangliosides. This concentration will be increase up to 3-fold from 10-weeks' gestation to 5 years of age [211]. GM1, GM3, and GD3 are three types of gangliosides in human milk that bind pathogens without causing inflammatory reactions [212]. A study by Thakkar et al. (2013) demonstrated that the concentration of gangliosides in human milk was significantly higher at 120 days postpartum in male infants [213].
	α1-antitrypsin	The concentration of α1-antitrypsin from 190 human milk samples of 94 maternal mothers between 1 and 160 days after delivery was found to decrease over time [179].
	Alpha-amylase	The decrease in alpha-amylase concentration could range up to 35% ($p < 0.001$) during the first 3 months [214].

Table 2. *Cont.*

Influence Factors	Immunological Properties	Alteration
Pregnancy length	Chemokines	A study by Michie et al. (1998) showed that there were no differences in chemokines concentration between human milk of mothers that delivered preterm or full term [215].
	G-CSF	The production of G-CSF (at the first 2 days of colostrum) of premature delivered mothers was lower than that of term mothers [216].
	TNFα	A study demonstrated that the concentration of TNFα in colostrum of mothers with premature babies was lower than the mothers with full term babies [205].
	Alpha-amylase	No differences in alpha-amylase concentration were found in the human milk produced for preterm infants or full-term infants [217].
	BSSL	There was no significant difference of characteristics of BSSL found in the human milk produced from mothers with preterm or full-term infants [218].
	Osteoprotegerin	The concentration of osteoprotegerin in the milk of mothers whose deliver their infant prematurely and full term was not significantly different [172].
Type of Breastfeeding	IL-7	A higher concentration of IL-7 was found in exclusively breastfed infants compared to those that received mixed feeding [204].
Nutritional Status of Infant	IGF-1	A study showed that high BMI of infant was negatively correlated with the concentration of IGF-1 in human milk [19].

4. Conclusions

Based on literature that were collected and reviewed, obesity and overweight of mothers appears have a clear influence on the concentration of 29 immunological properties in human milk. An alteration of any of these immunological properties could influence the function of the human milk. An increase or decrease of immunological properties in human milk are predicted to influence the health status of infants in both the short term and long term. Further research is needed to study the correlation of high BMI with the 31 yet unstudied or unreported immunological properties in human milk and the risk of specific diseases to the infants.

There are some limitations of this review. First, the range of years for publications of studies reviewed was not limited. Some current studies might have different results compared to older studies because of different number of participants, data collection methods, immunoassay methods and detection, and statistical analyses. Second, many studies did not classify the samples collected as to whether they were colostrum, hind milk, and fore milk. Differences in the composition of milk over the duration of lactation should have different concentration of immunological properties. Furthermore, these variations might influence the output of immunological actions related to their function as protectors and supporters for infant growth and development.

5. Practice Points

Maternal overweight and obesity has been found to affect 29 immunological properties of human milk. While a complete understanding of the implications of these changes to the infant are not known, they add to the body of evidence supporting healthy weight management:

1. Medical practitioners, dietitian/nutritionists, and nurses should educate the pregnant mothers with BMI > 30 kg/m^2 about the risk of obesity to mother and infant and suggest restricting excessive caloric intake outside of the recommended amounts needed for healthy infant growth. The health care professionals support the pregnant mother in obesity by providing weight management targets and the information about nutrient dense eating plans or menus.
2. After delivery, exclusive breastfeeding can be promoted as another component of energy expenditure which will help prevent further weight gain by the obese mother. Breastfeeding mothers should not follow a weight loss program until 6–8 weeks after delivery because it may influence the quality and quantity of milk production. The overweight and obese mothers should be encouraged and supported to exclusively breastfeed for at least for 6 months.
3. At 6–8 weeks postpartum, obese mothers should be encouraged to pursue healthy weight loss programs with nutrient dense eating plans after lactation is well established. Health professionals can emphasize that achieving a healthier BMI provides health benefits to both the

infants (by normalizing the immunological properties concentration in milk) and to the mothers (by decreasing the risk of metabolic diseases, such as, heart disease, diabetes mellitus, and cancer).

4. Even though composition of some immunological factors in milk changes in obesity, there is not enough evidence at this time to suggest that the changes in the milk should preclude breastfeeding the infant, as the benefits of breastfeeding are well established. In fact, obese mothers in the US, Europe, Brazil, Latin America, Singapore, India, South Africa, and Australia, that cannot breastfeed, can consult with international board certified lactation consultants (IBCLC) and pediatricians to consider obtaining donor milk from a milk bank in one of these countries.

6. Research Agenda

1. The association of maternal obesity with immunological properties reported as unstudied in this review need investigation.
2. The alteration of specific immunological properties and these effects on infant health condition (when breastfed by obese mothers) should be examined both in the short term and long term. A reduction of protective factors could have a negative impact on infant health.
3. The differences in infant health status after exclusive breastfeeding by obese mothers should also be explored for obese mothers that practice nonexclusive breastfeeding. Exclusive breastfeeding has been widely encouraged due to the benefits for mothers and infants. It is conceivable that alterations in the immunological properties due to obesity may affect the usual advantages associated with exclusive breastfeeding.

Author Contributions: Conceptualization, U.D.E. and A.D.F.; Data collection, U.D.E.; Supervision, A.D.F.; Writing—original draft, U.D.E.; Writing—review, A.D.F.; Writing—editing, U.D.E. and A.D.F.

Funding: This research received no external funding.

Acknowledgments: This review was supported by Indonesia Endowment Fund for Education (LPDP) and Indiana University Bloomington.

Conflicts of Interest: The authors declare no conflict of interest.

References

1. Chen, C.; Xu, X.; Yan, Y. Estimated global overweight and obesity burden in pregnant women based on panel data model. *PLoS ONE* **2018**, *13*, 1–14. [CrossRef] [PubMed]
2. Singh, G.K.; Siahpush, M.; Hiatt, R.A.; Timsina, L.R. Dramatic Increases in Obesity and Overweight Prevalence and Body Mass Index Among Ethnic-Immigrant and Social Class Groups in the United States, 1976–2008. *J. Community Health* **2010**, *36*, 94–110. [CrossRef] [PubMed]
3. US Department of Health and Human Services. *The National Health Interview Survey, Questionnaires, Datasets, and Related Documentation: 1976–2014 Public Use Data Files*; National Center for Health Statistics: Hyattsville, MD, USA, 2015.
4. Hamilton, B.E.; Martin, J.A.; Michelle, J.L.; Osterman, M.H.S.; Curtin, M.A.; Matthews, M.S. Births: Final data for 2014. *Natl. Vital Stat. Rep.* **2015**, *64*, 1–64. [PubMed]
5. Fisher, S.C.; Kim, S.Y.; Sharma, A.J.; Rochat, R.; Morrow, B. Is obesity still increasing among pregnant women? Prepregnancy obesity trends in 20 states, 2003–2009. *Prev. Med.* **2013**, *56*, 372–378. [CrossRef] [PubMed]
6. Deputy, N.P.; Dub, B.; Sharma, A.J. Prevalence and Trends in Prepregnancy Normal Weight—48 States, New York City, and District of Columbia, 2011–2015. *MMWR Morb. Mortal. Wkly. Rep.* **2018**, *66*, 1402–1407. [CrossRef] [PubMed]
7. Fields, D.A.; Schneider, C.R.; Pavela, G. A narrative review of the associations between six bioactive components in breast milk and infant adiposity. *Obesity (Silver Spring)* **2016**, *24*, 1213–1221. [CrossRef] [PubMed]
8. Whitaker, K.M.; Marino, R.C.; Haapala, J.L.; Foster, L.; Smith, K.D.; Teague, A.M.; Jacobs, D.; Fontaine, P.L.; McGovern, P.M.; Schoenfuss, T.C.; et al. Associations of Maternal Weight Status Before, During, and After Pregnancy with Inflammatory Markers in Breast Milk. *Obesity (Silver Spring)* **2017**, *25*, 2092–2099. [CrossRef]

9. De Luca, A.; Frasquet-Darrieux, M.; Gaud, M.A.; Christin, P.; Boquien, C.Y.; Millet, C.; Herviou, M.; Darmaun, D.; Robins, R.J.; Ingrand, P.; et al. Higher Leptin but Not Human Milk Macronutrient Concentration Distinguishes Normal-Weight from Obese Mothers at 1-Month Postpartum. *PLoS ONE* **2016**, *11*, e0168568. [CrossRef]

10. Lemas, D.J.; Young, B.E.; Baker, P.R.; Tomczik, A.C.; Soderborg, T.K.; Hernandez, T.L.; A de la Houssaye, B.; Robertson, C.E.; Rudolph, M.C.; Patinkin, D.I.Z.W.; et al. Alterations in human milk leptin and insulin are associated with early changes in the infant intestinal microbiome. *Am. J. Clin. Nutr.* **2016**, *103*, 1291–1300. [CrossRef]

11. Uysal, F.K.; Onal, E.E.; Aral, Y.Z.; Adam, B.; Dilmen, U.; Ardicolu, Y. Breast milk leptin: Its relationship to maternal and infant adiposity. *Clin. Nutr.* **2002**, *21*, 157–160. [CrossRef]

12. Andreas, N.J.; Hyde, M.J.; Gale, C.; Parkinson, J.R.; Jeffries, S.; Holmes, E.; Modi, N. Effect of maternal body mass index on hormones in breast milk: A systematic review. *PLoS ONE* **2014**, *9*, 1–25. [CrossRef] [PubMed]

13. Quinn, E.A.; Largado, F.; Borja, J.B.; Kuzawa, C.W. Maternal Characteristics Associated with Milk Leptin Content in a Sample of Filipino Women and Associations with Infant Weight for Age. *J. Hum. Lact.* **2014**, *31*, 273–281. [CrossRef] [PubMed]

14. Eilers, E.; Ziska, T.; Harder, T.; Plagemann, A.; Obladen, M.; Loui, A. Leptin determination in colostrum and early human milk from mothers of preterm and term infants. *Early Hum. Dev.* **2011**, *87*, 415–419. [CrossRef] [PubMed]

15. Panagos, P.G.; Vishwanathan, R.; Penfield-Cyr, A.; Matthan, N.R.; Shivappa, N.; Wirth, M.D.; Hebert, J.R.; Sen, S. Breastmilk from obese mothers has pro-inflammatory properties and decreased neuroprotective factors. *J. Perinatol.* **2016**, *36*, 284–290. [CrossRef] [PubMed]

16. Brunner, S.; Schmid, D.; Zang, K.; Much, D.; Knoeferl, B.; Kratzsch, J.; Amann-Gassner, U.; Bader, B.L.; Hauner, H. Breast milk leptin and adiponectin in relation to infant body composition up to 2 years. *Pediatr. Obes.* **2015**, *10*, 67–73. [CrossRef]

17. Grunewald, M.; Hellmuth, C.; Demmelmair, H.; Koletzko, B. Excessive weight gain during full breast-feeding. *Ann. Nutr. Metab.* **2014**, *64*, 271–275. [CrossRef] [PubMed]

18. Fields, D.A.; George, B.; Williams, M.; Whitaker, K.; Allison, D.B.; Teague, A.; Demerath, E.W. Associations between human breast milk hormones and adipocytokines and infant growth and body composition in the first 6 months of life. *Pediatr. Obes.* **2017**, *12* (Suppl. 1), 78–85. [CrossRef]

19. Khodabakhshi, A.; Ghayour-Mobarhan, M.; Rooki, H.; Vakili, R.; Hashemy, S.I.; Mirhafez, S.R.; Shakeri, M.-T.; Kashanifar, R.; Pourbafarani, R.; Mirzaei, H.; et al. Comparative measurement of ghrelin, leptin, adiponectin, EGF and IGF-1 in breast milk of mothers with overweight/obese and normal-weight infants. *Eur. J. Clin. Nutr.* **2015**, *69*, 614–618. [CrossRef] [PubMed]

20. Lonnerdal, B. Bioactive proteins in breast milk. *J. Paediatr. Child Health.* **2013**, *49* (Suppl. 1), 1–7. [CrossRef]

21. Pang, W.W.; Hartmann, P.E. Initiation of human lactation: Secretory differentiation and secretory activation. *J. Mammary Gland Biol. Neoplasia.* **2007**, *12*, 211–221. [CrossRef]

22. Ballard, O.; Morrow, A.L. Human milk composition: Nutrients and bioactive factors. *Pediatr. Clin. N. Am.* **2013**, *60*, 49–74. [CrossRef] [PubMed]

23. Saarela, T.; Kokkonen, J.; Koivisto, M. Macronutrient and energy contents of human milk fractions during the first six months of lactation. *Acta Paediatr.* **2005**, *94*, 1176–1181. [CrossRef] [PubMed]

24. Nommsen, L.A.; Lovelady, C.A.; Heinig, M.J.; Lonnerdal, B.; Dewey, K.G. Determinants of energy, protein, lipid, and lactose concentrations in human milk during the first 12 mo of lactation- the DARLING Study13. *Am. J. Clin. Nutr.* **1991**, *53*, 457–465. [CrossRef] [PubMed]

25. Dewey, K.G.; Finley, D.A.; Lönnerdal, B. Breast milk volume and composition during late lactation (7–20 months). *J. Pediatric Gastroenterol. Nutr.* **1984**, *3*, 713–720. [CrossRef]

26. Cabrera-Rubio, R.; Collado, M.C.; Laitinen, K.; Salminen, S.; Isolauri, E.; Mira, A. The human milk microbiome changes over lactation and is shaped by maternal weight and mode of delivery. *Am. J. Clin. Nutr.* **2012**, *96*, 544–551. [CrossRef] [PubMed]

27. O'Reilly, J.R.; Reynolds, R.M. The risk of maternal obesity to the long-term health of the offspring. *Clin. Endocrinol. (Oxf.)* **2013**, *78*, 9–16. [CrossRef] [PubMed]

28. Dieterich, C.M.; Felice, J.P.; O'Sullivan, E.; Rasmussen, K.M. Breastfeeding and health outcomes for the mother-infant dyad. *Pediatr. Clin. N. Am.* **2013**, *60*, 31–48. [CrossRef] [PubMed]

29. Horta, B.L.; Victora, C.G. *Long-Term Effects of Breastfeeding: A Systematic Review*; World Health Organization: Geneva, Switzerland, 2013.

30. Steinhoff, M.C.; Omer, S.B. A review of fetal and infant protection associated with antenatal influenza immunization. *Am. J. Obstet. Gynecol.* **2012**, *207*, S21–S27. [CrossRef]

31. Valentine, C.J. Optimizing Humn Milk Fortification for the Preterm Infant. *Pnpg. Build. Block Life.* **2011**, *34*, 9–11.

32. Houghton, M.R.; Gracey, M.; Burke, V.; Bottrell, C.; Spargo, R.M. Breast milk lactoferrin levels in relation to maternal nutritional status. *J. Pediatr. Gastroenterolnutr.* **1985**, *4*, 230–233.

33. Islam, S.K.; Ahmed, L.; Khan, M.N.; Huque, S.; Begum, A.; Yunus, A.B. Immune components (IgA, IgM, IgG, immune cells) of colostrum of Bangladeshi mothers. *Pediatr. Int.* **2006**, *48*, 543–548. [CrossRef] [PubMed]

34. Twigger, A.J.; Hepworth, A.R.; Tat Lai, C.; Chetwynd, E.; Stuebe, A.M.; Blancafort, P.; Hartmann, P.E.; Geddes, D.T.; Kakulas, F. Gene expression in breastmilk cells is associated with maternal and infant characteristics. *Scientific Reports.* **2015**, *5*, 1–14. [CrossRef] [PubMed]

35. Collado, M.C.; Laitinen, K.; Salminen, S.; Isolauri, E. Maternal weight and excessive weight gain during pregnancy modify the immunomodulatory potential of breast milk. *Pediatr. Res.* **2012**, *72*, 77–85. [CrossRef] [PubMed]

36. Martin, L.J.; Woo, J.G.; Geraghty, S.R.; Altaye, M.; Davidson, B.S.; Banach, W.; Dolan, L.M.; Ruiz-Palacios, G.M.; Morrow, A.L. Adiponectin is present in human milk and is associated with maternal factors. *Am. J. Clin. Nutr.* **2006**, *83*, 1106–1111. [CrossRef] [PubMed]

37. Zhang, N.; Yuan, C.; Li, Z.; Li, J.; Li, X.; Li, C.; Li, R.; Wang, S.R. Meta-Analysis of the Relationship Between Obestatin and Ghrelin Levels and the Ghrelin: Obestatin Ratio with Respect to Obesity. *Am. J. Med. Sci.* **2011**, *341*, 48–55. [CrossRef] [PubMed]

38. Aydin, S.; Ozkan, Y.; Erman, F.; Gurates, B.; Kilic, N.; Colak, R.; Gundogan, T.; Catak, Z.; Bozkurt, M.; Akin, O.; et al. Presence of obestatin in breast milk: Relationship among obestatin, ghrelin, and leptin in lactating women. *Nutrition* **2008**, *24*, 689–693. [CrossRef] [PubMed]

39. Andreas, N.J.; Hyde, M.J.; Herbert, B.R.; Jeffries, S.; Santhakumaran, S.; Mandalia, S.; Holmes, E.; Modi, N. Impact of maternal BMI and sampling strategy on the concentration of leptin, insulin, ghrelin and resistin in breast milk across a single feed: A longitudinal cohort study. *BMJ Open.* **2016**, *6*, e010778. [CrossRef] [PubMed]

40. Savino, F.; Sorrenti, M.; Benetti, S.; Lupica, M.M.; Liguori, S.A.; Oggero, R. Resistin and leptin in breast milk and infants in early life. *Early Hum. Dev.* **2012**, *88*, 779–782. [CrossRef] [PubMed]

41. Miranda, R.; Saravia, N.G.; Ackerman, R.; Murphy, N.; Berman, S.; McMurray, D.N. Effect of maternal nutritional statuses on immunological substances in human colostrum and milk. *Am. J. Clin. Nutr.* **1983**, *37*, 632–640. [CrossRef]

42. Fujimori, M.; Franca, E.L.; Fiorin, V.; Morais, T.C.; Honorio-Franca, A.C.; de Abreu, L.C. Changes in the biochemical and immunological components of serum and colostrum of overweight and obese mothers. *BMC Pregnancy Childbirth.* **2015**, *15*, 166. [CrossRef]

43. Makela, J.; Linderborg, K.; Niinikoski, H.; Yang, B.; Lagstrom, H. Breast milk fatty acid composition differs between overweight and normal weight women: The STEPS Study. *Eur. J. Nutr.* **2013**, *52*, 727–735. [CrossRef] [PubMed]

44. Delzenne, N.M.; Cani, P.D. Interaction between obesity and the gut microbiota: Relevance in nutrition. *Annu. Rev. Nutr.* **2011**, *31*, 15–31. [CrossRef] [PubMed]

45. Azad, M.B.; Robertson, B.; Atakora, F.; Becker, A.B.; Subbarao, P.; Moraes, T.J.; Mandhane, P.J.; Turvey, S.E.; Lefebvre, D.L.; Sears, M.R.; et al. Human Milk Oligosaccharide Concentrations Are Associated with Multiple Fixed and Modifiable Maternal Characteristics, Environmental Factors, and Feeding Practices. *J. Nutr.* **2018**, *148*, 1733–1742. [CrossRef] [PubMed]

46. Cerdo, T.; Ruiz, A.; Jauregui, R.; Azaryah, H.; Torres-Espinola, F.J.; Garcia-Valdes, L.; Segura, M.T.; Suárez, A.; Campoy, C. Maternal obesity is associated with gut microbial metabolic potential in offspring during infancy. *J. Physiol. Biochem.* **2018**, *74*, 159–169. [CrossRef] [PubMed]

47. Sanchez-Poso, A.; Lopez-Morales, J.; Ixquierdo, A.; Gil, A. Protein composition of human milk in relation to mother's weight and socio-economic status. *Hum. Nutr. Clin. Nutr.* **1987**, *41*, 115–125.

48. Jevitt, C.; Hernandez, I.; Groer, M. Lactation complicated by overweight and obesity: Supporting the mother and newborn. *J. Midwifery Women's Health* **2007**, *52*, 606–613. [CrossRef]

49. Siqueiros-Cendon, T.; Arevalo-Gallegos, S.; Iglesias-Figueroa, B.F.; Garcia-Montoya, I.A.; Salazar-Martinez, J.; Rascon-Cruz, Q. Immunomodulatory effects of lactoferrin. *Acta Pharmacol. Sin.* **2014**, *35*, 557–566. [CrossRef]

50. Suzuki, Y.A.; Shin, K.; Lönnerdal, B. Molecular cloning and functional expression of a human intestinal lactoferrin receptor. *Biochemistry.* **2002**, *40*, 15771–15779. [CrossRef]

51. Liao, Y.; Jiang, R.; Lonnerdal, B. Biochemical and molecular impacts of lactoferrin on small intestinal growth and development during early life. *Biochem. Cell Biol.* **2012**, *90*, 476–484. [CrossRef]

52. Wakabayashi, H.; Yamauchi, K.; Takase, M. Inhibitory effects of bovine lactoferrin and lactoferricin B on Enterobacter sakazakii. *Biocontrol. Sci.* **2008**, *13*, 29–32. [CrossRef]

53. Arnold, R.R.; Brewer, M.; Gauthier, J.J. Bactericidal Activity of Human Lactoferrin: Sensitivity of a Variety of Microorganisms. *Infect. Immun.* **1980**, *28*, 893–898. [PubMed]

54. Newburg, D.S.; Peterson, J.A.; Ruiz-Palacios, G.M.; Matson, D.O.; Morrow, A.L.; Shults, J.; de Lourdes Guerrero, M.; Chaturvedi, P.; Newburg, S.O.; Scallan, C.D.; et al. Role of human-milk lactadherin in protectoin against symptomatic rotavirus infection. *Lancet.* **1998**, *351*, 1160–1164. [CrossRef]

55. Aziz, M.; Jacob, A.; Matsuda, A.; Wang, P. Review: Milk fat globule-EGF factor 8 expression, function and plausible signal transduction in resolving inflammation. *Apoptosis.* **2011**, *16*, 1077–1086. [CrossRef] [PubMed]

56. Shi, J.; Heegaard, C.W.; Rasmussen, J.T.; Gilbert, G.E. Lactadherin binds selectively to membranes containing phosphatidyl-L-serine and increased curvature. *Biochim. Biophys. Acta.* **2004**, *1667*, 82–90. [CrossRef] [PubMed]

57. Kusunoki, R.; Ishihara, S.; Aziz, M.; Oka, A.; Tada, Y.; Kinoshita, Y. Roles of milk fat globule-epidermal growth factor 8 in intestinal inflammation. *Digestion* **2012**, *85*, 103–107. [CrossRef] [PubMed]

58. Chogle, A.; Bu, H.F.; Wang, X.; Brown, J.B.; Chou, P.M.; Tan, X.D. Milk fat globule-EGF factor 8 is a critical protein for healing of dextran sodium sulfate-induced acute colitis in mice. *Mol. Med.* **2011**, *17*, 502–507. [CrossRef]

59. Peterson, J.; Hamosh, M.; Scallan, C.; Ceriani, R.; Henderson, T.; Mehta, N.; Armand, M.; Hamosh, P. Milk fat globule glycoproteins in human milk and in gastric aspirates of mother's milk-fed preterm infants. *Pediatr. Res.* **1998**, *44*, 499–506. [CrossRef] [PubMed]

60. Steele, W.F.; Morrison, M. Antistreptococcal activity of lactoperoxidase. *J. Bacteriol.* **1969**, *97*, 635–639.

61. Björck, L.; Rosén, C.G.; Marshall, V.; Reiter, B. Antibacterial activity of lactoperoxidase system in milk against pseudomonas and other gram negative bacteria. *Appl. Microbiol.* **1975**, *30*, 199–204.

62. Shin, K.; Tomita, M.; Lonnerdal, B. Identification of lactoperoxidase in mature human milk. *J. Nutr. Biochem.* **2000**, *11*, 94–102. [CrossRef]

63. Ellison, R.; Giehl, T.J. Killing of Gram-negative Bacteria by Lactofernn and Lysozyme. *J. Clin. Investig.* **1991**, *88*, 1080–1091. [CrossRef] [PubMed]

64. Mosca, F.; Gianni, M.L. Human milk: Composition and health benefits. *Pediatr. Med. Chir.* **2017**, *39*, 155. [CrossRef] [PubMed]

65. Bansil, R.; Turner, B.S. Mucin structure, aggregation, physiological functions and biomedical applications. *Curr. Opin. Colloid Interface Sci.* **2006**, *11*, 164–170. [CrossRef]

66. Liu, B.; Yu, Z.; Chen, C.; Kling, D.E.; Newburg, D.S. Human milk mucin 1 and mucin 4 inhibit Salmonella enterica serovar Typhimurium invasion of human intestinal epithelial cells in vitro. *J Nutr.* **2012**, *142*, 1504–1509. [CrossRef] [PubMed]

67. Kvistgaard, A.S.; Pallesen, L.T.; Arias, C.F.; López, S.; Petersen, T.E.; Heegaard, C.W.; Rasmussen, J.T. Inhibitory Effects of Human and Bovine Milk Constituents on Rotavirus Infections. *J. Dairy Sci.* **2004**, *87*, 4088–4096. [CrossRef]

68. Schroten, H.; Hanisch, F.G.; Plogmann, R.; Hacker, J.; Uhlenbruck, G.; Nobis-Bosch, R. Inhibition of Adhesion of S-Fimbriated Escherichia coli to Buccal Epithelial Cells by Human Milk Fat Globule Membrane Components- a Novel Aspect of the Protective Function of Mucins in the Nonimmunoglobulin Fraction. *Infect Immun.* **1992**, *60*, 2893–2899. [PubMed]

69. Smith, C.W.; Goldman, A.S. Interactions of lymphocytes and macrophages from human colostrum- Electron microscopic studies of the interacting lymphocyte. *J. Reticuloendothel.* **1970**, *8*, 91–104.

70. Zhou, L.; Yoshimura, Y.; Huang, Y.; Suzuki, R.; Yokoyama, M.; Okabe, M.; Shimamura, M. Two independent pathways of maternal cell transmission to offspring-through placenta during pregnancy and by breast-feeding after birth. *Immunology* **2000**, *101*, 570–580. [CrossRef]

71. Cabinian, A.; Sinsimer, D.; Tang, M.; Zumba, O.; Mehta, H.; Toma, A.; Sant'Angelo, D.; Laouar, Y.; Laouar, A. Transfer of Maternal Immune Cells by Breastfeeding: Maternal Cytotoxic T Lymphocytes Present in Breast Milk Localize in the Peyer's Patches of the Nursed Infant. *PLoS ONE* **2016**, *11*, e0156762. [CrossRef] [PubMed]

72. Minniti, F.; Comberiati, P.; Munblit, D.; Piacentini, G.L.; Antoniazzi, E.; Zanoni, L.; Boner, A.L.; Peroni, D.G. Breast-Milk Characteristics Protecting Against Allergy. *Endocr. Metab. Immune Disord. Drug Targets* **2014**, *14*, 9–15. [CrossRef]

73. Buescher, E.; Malinowska, I. Soluble Receptors and Cytokine Antagonists in Human Milk. *Pediatr. Res.* **1996**, *40*, 839–844. [CrossRef] [PubMed]

74. English, B.K.; Burchett, S.K.; English, J.D.; Ammann, A.J.; Wara, D.W.; Wilson, C.B. Production of lymphotoxin and tumor necrosis factor by human neonatal mononuclear cells. *Pediatr. Res.* **1988**, *24*, 717–722. [CrossRef] [PubMed]

75. Ichikawa, M.; Sugita, M.; Takahashi, M.; Satomi, M.; Takeshita, T.; Araki, T.; Takahashi, H. Breast milk macrophages spontaneously produce granulocyte—Macrophage colony-stimulating factor and differentiate into dendritic cells in the presence of exogenous interleukin-4 alone. *Immunology* **2003**, *108*, 189–195. [CrossRef] [PubMed]

76. Yagi, Y.; Watanabe, E.; Watari, E.; Shinya, E.; Satomi, M.; Takeshita, T.; Takahashi, H. Inhibition of DC-SIGN-mediated transmission of human immunodeficiency virus type 1 by Toll-like receptor 3 signalling in breast milk macrophages. *Immunology* **2010**, *130*, 597–607. [CrossRef]

77. Jackson, K.M.; Nazar, A.M. Breastfeeding, the Immune Response, and Long-term Health. *J. Am. Osteopath Assoc.* **2006**, *106*, 203–207. [PubMed]

78. Grazioso, C.F.; Buescher, E.S. Inhibition of Neutrophil Function by Human Milk. *Cell Immunol.* **1996**, *168*, 125–132. [CrossRef] [PubMed]

79. Hassiotou, F.; Beltran, A.; Chetwynd, E.; Stuebe, A.M.; Twigger, A.J.; Metzger, P.; Trengove, N.; Lai, C.T.; Filgueira, L.; Blancafort, P.; et al. Breastmilk is a novel source of stem cells with multilineage differentiation potential. *Stem Cells* **2012**, *30*, 2164–2174. [CrossRef] [PubMed]

80. Kakulas, F. Breast milk: A source of stem cells and protective cells for the infant. *Infant* **2015**, *11*, 3.

81. Keller, T.; Korber, F.; Oberthuer, A.; Schafmeyer, L.; Mehler, K.; Kuhr, K.; Kribs, A. Intranasal breast milk for premature infants with severe intraventricular hemorrhage-an observation. *Eur. J. Pediatr.* **2019**, *178*, 199–206. [CrossRef] [PubMed]

82. Cacho, N.T.; Lawrence, R.M. Innate Immunity and Breast Milk. *Front. Immunol.* **2017**, *8*, 584. [CrossRef] [PubMed]

83. Avalos, B. Molecular analysis of the granulocyte colony-stimulating factor receptor. *Blood* **1996**, *88*, 761–777. [PubMed]

84. Wallace, J.M.; Ferguson, S.J.; Loane, P.; Kell, M.; Millar, S.; Gillmore, W.S. Cytokines in human breast milk. *Br. J. Biomed. Sci.* **1997**, *54*, 85–87. [PubMed]

85. Calandra, T.; Bernhagen, J.; Mitchell, R.A.; Bucala, R. The Macrophage Is an Important and Previously Unrecognized Source of Macrophage Migration Inhibitory Factor. *J. Exp. Med.* **1994**, *179*, 1895–1902. [CrossRef] [PubMed]

86. Bacher, M.; Metz, C.N.; Calandra, T.; Mayer, K.; Chesney, J.; Lohoff, M.; Gemsa, D.; Donnelly, T.; Bucala, R. An essential regulatory role for macrophage migration inhibitory factor in T-cell activation. *Proc. Natl. Acad. Sci. USA* **1996**, *93*, 7849–7854. [CrossRef] [PubMed]

87. Magi, B.; Bini, L.; Liberatori, S.; Marzocchi, B.; Raggiaschi, R.; Arcuri, F.; Tripodi, S.A.; Cintorino, M.; Tosi, P.; Pallini, V. Charge heterogeneity of macrophage migration inhibitory factor (MIF) in human liver and breast tissue. *Electrophoresis* **1998**, *19*, 2010–2013. [CrossRef]

88. Lonnerdal, B. Nutritional and physiologic significance of human milk proteins. *Am. J. Clin. Nutr.* **2003**, *77*, 1537S–1543S. [CrossRef] [PubMed]

89. Brenmoehl, J.; Ohde, D.; Wirthgen, E.; Hoeflich, A. Cytokines in milk and the role of TGF-beta. *Best Pract. Res. Clin. Endocrinol. Metab.* **2018**, *32*, 47–56. [CrossRef] [PubMed]

90. Basolo, F.; Fiore, L.; Fontanini, G.; Giulio, P.; Simonetta, C.; Falcone, V. Expression of response to interleukin 6 (IL6) in human mammary tumors. *Cancer Res.* **1996**, *56*, 3118–3122. [PubMed]

91. Meki, A.-R.; Saleem, T.H.; Al-Ghazali, M.H.; Sayed, A.A. Interleukins -6, -8 and -10 and tumor necrosis factor-alpha and its soluble receptor I in human milk at different periods of lactation. *Nutr. Res.* **2003**, *23*, 845–855. [CrossRef]

92. Mizuno, K.; Hatsuno, M.; Aikawa, K.; Takeichi, H.; Himi, T.; Kaneko, A.; Kodaira, K.; Takahashi, H.; Itabashi, K. Mastitis Is Associated with IL-6 Levels and Milk Fat Globule Size in Breast Milk. *J. Hum. Lact.* **2012**, *28*, 529–534. [CrossRef]

93. Bottcher, M.F.; Jenmalm, M.C.; Garofalo, R.P.; Bjorksten, B. Cytokines in breast milk from allergic and nonallergic mothers. *Pediatr. Res.* **2000**, *47*, 157–162. [CrossRef] [PubMed]

94. Aspinall, R.; Prentice, A.M.; Ngom, P.T. Interleukin 7 from maternal milk crosses the intestinal barrier and modulates T-cell development in offspring. *PLoS ONE* **2011**, *6*, e20812. [CrossRef] [PubMed]

95. Wagner, C.L.; Taylor, S.N.; Johnson, D. Host factors in amniotic fluid and breast milk that contribute to gut maturation. *Clin. Rev. Allergy Immunol.* **2008**, *34*, 191–204. [CrossRef] [PubMed]

96. Higashiyama, S.; Abraham, J.A.; Miller, J.; Fiddes, J.C.; Klagsbrun, M. A heparin-binding growth factor secreted by macrophage-like cells that is related to EGF. *Science* **1991**, *251*, 936–939. [CrossRef] [PubMed]

97. Michalsky, M.P.; Lara-Marquez, M.; Chun, L.; Besner, G.E. Heparin-binding EGF-like growth factor is present in human amniotic fluid and breast milk. *J. Pediatr. Surg.* **2002**, *37*, 1–6. [CrossRef] [PubMed]

98. Pillai, S.B.; Turman, M.A.; Besner, G.E. Heparin-Binding EGF-Like Growth Factor Is Cytoprotective for Intestinal Epithelial Cells Exposed to Hypoxia. *J. Pediatr. Surg.* **1998**, *33*, 973–978. [CrossRef]

99. Michalsky, M.P.; Kuhn, A.; Mehta, V.; Besner, G.E. Heparin-binding EGF-like growth factor decreases apoptosis in intestinal epithelial cells in vitro. *J. Pediatr. Surg.* **2001**, *36*, 1130–1135. [CrossRef] [PubMed]

100. Xia, G.; Martin, A.E.; Besner, G.E. Heparin-binding EGF-like growth factor downregulates expression of adhesion molecules and infiltration of inflammatory cells after intestinal ischemia/reperfusion injury. *J. Pediatr. Surg.* **2003**, *38*, 434–439. [CrossRef] [PubMed]

101. Kuhn, M.A.; Xia, G.; Mehta, V.B.; Glenn, S.; Michalsky, M.P. Antioxidants & redox signaling.-Heparin-binding EGF-like growth factor (HB-EGF) decreases oxygen free radical production in vitro and in vivo. *Antioxid. Redox Signal.* **2002**, *4*, 639–646. [PubMed]

102. LeBon, T.R.; Jacobs, S.; Cuatrecasas, P.; Kathuria, S.; Fujita-Yamaguchi, Y. Purification of Insulin-like Growth Factor I Receptor from Human Placental Membranes. *J. Biol. Chem.* **1986**, *261*, 7685–7689.

103. Larnkjaer, A.; Ong, K.K.; Carlsen, E.M.; Ejlerskov, K.T.; Molgaard, C.; Michaelsen, K.F. The Influence of Maternal Obesity and Breastfeeding on Infant Appetite- and Growth-Related Hormone Concentrations: The SKOT Cohort Studies. *Horm. Res. Paediatr.* **2018**, *90*, 28–38. [CrossRef] [PubMed]

104. Fonseca, L.T.; Senna, D.C.; Eckert, G.U.; Silveira, R.C.; Procianoy, R.S. Association between human breast milk and retinopathy of prematurity. *Arq. Bras. Oftalmol.* **2018**, *81*, 102–109. [CrossRef] [PubMed]

105. Bradshaw, R.A. Nerve growth factor. *Annu. Rev. Biochem.* **1978**, *47*, 191–216. [CrossRef] [PubMed]

106. Tam, S.Y.; Tsai, M.; Yamaguchi, M.; Yano, K.; Butterfield, J.H.; Galli, S.J. Expression of Functional TrkA Receptor Tyrosine Kinase in the HMC-1 Human Mast Cell Line and in Human Mast Cells. *Blood* **1997**, *90*, 1807–1820. [PubMed]

107. Kobata, R.; Tsukahara, H.; Ohshima, Y.; Ohta, N.; Tokuriki, S.; Tamura, S.; Mayumi, M. High levels of growth factors in human breast milk. *Early Hum. Dev.* **2008**, *84*, 67–69. [CrossRef] [PubMed]

108. Quinn, T.; Peters, K.G.; de Vries, C.; Ferrara, N.; Williams, L.T. Fetal liver kinase 1 is a receptor for vascular endothelial growth factor and is selectively expressed in vascular endothelium. *Proc. Natl. Acad. Sci. USA* **1993**, *90*, 7533–7537. [CrossRef] [PubMed]

109. Weyermann, M.; Brenner, H.; Rothenbacher, D.A. Adipokines in human milk and risk of overweight in early childhood: A Prospective Cohort Study. *Epidemiology* **2007**, *18*, 722–729. [CrossRef] [PubMed]

110. Catli, G.; Dundar, N.O.; Dundar, B.N. Adipokines in breast milk: An update. *J. Clin. Res. Pediatr. Endocrinol.* **2014**, *6*, 192–201. [CrossRef] [PubMed]

111. Savino, F.; Petrucci, E.; Nanni, G. Adiponectin: An intriguing hormone for paediatricians. *Acta Paediatr.* **2008**, *97*, 701–705. [CrossRef] [PubMed]

112. Woo, J.G.; Guerrero, M.L.; Altaye, M.; Ruiz-Palacios, G.M.; Martin, L.J.; Dubert-Ferrandon, A.; Newburg, D.S.; Morrow, A.L. Human milk adiponectin is associated with infant growth in two independent cohorts. *Breastfeed. Med.* **2009**, *4*, 101–109. [CrossRef] [PubMed]

113. Konturek, S.J.; Radecki, T.; Konturek, D.; Dimitreascu, T. Effect of Calcitonin on Gastric and Pancreatic Secretion and Peptic Ulcer Formation in Cats. *Dig. Dis. Sci.* **1974**, *19*, 235–241. [CrossRef]

114. Bueno, L.; Ferre, J.P.; Fioramonti, J.; Honde, C. Effects of intracerebroventricular administration of neurotensin, substance P and calcitonin on gastrointestinal motility in normal and vagotomized rats. *Regul. Pept.* **1983**, *6*, 197–205. [CrossRef]

115. Appelgren, B.H.; Arver, S.; Sagulin, G.B. Effect of intracerebroventricular calcitonin on renal hydromineral excretion in sheep. *Am. J. Physiol. Regul. Integr. Comp. Physiol.* **1986**, *250*, R980–R983. [CrossRef] [PubMed]
116. Arver, S.; Bucht, E.; Sjoberg, E. Calcitonin-like immunoreactivity in human milk, longitudinal alterations andl divalent cations. *Acta Physiol. Scand.* **1984**, *122*, 461–464. [CrossRef] [PubMed]
117. Root, A.W. *Disorders of Calcium and Phosphorus Homeostasis in the Newborn and Infant*, 4th ed.; Elseviers: Amsterdam, The Netherlands, 2014.
118. Kojima, M.; Hosoda, H.; Date, Y.; Nakazato, M.; Matsuo, H.; Kangawa, K. Ghrelin is a growth-hormone-releasing acylated peptide from stomach. *Nature* **1999**, *402*, 652–660. [CrossRef] [PubMed]
119. van der Lely, A.J.; Tschop, M.; Heiman, M.L.; Ghigo, E. Biological, physiological, pathophysiological, and pharmacological aspects of ghrelin. *Endocr. Rev.* **2004**, *25*, 426–457. [CrossRef] [PubMed]
120. Cesur, G.; Ozguner, F.; Yilmaz, N.; Dundar, B. The relationship between ghrelin and adiponectin levels in breast milk and infant serum and growth of infants during early postnatal life. *J. Physiol. Sci.* **2012**, *62*, 185–190. [CrossRef] [PubMed]
121. Ley, S.H.; Hanley, A.J.; Sermer, M.; Zinman, B.; O'Connor, D.L. Associations of prenatal metabolic abnormalities with insulin and adiponectin concentrations in human milk. *Am. J. Clin. Nutr.* **2012**, *95*, 867–874. [CrossRef] [PubMed]
122. Butte, N.F.; Garza, C.; Burr, R.; Goldman, A.S.; Kennedy, K.; Kitzmiller, J.L. Milk composition of insulin-dependent diabetic women. *J. Pediatr. Gastroenterol. Nutr.* **1987**, *6*, 936–941. [CrossRef] [PubMed]
123. Jovanovic-Peterson, L.; Fuhrmann, K.; Hedden, K.; Walker, L.; Peterson, C.M. Maternal milk and plasma glucose and insulin levels: Studies in normal and diabetic subjects. *J. Am. Coll. Nutr.* **1989**, *8*, 125–131. [CrossRef] [PubMed]
124. Plagemann, A.; Harder, T.; Franke, K.; Kohlhoff, R. Long-term impact of neonatal breast feeding on body weight and glucose tolerance in children of diabetic mothers. *Diabetes Care* **2002**, *25*, 16–22. [CrossRef] [PubMed]
125. Zhang, Y.; Proenca, R.; Maffei, M.; Barone, M.; Leopold, L.; Friedman, J.M. Positional cloning of the mouse obese gene and its human homologue. *Nature (London)* **1994**, *372*, 425–432. [CrossRef] [PubMed]
126. Friedman, J.M. Leptin at 14 y of age: An ongoing story. *Am. J. Clin. Nutr.* **2009**, *89*, 973S–979S. [CrossRef] [PubMed]
127. Lacquaniti, A.; Donato, V.; Chirico, V.; Buemi, A.; Buemi, M. Obestatin: An interesting but controversial gut hormone. *Ann. Nutr. Metab.* **2011**, *59*, 193–199. [CrossRef] [PubMed]
128. Zhang, J.V.; Ren, P.G.; Avsian-Kretchmer, O.; Luo, C.W.; Rauch, R.; Klein, C.; Hsueh, A.J. Obestatin, a peptide encoded by the ghrelin gene, opposes ghrelin's effects on food intake. *Science* **2005**, *310*, 996–999. [CrossRef]
129. Savino, F.; Liguori, S.A.; Fissore, M.F.; Oggero, R. Breast milk hormones and their protective effect on obesity. *Int. J. Pediatr. Endocrinol.* **2009**, *2009*, 327505. [CrossRef] [PubMed]
130. Ilcol, Y.O.; Hizli, Z.B.; Eroz, E. Resistin is present in human breast milk and it correlates with maternal hormonal status and serum level of C-reactive protein. *Clin. Chem. Lab. Med.* **2008**, *46*, 118–124. [CrossRef] [PubMed]
131. Marchini, G.; Winberg, J.; Uvnas-Mobergi, K. Plasma concentrations of gastrin and somatostatin after breast feeding in 4 day old infants. *Arch. Dis. Child.* **1988**, *63*, 1218–1221. [CrossRef]
132. Hoist, N.; Jenssen, T.G.; Burhol, P.G. A characterization of immunoreactive somastostatin in human milk. *J. Pediatr. Gastroenterol. Nutr.* **1990**, *10*, 47–52.
133. Brandtzaeg, P.; Kiyono, H.; Pabst, R.; Russell, M.W. Terminology: Nomenclature of mucosa-associated lymphoid tissue. *Mucosal. Immunol.* **2008**, *1*, 31–37. [CrossRef] [PubMed]
134. Palmeira, P.; Costa-Carvalho, B.T.; Arslanian, C.; Pontes, G.N.; Nagao, A.T.; Carneiro-Sampaio, M.M. Transfer of antibodies across the placenta and in breast milk from mothers on intravenous immunoglobulin. *Pediatr. Allergy Immunol.* **2009**, *20*, 528–535. [CrossRef] [PubMed]
135. Carbonare, S.B.; Silva, M.L.M.; Palmeira, P.; Carneiro-Sampaio, M.M. Human colostrum IgA antibodies reacting to enteropathogenic Escherichia coli antigens and their persistence in the faeces of a breastfed infant. *J. Diarrhoeal. Dis. Res.* **1997**, *15*, 53–58. [PubMed]
136. Hanson, L.A.; Korotkova, M.; Lundin, S.; Håversen, L.; Silfverdal, S.A.; Mattsby-Baltzer, I.; Strandvik, B.; Telemo, E. The transfer of immunity from mother to child. *N. Y. Acad. Sci.* **2003**, *987*, 199–206. [CrossRef]
137. Kling, P.J.; Taing, K.M.; Dvorak, B.; Woodward, S.S.; Philipps, A.F. Insulin-like growth factor-I stimulates erythropoiesis when administered enterally. *Growth Factors* **2006**, *24*, 218–223. [CrossRef] [PubMed]

138. Blum, J.W.; Baumrucker, C.R. Colostral and milk insulin-like growth factors and related substances- Mammary gland and neonatal (intestinal and systemic) targets. *Domest. Anim. Endocrinol.* **2002**, *23*, 101–110. [CrossRef]

139. Grote, V.; Verduci, E.; Scaglioni, S.; Vecchi, F.; Contarini, G.; Giovannini, M.; Koletzko, B.; Agostoni, C. Breast milk composition and infant nutrient intakes during the first 12 months of life. *Eur. J. Clin. Nutr.* **2016**, *70*, 250–256. [CrossRef] [PubMed]

140. Milani, C.; Duranti, S.; Bottacini, F.; Casey, E.; Turroni, F.; Mahony, J.; Belzer, C.; Delgado Palacio, S.; Arboleya Montes, S.; Mancabelli, L.; et al. The first microbial colonizers of the human gut: Composition, activities, and health implications of the infant gut microbiota. *Microbiol. Mol. Biol. Rev.* **2017**, *81*, e00036-17. [CrossRef]

141. German, J.B.; Dillard, C.J. Composition, structure and absorption of milk lipids: A source of energy, fat-soluble nutrients and bioactive molecules. *Crit. Rev. Food Sci. Nutr.* **2006**, *46*, 57–92. [CrossRef]

142. Bergsson, G.; Arnfinnsson, J.; Karlsson, S.M.; Steingrimsson, O.; Thormar, H. In Vitro Inactivation of Chlamydia trachomatis by Fatty Acids and Monoglycerides. *Antimicrob. Agents Chemother.* **1998**, *42*, 2290–2294. [CrossRef]

143. Gibson, R.A.; Chen, W.; Makrides, M. Randomized trials with polyunsaturated fatty acid interventions in preterm and term infants: Functional and clinical outcomes. *Lipids* **2001**, *36*, 873–883. [CrossRef]

144. Lapillonne, A.; Carlson, S.E. Polyunsaturated fatty acids and infant growth. *Lipids* **2001**, *36*, 901–911. [CrossRef] [PubMed]

145. Makrides, M.; Gibson, R.A.; Udell, T.; Ried, K.; International LUPUFA Investigators. Supplementation of infant formula with long-chain polyunsaturated fatty acids does not influence the growth of term infants. *Am. J. Clin. Nutr.* **2005**, *81*, 1094–1101. [PubMed]

146. Savino, F.; Benetti, S.; Lupica, M.M.; Petrucci, E.; Palumeri, E.; Cordero di Montezemolo, L. Ghrelin and obestatin in infants, lactating mothers and breast milk. *Horm. Res. Paediatr.* **2012**, *78*, 297–303. [CrossRef] [PubMed]

147. Walker, W.A.; Iyengar, R.S. Breast milk, microbiota, and intestinal immune homeostasis. *Pediatr. Res.* **2015**, *77*, 220–228. [CrossRef] [PubMed]

148. Toscano, M.; De Grandi, R.; Grossi, E.; Drago, L. Role of the Human Breast Milk-Associated Microbiota on the Newborns' Immune System: A Mini Review. *Front. Microbiol.* **2017**, *8*, 2100. [CrossRef] [PubMed]

149. Bode, L. Human milk oligosaccharides: Every baby needs a sugar mama. *Glycobiology* **2012**, *22*, 1147–1162. [CrossRef] [PubMed]

150. Rodriguez, J.M. The origin of human milk bacteria: Is there a bacterial entero-mammary pathway during late pregnancy and lactation? *Adv. Nutr.* **2014**, *5*, 779–784. [CrossRef] [PubMed]

151. Kozyrskyj, A.L.; Kalu, R.; Koleva, P.T.; Bridgman, S.L. Fetal programming of overweight through the microbiome: Boys are disproportionately affected. *J. Dev. Orig. Health Dis.* **2016**, *7*, 25–34. [CrossRef] [PubMed]

152. Gregora, M. Lactobacillus Species in Breast Milk. In *Probiotics—Current Knowledge and Future Prospects*; IntechOpen: London, UK, 2018.

153. Turnbaugh, P.J.; Hamady, M.; Yatsunenko, T.; Cantarel, B.L.; Duncan, A.; Ley, R.E.; Sogin, M.L.; Jones, W.J.; Roe, B.A.; Affourtit, J.P.; et al. A core gut microbiome in obese and lean twins. *Nature* **2009**, *457*, 480–484. [CrossRef] [PubMed]

154. DiBaise, J.K.; Frank, D.N.; Mathur, R. Impact of the Gut Microbiota on the Development of Obesity: Current Concepts. *Am. J. Gastroenterol. Suppl.* **2012**, *1*, 22–27. [CrossRef]

155. Karvonen, A.M.; Hyvarinen, A.; Rintala, H.; Korppi, M.; Taubel, M.; Doekes, G.; Gehring, U.; Renz, H.; Pfefferle, P.I.; Genuneit, J.; et al. Quantity and diversity of environmental microbial exposure and development of asthma: A birth cohort study. *Allergy* **2014**, *69*, 1092–1101. [CrossRef] [PubMed]

156. Kostic, A.D.; Gevers, D.; Siljander, H.; Vatanen, T.; Hyötyläinen, T.; Hämäläinen, A.-M.; Peet, A.; Tillmann, V.; Pöhö, P.; Mattila, I.; et al. The dynamics of the human infant gut microbiome in development and in progression toward type 1 diabetes. *Cell Host Microbe* **2015**, *17*, 260–273. [CrossRef] [PubMed]

157. Frank, D.N.; Amand, A.L.S.; Feldman, R.A.; Boedeker, E.C.; Harpaz, N.; Pace, N.R. Molecular-phylogenetic characterization of microbial community imbalances in human inflammatory bowel diseases. *Proc. Natl. Acad. Sci. USA* **2007**, *104*, 13780–13785. [CrossRef] [PubMed]

158. Kalliomäki, M.; Carmen Collado, M.; Salminen, S.; Isolauri, E. Early differences in fecal microbiota composition in children may predict overweight. *Am. J. Clin. Nutr.* **2012**, *96*, 544–551. [CrossRef] [PubMed]

159. Thorell, L. Nucleotides in human milk: Sources and metabolism by the newborn infant. *Pediatr. Res.* **1996**, *40*, 845–852. [CrossRef] [PubMed]

160. Uauy, R. *Dietary Nucleotides and Requirements in Early Life: Text-Book of Gastroenterology and Nutrition in Infancy*; Raven Press Ltd.: New York, NY, USA, 1989.

161. Quan, R.; Barness, L.A.; Uauy, R. Do Infants Need Nucleotide Supplemented Formula for Optimal Nutrition? *J. Pediatr. Gastroenterol. Nutr.* **1990**, *11*, 429–433.

162. Carver, J.D.; Walker, W.A. The role of nucleotides in human nutrition. *J. Nutr. Biochem.* **1995**, *6*, 58–72. [CrossRef]

163. Gil, A.; Uauy, A. Nucleotides and related compounds in human and bovine milks. In *Handbook of Milk Composition*; Jensen, R.G., Ed.; Academic Press: San Diego, CA, USA, 1995.

164. Plaza-Diaz, J.; Fontana, L.; Gil, A. Human Milk Oligosaccharides and Immune System Development. *Nutrients* **2018**, *10*, 1038. [CrossRef] [PubMed]

165. Newburg, D.S.; Walker, W.A. Protection of the neonate by the innate immune system of developing gut and of human milk. *Pediatr. Res.* **2007**, *61*, 2–8. [CrossRef]

166. Newburg, D.S.; He, Y. Neonatal gut microbiota and human milk glycans cooperate to attenuate infection and inflammation. *Clin. Obstet. Gynecol.* **2015**, *58*, 814–826. [CrossRef]

167. He, Y.; Liu, S.; Leone, S.; Newburg, D.S. Human colostrum oligosaccharides modulate major immunologic pathways of immature human intestine. *Mucosal. Immunol.* **2014**, *7*, 1326–1339. [CrossRef] [PubMed]

168. Newburg, D.S.; Morelli, L. Human milk and infant intestinal mucosal glycans guide succession of the neonatal intestinal microbiota. *Pediatr. Res.* **2015**, *77*, 115–120. [CrossRef] [PubMed]

169. Alderete, T.L.; Autran, C.; Brekke, B.E.; Knight, R.; Bode, L.; Goran, M.I.; Fields, D.A. Associations between human milk oligosaccharides and infant body composition in the first 6 mo of life. *Am. J. Clin. Nutr.* **2015**, *102*, 1381–1388. [CrossRef]

170. Coppa, G.V.; Gabrielli, O.; Zampini, L.; Galeazzi, T.; Maccari, F.; Buzzega, D.; Galeotti, F.; Bertino, E.; Volpi, N. Glycosaminoglycan content in term and preterm milk during the first month of lactation. *Neonatology* **2012**, *101*, 74–76. [CrossRef] [PubMed]

171. Blewett, H.J.H.; Cicalo, M.C.; Holland, C.D.; Field, C.J. The Immunological Components of Human Milk. In *Advances in Food and Nutrition Research*; Academic Press: Cambridge, MA, USA, 2008; pp. 45–80.

172. Vidal, K.; Van Den Broek, P.; Lorget, F.; Donnet-Hughes, A. Osteoprotegerin in Human Milk: A Potential Role in the Regulation of Bone Metabolism and Immune Development. *Pediatr. Res.* **2004**, *55*, 1001–1008. [CrossRef]

173. Vidal, K.; Serrant, P.; Schlosser, B.; van den Broek, P.; Lorget, F.; Donnet-Hughes, A. Osteoprotegerin production by human intestinal epithelial cells- a potential regulator of mucosal immune responses. *Am. J. Physiol. Gastrointest. Liver Physiol.* **2004**, *287*, G836–G844. [CrossRef]

174. Phillips, N.I.; Jennes, R. Isolation and properties of human alpha-lactalbumin. *Biochim. Biophys. Acta* **1971**, *229*, 407–410. [CrossRef]

175. Permyakov, E.; Berliner, L.J. Alpha-Lactalbumin structure and function. *Fed. Eur. Biochem. Soc.* **2000**, *473*, 269–274. [CrossRef]

176. Heine, W.E.; Klein, P.D.; Reeds, P.J. The importance of alpha-lactalbumin in infant nutrition. *J. Nutr.* **1991**, *121*, 277–283. [CrossRef]

177. Kunz, C.; Lönnerdal, B. Re-evaluation of the whey protein:casein ratio of human milk. *Acta Paediatr.* **1992**, *81*, 107–112. [CrossRef]

178. Lönnerdal, B.; Forsum, E.; Gebre-Medhin, M.; Hambraeus, L. Breast milk composition in Ethiopian and Swedish mothers. II. Lactose, nitrogen, and protein contents. *Am. J. Clin. Nutr.* **1976**, *9*, 1134–1141. [CrossRef] [PubMed]

179. Lindberg, T.; Ohlsson, K.; Weström, B. Protease inhibitors and their relation to proteases in human milk. *Pediatr. Res* **1982**, *16*, 479–483. [CrossRef] [PubMed]

180. Davidson, L.A.; Lönnerdal, B. Persistence of human milk proteins in the breast fed infant. *Acta Paediatr. Scand.* **1987**, *76*, 733–740. [CrossRef] [PubMed]

181. Davidson, L.A.; Lönnerdal, B. Fecal alpha1 antitrypsin in breast fed infants is derived from human milk and is not indicative of enteric protein loss. *Acta Paediatr. Scand.* **1990**, *79*, 137–141. [CrossRef] [PubMed]

182. Lonnerdal, B.; Lien, E.L. Nutritional and Physiologic Significance of Alpha Lactalbumin in Infants. *Nutr. Rev.* **2003**, *61*, 295–305. [CrossRef] [PubMed]

183. Lonnerdal, B. Bioactive proteins in human milk: Mechanisms of action. *J. Pediatr.* **2010**, *156*, S26–S30. [CrossRef] [PubMed]

184. Chowanadisai, W.; Lonnerdal, B. α(1)-antitrypsin and antichymotrypsin in human milk: Origin, concentrations, and stability. *Am. J. Clin. Nutr.* **2002**, *76*, 828–833. [CrossRef] [PubMed]

185. Lindberg, T.; Skude, G. Amylase in human milk. *Pediatrics* **1982**, *70*, 235–238.

186. Heitlinger, L.A.; Lee, P.C.; Dillon, W.P.; Lebenthal, E. Mammary amylase: A possible alternate pathway of carbohydrate digestion in infancy. *Pediatr. Res.* **1979**, *13*, 969–972. [CrossRef]

187. Andersson, Y.; Savman, K.; Blackberg, L.; Hernell, O. Pasteurization of mother's own milk reduces fat absorption and growth in preterm infants. *Acta Paediatr.* **2007**, *96*, 1445–1449. [CrossRef]

188. Greenberg, R.; Groves, M.L.; Dower, H.J. Human beta-casein. Amino acid sequence and identification of phosphorylation sites. *J. Biol. Chem.* **1984**, *259*, 5128–5132.

189. Plaisancie, P.; Claustre, J.; Estienne, M.; Henry, G.; Boutrou, R.; Paquet, A.; Léonil, J. A novel bioactive peptide from yoghurts modulates expression of the gel-forming MUC2 mucin as well as population of goblet cells and Paneth cells along the small intestine. *J. Nutr. Biochem.* **2013**, *24*, 213–221. [CrossRef] [PubMed]

190. Haschke, F.; Haiden, N.; Thakkar, S.K. Nutritive and Bioactive Proteins in Breastmilk. *Ann. Nutr. Metab.* **2016**, *69* (Suppl. 2), 17–26. [CrossRef]

191. Sproston, N.R.; Ashworth, J.J. Role of C-Reactive Protein at Sites of inflammation and infection. *Front. Immunol.* **2018**, *9*, 1–11. [CrossRef] [PubMed]

192. Williams, M.J.; Williams, S.M.; Poulton, R. Breast feeding is related to C reactive protein concentration in adult women. *J. Epidemiol. Community Health* **2006**, *60*, 146–148. [CrossRef] [PubMed]

193. Henderson, G.B. Folate-Binding Proteins. *Annu. Rev. Nutr.* **1990**, *10*, 319–335. [CrossRef]

194. Antony, A.C.; Utley, C.S.; Marcell, P.D.; Kolhouse, J.F. Isolation, characterization, and comparison of the solubilized particulate and soluble folate binding proteins from human milk. *J. Biol. Chem.* **1982**, *257*, 10081–10089. [PubMed]

195. Said, H.M.; Horne, D.W.; Wagner, C. Effect of human milk folate binding protein on folate intestinal transport. *Arch. Biochem. Biophys.* **1986**, *251*, 114–120. [CrossRef]

196. Adkins, Y.; Lönnerdal, B. Potential host-defense role of a human milk vitamin B-12–binding protein, haptocorrin, in the gastrointestinal tract of breastfed infants, as assessed with porcine haptocorrin in vitro. *Am. J. Clin. Nutr.* **2003**, *77*, 1234–1240. [CrossRef] [PubMed]

197. Adkins, Y.; Lönnerdal, B. Mechanisms of vitamin B12 absorption in breast-fed infants. *J. Pediatr. Gastroenterol. Nutr.* **2002**, *35*, 192–198. [CrossRef] [PubMed]

198. Chung, M.Y. Factors affecting human milk composition. *Pediatr. Neonatol.* **2014**, *55*, 421–422. [CrossRef] [PubMed]

199. Hennart, P.F.; Brasseur, D.J.; Delogne-Desnoeck, J.B.; Dramaix, M.M.; Robyn, C.E. Lysozyme, lactoferrin, and secretory immunoglobulin A content in breast milk: Influence of duration of lactation, nutrition status, prolactin status, and parity of mother. *Am. J. Clin. Nutr.* **1991**, *53*, 32–39. [CrossRef] [PubMed]

200. Chang, S.J. Antimicrobial proteins of maternal and cord sera and milk in relation to maternal nutritional status. *Am. J. Clin. Nutr.* **1990**, *51*, 183–187. [CrossRef] [PubMed]

201. Wirt, D.P.; Adkins, L.T.; Palkowetz, K.H.; Schmalstieg, F.C.; Goldman, A.S. Activated and memory T lymphocytes in human milk. *Cytometry* **1992**, *13*, 282–290. [PubMed]

202. Hassiotou, F.; Geddes, D.T. Immune cell-mediated protection of the mammary gland and the infant during breastfeeding. *Adv Nutr.* **2015**, *6*, 267–275. [CrossRef] [PubMed]

203. Vigh, E.; Bodis, J.; Garai, J. Longitudinal changes in macrophage migration inhibitory factor in breast milk during the first three months of lactation. *J. Reprod. Immunol.* **2011**, *89*, 92–94. [CrossRef]

204. Hossny, E.M.; El-Ghoneimy, D.H.; El-Owaidy, R.H.; Mansour, M.G.; Hamza, M.T.; El-Said, A.F. Breast milk interleukin-7 and thymic gland development in infancy. *Eur. J. Nutr.* **2019**. [CrossRef]

205. Ustundag, B.; Yilmaz, E.; Dogan, Y.; Akarsu, S.; Canatan, H.; Halifeoglu, I.; Cikim, G.; Aygun, A.D. Levels of cytokines (IL-1beta, IL-2, IL-6, IL-8, TNF-α) and trace elements (Zn, Cu) in breast milk from mothers of preterm and term infants. *Mediat. Inflamm.* **2005**, *2005*, 331–336. [CrossRef]

206. Eriksson, U.; Duc, G.; Froesch, E.R.; Zapf, J. Insulin-like growth factors (IGF) I and I1 and IGF binding proteins (IGFBPs) in human colostrum:transitory milk during the first week postpartum. *Biochem. Biophys. Res. Commun.* **1993**, *196*, 267–273. [CrossRef]

Nutrients **2019**, *11*, 1284

207. Ai, Z.; Yumei, Z.; Titi, Y.; Qinghai, S.; Xiaohong, K.; Peiyu, W. The concentrations of some hormones and growth factors in bovine and human colostrums: Short communication. *Int. J. Dairy Technol.* **2012**, *65*, 507–510. [CrossRef]

208. Savino, F.; Fissore, M.; Grassino, E.; Nanni, G.; Oggero, R.; Silvestro, L. Ghrelin, leptin and IGF-I levels in breast-fed and formula-fed infants in the first years of life. *Acta Paediatr.* **2005**, *94*, 531–537. [CrossRef] [PubMed]

209. James, R.J.; Drewett, R.F.; Cheetham, T.D. Low cord ghrelin levels in term infants are associated with slow weight gain over the first 3 months of life. *J. Clin. Endocrinol. Metab.* **2004**, *89*, 3847–3850. [CrossRef] [PubMed]

210. Bitman, J.; Freed, L.M.; Neville, M.C.; Wood, D.L.; Hamosh, P.; Hamosh, M. Lipid composition of prepartum human mammary secretion and postpartum milk. *J. Pediatr. Gastroenterol. Nutr.* **1986**, *5*, 608–615. [CrossRef] [PubMed]

211. Svennerholm, L.; Bostrom, K.; Fredman, P.; Mansson, J.E.; Rosengren, B.; Rynmark, B.M. Human brain gangliosides: Developmental changes from early fetal stage to advanced age. *Biochim. Biophys. Acta* **1989**, *1005*, 109–117. [CrossRef]

212. Liu, B.; Newburg, D.S. Human milk glycoproteins protect infants against human pathogens. *Breastfeed. Med.* **2013**, *8*, 354–362. [CrossRef] [PubMed]

213. Thakkar, S.K.; Giuffrida, F.; Cristina, C.H.; De Castro, C.A.; Mukherjee, R.; Tran, L.-A.; Steenhout, P.; Lee, L.Y.; Destaillats, F. Dynamics of human milk nutrient composition of women from Singapore with a special focus on lipids. *Am. J. Hum. Biol.* **2013**, *25*, 770–779. [CrossRef] [PubMed]

214. Dewit, O.; Dibba, B.; Prentice, A. Breast-milk amylase activity in English and Gambian mothers: Effects of prolonged lactation, maternal parity, and individual variations. *Pediatr. Res* **1990**, *28*, 502–506. [CrossRef] [PubMed]

215. Michie, C.A.; Tantscher, E.; Schall, T.; Rot, A. Physiological secretion of chemokines in human breast milk. *Eur. Cytokine Netw.* **1998**, *9*, 123–129. [PubMed]

216. Calhoun, D.A.; Lunoe, M.; Du, Y.; Christensen, R.D. Granulocyte colony stimulating factor is present in human milk and its receptor is present in human fetal intestine. *Pediatrics* **2000**, *105*, 1–6. [CrossRef]

217. Hegardt, P.; Lindberg, T.; Börjesson, J.; Skude, G. Amylase in human milk from mothers of preterm and term infants. *J. Pediatr. Gastroenterol. Nutr.* **1984**, *3*, 563–566. [CrossRef]

218. Freed, L.M.; York, C.M.; Hamosh, P.; Mehta, N.R.; Hamosh, M. Bile Salt-Stimulated Lipase of Human Milk Characteristics of the Enzyme in the Milk of Mothers of Premature and Full-Term Infants. *J. Pediatr. Gastroenterol. Nutr.* **1987**, *6*, 598–604. [CrossRef] [PubMed]

MDPI

St. Alban-Anlage 66

4052 Basel

Switzerland

Tel. +41 61 683 77 34

Fax +41 61 302 89 18

www.mdpi.com

Nutrients Editorial Office

E-mail: nutrients@mdpi.com

www.mdpi.com/journal/nutrients